T0305513

Advances in Water Management Under Climate Change

Due to increasing population, decreased cultivable land, and mounting scarcity of water, it is essential to optimize the use of available resources. Climate change is occurring across the world but its effect may be local or region-specific, including localized watershed management. In order to minimize these effects, governments and environmental agencies encourage the adoption of "climate-smart" agricultural technologies, which involve implementing plans, programs, and projects to sustain and enhance watersheds. Natural ecosystems, in their altered states, have always been relied upon to support the continuity of agricultural production and ecosystem services, such as flood and erosion control, mediation of water quality, stream flow regulation, microclimate regulation, and biodiversity in its various forms. According to the Food and Agriculture Organization of the United Nations, the adoption of these sustainable water management practices has resulted in savings of water and energy as well as a reduction of carbon emissions, decreased erosion, increased organic matter content and biotic activity in soils, increased crop water availability and thus resilience to drought, improved recharge of aquifers, and reduced impact of the variability in weather due to climate change. *Advances in Water Management Under Climate Change* examines all of these issues and provides best practices for sustainability.

Features:

- Presents the latest research in hydrology, hydraulics, water resources engineering, and agricultural best practices;
- Examines water management practices to best address and ideally mitigate climate change;
- Explains the nexus of agriculture, micro irrigation, AI applications in water management, and the impact of climate change on water resources;
- Includes practical examples to present practical insights on water management for climate change mitigation.

Advances in Water Management Under Climate Change

Edited by
Mukesh Kumar, Rohitashw Kumar,
and Vijay P. Singh

CRC Press
Taylor & Francis Group
Boca Raton London New York

CRC Press is an imprint of the
Taylor & Francis Group, an **informa** business

Designed cover image: Shutterstock

First edition published 2024
by CRC Press
6000 Broken Sound Parkway NW, Suite 300, Boca Raton, FL 33487-2742

and by CRC Press
4 Park Square, Milton Park, Abingdon, Oxon, OX14 4RN

CRC Press is an imprint of Taylor & Francis Group, LLC

ISBN: 978-1-032-39851-8 (hbk)
ISBN: 978-1-032-39853-2 (pbk)
ISBN: 978-1-003-35167-2 (ebk)

DOI: 10.1201/9781003351672

Typeset in Times
by KnowledgeWorks Global Ltd.

Mukesh Kumar: My late mother (Smt. Rajwati Devi) and my father (Sh. Om Prakash), my mother-in-law (Late Smt. Dropa Rani), father-in-law (Sh. Hoti Lal), my wife (Dr. Bhavnesh Jaint) and my lovely sons (Agrim and Arnav).

Rohitashw Kumar: My late parents (Smt. Rajwan Devi and Sh. Chhalu Ram), my wife (Mrs Reshma Devi), daughter (Meenu) and son (Vineet Kumar).

Vijay P. Singh: Wife Anita who is no more; son Vinay; daughter Arti; daughter-in-law Sonali; son-in-law Vamsi; grandsons Ronin, Kayden, and Davin.

Contents

Preface

Life cannot be sustained without water, as it is the most important natural resource required for the survival of humans, animals, plants, and other living organisms. The role of water in socio-economic and human development cannot be over-emphasized, as it is vital for each and every aspect of people's lives and livelihoods. The ever-increasing population and urbanization are creating mounting pressure on this vital natural resource, resulting in the scarcity of water. Water is one of the potential inputs for the agriculture sector. Rainfall is the main source of fresh water on the earth which is highly erratic in nature and varies with time and space resulting in the variation in spatial and temporal distribution of runoff, groundwater resources, etc. leading to limited availability of water resources in different parts of the country for agriculture and ultimately low yield and inferior quality of produce.

About two-thirds of cultivated area in India is still rainfed and more than 50% of the country's population is engaged in agriculture and is dependent for livelihood. Rainfed agriculture has great potential in providing agricultural production, livelihood security, and food and nutritional security. Climate change not only affects the water availability for the agricultural sector but also for other sectors. Water is also a critical resource for social and economic development. Due to the sudden increase in climatic extreme conditions, the quality of water is getting deteriorated and the quantity decreases, thus resulting in the scarcity of water and leading to droughts. On the other hand, its excessive quantity causes floods. In recent decades, climate change has been linked to changes in the global hydrological cycle, rising atmospheric water content, and shifting rainfall patterns.

Agriculture not only plays an important role in development but also constitutes the backbone of economic development of many South Asian countries. The growth in agricultural sector and productivity enhancement are vital for food and nutritional security which is being threatened by global warming and climate change. The issue of climate change is receiving a lot of attention at national and international levels, as the impact of climate change is being felt almost everywhere.

The water table and water quality are decreasing in many parts of the world particularly in developing nations. There is a need to take steps for saving water and environment using new and innovative on-farm techniques which not only increase agricultural production, soil productivity, and farmers' income but also protect the environment. Advanced and efficient methods of irrigation coupled with IoT play an important role not only in the efficient use of water in agriculture by growing more crop per drop of water but also help in mitigating the impact of climate change. As per Sustainable Development Goal (SDG)-6, access to fresh water in sufficient quantity and quality is required to fulfill all aspects of life and sustainable development. The human rights to water and sanitation are widely recognized all over the world. Water resources are embedded in all forms of development, like food security, health promotion, and poverty reduction in sustaining economic growth in agricultural industry and energy generation and in maintaining healthy ecosystem. In the present-day policy debates, water and climate change occupy a central position and it is being increasingly argued that the growth and development of a society in particular and nation as a whole are determined by the sensitization of each and everyone about water and climate change. This book discusses the significance of water resources, water harvesting, water conservation, effective ways, issues and challenges related to water and climate change, policy and government initiatives to deal with the question of sustainable management of water resources, and the issue of climate change in the 21st century.

This book provides a discussion of important aspects of water and its management, issues and challenges, concepts, scope, and guiding principles along with the objectives of integrated farming system water management, climate-resilient agriculture along with the impact of climate change on agriculture and food security, impact of climate change on irrigation demand and groundwater depletion and impact of irrigation on greenhouse gas emissions, climate-resilient agriculture to double the farmer's income, drip irrigation systems along with significant advancements in design, technology, and management, crop water demand, evapotranspiration, importance, need and applications of micro irrigation in hilly and sloppy areas, artificial intelligence in water management, soil carbon sequestration and conservation agriculture, and role of livestock in watershed management and its effect on environment. The book provides information to deal with different aspects of water management. It also discusses the issue of groundwater management, climate change, carbon sequestration, micro irrigation, protected cultivation, and artificial intelligence. The book consists of 18 chapters.

Chapter 1 discusses the importance, status, and classification of water resources in India as well as in the world. The chapter also describes types of water, alternatives to water, techniques of water saving, rooftop rainwater harvesting, complexities of agricultural water management along with management of water resources in the agricultural sector, and the impact of climate change on water resources.

Chapter 2 is related to water management, issues, and challenges and gives a broad overview of water management and water safety and security, which is a major concern nowadays not only in the country but also all over the globe. The chapter also discusses the importance of water harvesting in the present-day scenario along with major schemes and initiatives of the government of India to deal with the question of sustainable management of water resources in the 21st century, importance of water education, etc.

Water resources status in India is highlighted in Chapter 3, along with the impact of climate change on irrigation demand and groundwater depletion and the impact of irrigation on greenhouse gas emissions. The chapter also highlights the needs, advantages, and design criteria of micro-irrigation technology, and estimation of water requirement. It describes systems for reducing flooding and emissions during the crop-growing season falling into four categories. The chapter also describes conservation tillage, mulching, site-specific nutrient management, and nanotechnology.

Chapter 4 discusses water resources potential along with groundwater development and utilization prospects in India which include the use of groundwater in agriculture and food security, impact of climate change on groundwater resources and groundwater depletion and water quality issues in India. It also explains the WHO drinking water quality criteria.

Concepts, guiding principles, and objectives of integrated farming schemes (IFS) are discussed in Chapter 5 which also describes the climate change nexus with agriculture along with climate resilience and sustainability through IFS.

Chapter 6 explains climate-resilient agriculture along with the impact of climate change on agriculture and food security. Strategies to achieve the objectives and key characteristics of climate-resilient agriculture are also discussed. The chapter also throws light on climate-resilient agricultural practices along with important steps in climate-resilient agriculture. Some action-oriented case studies are also included in the chapter.

Chapter 7 deals with climate-resilient agriculture to double the farmer's income. This chapter describes strategies and technologies for climate change mitigation, climate change and agriculture along with the importance of crop diversification. Different interventions under different modules of the NICRA project which have helped in doubling the farmers' income in Chamba (HP) are discussed as case studies in the chapter.

The need for increasing farmers' income is discussed in Chapter 8 under integrated farming system – a dynamic approach toward increasing the income of small and marginal farmers. This chapter discusses the challenges faced by small and marginal farmers in India along with the restraining forces in doubling farmer income and problems of present-day agriculture. It also describes the farming system and its goal. The concept of integrated farming system (IFS) along with its need, components, benefits, and agronomic practices is also included. The chapter also includes criteria for the selection of enterprises, enterprises linked through IFS under different conditions, and beneficial IFS combination under varied ecosystems along with the present status of farming system research.

Chapter 9 deals with the importance of drip irrigation systems along with significant advancements in design, technology, and management over the last decade. Descriptions of empirical models along with drip irrigation water interface, etc. are described in the chapter.

Chapter 10 describes protected cultivation as an alternative to growing high-value nutritional horticultural crops throughout the year on a sustainable basis and food security in India. Pressurized irrigation system networks along with different types of pressurized irrigation systems and essential elements of a drip irrigation system are discussed. The soil-water-plant-climate relationship in relation to drip irrigation systems along with crop water requirement is explained. The chapter also gives an idea of fertigation for protected cultivation technology, automation of pressurized irrigation system, and design of drip irrigation system along with the maintenance of drip fertigation system for protected cultivation.

Chapter 11 describes the soil carbon sequestration and conservation agriculture as a way forward for soil carbon sequestration. Potential impacts of climate change on soil properties and food security. The chapter also discusses the effects of conservation agriculture on SOC and challenges to the adoption of conservation agriculture.

Chapter 12 describes water productivity of temperate fruits under climate change scenario. It also describes the effect of modern irrigation methods like drip irrigation and their significance on the plant's growth and yield parameters of temperate fruits along with water use efficiency, water stress, and maturity time.

Chapter 13 discusses the livestock wealth in India and its role in country's economy. It describes the role of livestock in watershed management and its effect on the environment, as well as the role of livestock in livelihood security and watershed health.

Chapter 14 explains the role of artificial intelligence in water management and typology of AI applications. The chapter also throws light on artificial intelligence techniques and their application to hydrological modeling and water resources, simulation optimization, and management.

Crop water demand is determined, based on evapotranspiration and climatic factors affecting evapotranspiration. Chapter 15 describes the significance of evapotranspiration in context of climate change along with reference evapotranspiration (ETo) models. The influence of global warming on the water cycle revealed through satellite data is also presented in the chapter.

Chapter 16 describes the role, importance, need, and applications of micro irrigation in hilly and sloppy areas for doubling farm income. It explains the potential and future prospects of micro irrigation in India along with the need and scope of gravity-fed drip irrigation. The chapter also includes moisture depletion along with the role and responsibilities of the National Mission on Micro Irrigation (NMMI) in India.

Chapter 17 describes about micro irrigation which leads to significant economic and social benefits to Indian agriculture. This technique is effective for improving water use efficiency. It is found to be significant to greater extent for horticulture, ornamental, and landscape and has been used in large variety of meteorological conditions. Micro irrigation is advantageous in saving water and energy, increases yield, improves fertilizer application, decreases the rate of salinization, reduces weeds and diseases and reduces the use of labors. Advancement in dripper or emitter techniques and development of low-cost filters has facilitated to expansion of land under micro irrigation. Using the micro-irrigation system will be therefore of vital significance in terms of conserving water and sustainable management of soil and water resources.

Chapter 18 describes the ground water recharge techniques. The rate of water consumption is outpacing that of replenishment. The groundwater problem in India is particularly severe. Water conservation through recharge and resource management by artificial recharge is critical for solving the problems of the 21st century. To boost our water resources, artificial groundwater recharge techniques can be utilized. Various approaches can be utilized depending on the meteorological and geographic conditions. In order to highlight a variety of approaches to recharge groundwater, we have reviewed and summarized numerous studies in this chapter. One approach for resolving groundwater issues that is long-lasting is artificially recharging aquifers. The main objective of this technology is to safeguard or improve groundwater resources across India. This comprises temporary groundwater abstraction regulation, floodwater conservation or disposal, saltwater intrusion mitigation, water storage to reduce pumping and pipe expenses, and water quality enhancement through dilution by combining with pre-existing groundwater. The aquifer map will give details on the aquifer system's size and shape, its susceptibility to exploitation, contamination, and seawater intrusion, as well as areas appropriate for artificial recharge.

It is emphasized that we have just begun to scratch the surface with some of the recent advances in water management under climate change. This book discusses advanced technologies for tackling the issue of water and climate change along with applications in agriculture in different agro-climatic regions in the globe. It is hoped that the book will be useful to increase agricultural production, productivity, profitability, and suitability with the mitigation of the impact of climate change.

Mukesh Kumar
IGNOU, New Delhi, India

Rohitashw Kumar
SKUAST-Kashmir, Srinagar, India

Vijay P. Singh
A&M University, College Station, Texas, USA

Acknowledgments

I sincerely acknowledge the efforts of Professor Vijay Pal Singh, distinguished Professor, Department of Biological and Agricultural Engineering, Texas A&M University, College Station, Texas, USA, for his support and guidance to bring this manuscript to final refined shape, constructive criticism, and utmost cooperation at every stage during this work. He is always an inspiration for us.

I am grateful to Dr. Mukesh Kumar, who put in great efforts and worked day and night to bring this manuscript to a refined shape. I am also highly thankful to all contributors to different chapters of this book.

I am grateful to my students, Dr. Sakeel Ahmad Bhat, Er. Munjid Maryam, Er. Zeenat Farooq, Er. Tanzeel Khan, Er. Dinesh Vishkarma, Er. Faizan Masoodi, Er, Noureen, Er. Mahrukh, Khilat Shabir, Munneza farooq, and Riaze Ashraf, for providing all necessary help to write this book.

I am highly obliged to Hon'ble Vice Chancellor Prof. N A Ganai for his support and affection and encouragement always for innovation. I wish to extend my sincere thanks to College of Agricultural Engineering and Technology, SKUAST-Kashmir for their support and encouragement.

I express my regards and reverence to my late parents as their contribution to whatever I have achieved till date is beyond expression. It was their love, affection and blessed care that has helped me to move ahead in my difficult times and complete my work successfully. I thank my family members, who have been a source of inspiration always. I thankfully acknowledge the contribution of all my teachers since schooldays, for showing me the right path at different steps of life.

I consider it a proud privilege to express my heartfelt gratitude to ICAR – All India Coordinated Research Project on Plastic Engineering in Agriculture Structures & Environment Management, for providing all facilities to carry out this project at SKUAST-Kashmir, Srinagar.

I sincerely acknowledge with love, the patience, and support of my wonderful wife Reshu. She has loved and cared for me without ever asking anything in return and I am thankful to God for blessing me with her. She has spent the best and the worst times with me but her faith in my decisions and my abilities has never wavered. I would also like to thank my beloved daughter Meenu and son Vineet for making my home lovely with their sweet activities.

Finally, I bow my head before the almighty God, whose divine grace gave me the required courage, strength, and perseverance to overcome various obstacles that stood in my way.

Rohitashw Kumar
Associate Dean,
College of Agricultural Engineering and Technology,
SKUAST – Kashmir, Srinagar

Contributors

Tarun Ameta
ICAR-IARI
Pusa, New Delhi, India

M. H. Chavda
Sardarkrushinagar Dantiwada
   Agricultural University
Deesa, Gujarat, India

Deepshikha
Sher-e-Kashmir University of
   Agricultural Sciences and
   Technology of Kashmir
Shalimar Campus, Srinagar, Jammu
   and Kashmir, India

Muneeza Farooq
Sher-e-Kashmir University of
   Agricultural Sciences and
   Technology of Kashmir
Srinagar, Jammu and Kashmir, India

Kishor Pandurag Gavhane
ICAR-IARI
Pusa, New Delhi, India

Murtaza Hasan
ICAR-IARI
Pusa, New Delhi, India

Salim Heddam
University 20 Août 1955
Skikda, Algeria

Naushad Khan
C. S. Azad University of Agriculture
   and Technology
Kanpur, Uttar Pradesh, India

Amit Kumar
Sher-e-Kashmir University of
   Agricultural Sciences and
   Technology of Kashmir
Srinagar, Jammu and Kashmir, India

Love Kumar
ICAR-IARI
Pusa, New Delhi, India

Mukesh Kumar
IGNOU
New Delhi, India

Rohitashw Kumar
Sher-e-Kashmir University of
   Agricultural Sciences and
   Technology of Kashmir
Srinagar, Jammu and Kashmir, India

Sanjeev Kumar
Indian Agriculture Statistics Research
   Institutes
New Delhi, India

Kaberi Mahanta
Assam Agricultural University
Jorhat, Assam, India

S. H. Malve
Sardarkrushinagar Dantiwada
   Agricultural University
Deesa, Gujarat, India

Indra Mani
ICAR-IARI
Pusa, New Delhi, India

Munjid Maryam
Sher-e-Kashmir University of
   Agricultural Sciences and
   Technology of Kashmir
Srinagar, Jammu and Kashmir,
   India

Ananya Mishra
Sher-e-Kashmir University of
   Agricultural Sciences and
   Technology of Kashmir
Srinagar, Jammu and Kashmir, India

B. A. Pandit
Sher-e-Kashmir University of
  Agricultural Sciences and
  Technology of Kashmir
Srinagar, Jammu and Kashmir, India

Vinayak Paradkar
ICAR-IARI
Pusa, New Delhi, India

G. T. Patle
College of Agricultural Engineering
  and Post-Harvest Technology
CAU
Gangtok, Sikkim, India

Rajeev Raina
Dr. Y. S. Parmar University of
  Horticulture and Forestry
Nauni, Himachal Pradesh, India

D. J. Rajkhowa
ICAR Research Complex for NEH
  Region, Nagaland Centre
Medziphema, Nagaland, India

Rishi Richa
Sher-e-Kashmir University of
  Agricultural Sciences and
  Technology of Kashmir
Srinagar, Jammu and Kashmir, India

Atish Sagar
ICAR-IARI
Pusa, New Delhi, India

Khilat Shabir
Sher-e-Kashmir University of
  Agricultural Sciences and
  Technology of Kashmir
Srinagar, Jammu and Kashmir, India

Nirmal Sharma
Sher-e-Kashmir University of
  Agricultural Sciences and
  Technology of Jammu
Jammu, Jammu and Kashmir, India

Ritu Sharma
MHU
Karnal, Haryana, India

Karan Singh
Sher-e-Kashmir University of
  Agricultural Sciences and
  Technology of Kashmir
Srinagar, Jammu and Kashmir,
  India

Kehar Singh Thakur
Dr. Y. S. Parmar University of
  Horticulture and Forestry
Nauni, Himachal Pradesh, India

Y. B. Vala
Sardarkrushinagar Dantiwada
  Agricultural University
Deesa, Gujarat, India

Pradeep Kumar Verma
C. S. Azad University of Agriculture
  and Technology
Kanpur, Uttar Pradesh, India

P. Vijayakumar
Indira Gandhi National Open
  University (IGNOU)
Maidan Garhi, New Delhi, India

Dinesh Kumar Vishwakarma
G. B. Pant University of Agriculture
  and Technology
Pantnagar, Uttarakhand, India

Tajamul Farooq Wani
Sher-e-Kashmir University of
  Agricultural Sciences and
  Technology of Kashmir
Srinagar, Jammu and Kashmir,
  India

Ashish Kumar Yadav
C. S. Azad University of Agriculture
  and Technology
Kanpur, Uttar Pradesh, India

About the Authors

Dr. Mukesh Kumar has graduated in Agricultural Engineering with a Masters in Soil and Water Engineering from Punjab Agricultural University Ludhiana. He received his Ph.D. in Agricultural Engineering with the specialization in Soil and Water Conservation Engineering from Indian Agricultural Research Institute. Dr. Kumar has been working as Assistant Professor in School of Agriculture, Indira Gandhi National Open University, New Delhi, since 2006. He has associated in several research projects funded by the ICAR, different Ministries. He has published more than 40 research papers in peer reviewed journals and participated in 32 national and international conferences. He has also contributed to more than 22 book chapters and published two extension manuals. He has organized a number of national and international conferences as the organizing secretary, or a member of the organizing committee. He has developed number of educational programmes in the field of water management and watershed management at a national level. He is a life member of professional societies such as the Indian Society of Soil and Water Conservation, New Delhi; the Indian Society Agricultural Engineering, New Delhi; and the Indian Association of Hydrologists, Roorkee, Uttrakhand, India. He is involved as joint secretary and treasurer in the Indian Society of Soil and Water Conservation, New Delhi. He has received a Ph.D. Student Incentive Award (2012), Gold Medal (2014), Special Research Award (2016) and National Fellow (2019) from the Soil Conservation Society of India. The Fertilizer Association of India has awarded him with the Golden Jubilee Award for Doctoral Research in Fertilizer Uses in 2013 for outstanding research work on the dynamics of nutrient movements in soil under fertigation during Ph.D. He was also honored with the University Silver Medal (2015) by the Indira Gandhi National Open University, New Delhi, India.

Dr. Rohitashw Kumar (B.E., M.E., Ph.D.) is Associate Dean and Professor at the College of Agricultural Engineering and Technology, Sher-e-Kashmir University of Agricultural Sciences and Technology of Kashmir, Srinagar, India. He is also Professor Water Chair (Sheikkul Alam Shiekh Nuruddin Water Chair), Ministry of Jal Shakti, Government of India, at the National Institute of Technology, Srinagar (J&K). He is also Professor and Head, Division of Irrigation and Drainage

Engineering. He obtained his Ph.D. degree in Water Resources Engineering from NIT, Hamirpur, and Master of Engineering Degree in Irrigation Water Management Engineering from MPUAT, Udaipur. He received a leadership award in 2020, special research award in 2017 and a Student Incentive Award (2015) (Ph.D. Research) from the Soil Conservation Society of India, New Delhi. He also got the first prize in India for best M.Tech thesis in Agricultural Engineering in 2001. He has published over 100 papers in peer-reviewed journals, more than 25 popular articles, four books, two practical manuals, and 25 book chapters. He has guided one Ph.D. and 14 M.Tech. students in soil and water engineering. He has handled more than 12 research projects as a principal or co-principal investigator and since 2011 he has been Principal Investigator of ICAR - All India Coordinated Research Project on Plastic Engineering in Agriculture Structural and Environment Management.

Professor Vijay P. Singh is a Distinguished Professor, a Regents Professor, and the inaugural holder of the Caroline and William N. Lehrer Distinguished Chair in Water Engineering at the Texas A&M University. His research interests include surface-water hydrology, groundwater hydrology, hydraulics, irrigation engineering, environmental quality, water resources, water-food-energy nexus, climate change impacts, entropy theory, copula theory, and mathematical modeling. He graduated with a B.Sc. in Engineering and Technology with emphasis on Soil and Water Conservation Engineering in 1967 from the U.P. Agricultural University, India. He earned an M.S. in Engineering with specialization in Hydrology in 1970 from the University of Guelph, Canada; a Ph.D. in Civil Engineering with specialization in Hydrology and Water Resources in 1974 from the Colorado State University, Fort Collins, USA, and a D.Sc. in Environmental and Water Resources Engineering in 1998 from the University of the Witwatersrand, Johannesburg, South Africa. He has published extensively on a wide range of topics. His publications include more than 1,365 journal articles, 32 books, 80 edited books, 305 book chapters, and 315 conference proceedings papers. For his seminar contributions he has received more than 100 national and international awards, including three honorary doctorates. Currently he serves as Past President of the American Academy of Water Resources Engineers, the American Society of Civil Engineers (ASCE), and previously he served as President of the American Institute of Hydrology and Cahir, Watershed Council, ASCE. He is Editor-in-Chief of two book series, three journals, and serves on the editorial boards of more than 25 journals. He has served as Editor-in-Chief of three other journals. He is a Distinguished Member of the American Society of Civil Engineers, an Honorary Member of the American Water Resources Association, an Honorary Member of International Water Resource Association, and a Distinguished Fellow of the Association of Global Groundwater Scientists. He is a fellow of five professional societies. He is also a fellow or member of 11 national or international engineering or science academies.

1 Water Resource Management
An Approach to Sustainable Water Management

Ananya Mishra and Rohitashw Kumar

1.1 INTRODUCTION

Water is the most important and sustainable natural reserve, which is in serious distress due to its increasing demand though supply decreases. The requirement and demand for water have increased dramatically in cities as a result of increased industrialization, globalization, urban growth, and other such phenomena, which has been exacerbated by the increased population water demand resulting in various types of wastes and sewerages water. Governments and scientists have recognized the value of maximizing of various resource use while minimizing waste and losses. The available water resources in saline oceans contain 97 per cent of the world's water. Two-thirds of the remaining 3 per cent have deposited as ice masses in the polar region and in the form of ice in the mountains. As a corollary, fresh running water accounts for 1 per cent of the Earth's water, of which 98 per cent is groundwater (Jafari *et al.*, 2018a). In a global sense, India's water supply outlook owns just 4 per cent of the world's water supplies while housing 16 per cent of the global population, meaning India's per capita water available is very limited. The amount of accessible water here is approximately 6 trillion m^3, which is the world's highest for a nation of equivalent extent (Jafari *et al.*, 2018a). Water exists in the oceans, atmosphere, between the soil, and fragments of the Earth crust. Renewable radiation drives the movement of water as it transmits between the lithosphere and the environment via evaporation and precipitation. However, owing to wastage and inadequate water conservation schemes, as well as uneven rainfall distribution, a significant portion of the world suffers from water shortages during the year's dry period (Mushtaq *et al.*, 2020). For the past three decades, there has been an increase in water erosion. Not only has the rate and severity of water erosion increased but also the area of water erosion areas, and as a result of increased flooding. In about an area of 125 million hectares of watersheds (62 per cent of the total country), water erosion has been higher than the natural rate, and sedimentation has also increased (Jafari *et al.*, 2018b).

DOI: 10.1201/9781003351672-1

1.2   WATER'S DISTINCTIVE QUALITIES

Water is distinguished by the following characteristics:

a. It persists as a liquid over a broad temperature spectrum (0–100 °C).
b. It has high specific heat due to which it warms and cools gradually, avoiding temperature tremors in marine organisms.
c. It has the capacity to alter its state of nature; it can exist in the solid, liquid, or gaseous form.
d. It needs a significant amount of energy to vaporize as a consequence of high latent heat of vaporization. Water vaporizes and has a significant cooling effect on evaporation.
e. It is an exceptional solvent and therefore a sturdy carrier of plant nutrients.
f. It has high surface tension and cohesion properties.

Management of water sustainably has been a major problem around the world, including in Iran in the recent decade (Yazdanpanah *et al.*, 2015). Addressing to this problem, a wide range of strategies have been suggested, which can be roughly divided into two major groups: solution for demand-intended management and supply-intended management (Yazdanpanah *et al.*, 2015). As a result, there has been a global paradigm towards a demand-directed strategy (Beal *et al.*, 2013).

1.3   CLASSIFICATION OF WATER RESOURCES

Available water resources are divided into two categories (Fig. 1.1):

1. Surface water resources
2. Ground water resources

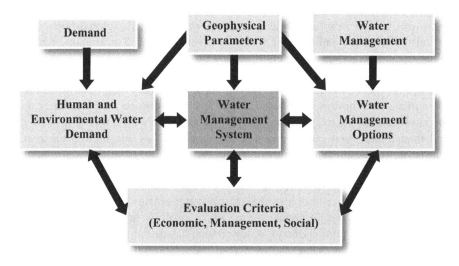

FIGURE 1.1   Conceptual diagram of water management (Parwal, 2015).

1.3.1 SURFACE WATER

Surface water is the naturally accessible water present on the Earth surface in the form of rivers, rivulets, streams, wetlands, and lakes. Water available from precipitation provides a considerable volume of water to available surface water, which is primarily used for commercial irrigation, farming, navigation purpose, and community water supply purposes. It may be divided into two types:

* Lotic (flowing) water
* Lentic (still) water

1.3.2 LOTIC WATERS

The term lotic refers to flowing water sources, such as streams, ponds, runoff, and springs, and is derived from the word "lavo" which means "to wash". Surface water bodies in the lotic region are mostly riverine and are dominated by advective movement. The fundamental goal of lotic water system is assumed to be conveyed surplus rainwater back to the oceans because the water in such system moves continuously. The lotic water system consists of ecosystems within them and is significant for a multitude of reason, including organic matter, nutrients, energy, and other compounds, absorbing constituents from the air and land transforming them, and/or transporting them to the sea.

The speed of the water flow, which varies from low current flow to turbulent rapids, is an important component in determining the complete ecology of these systems. Flow divergence occurs when it fluctuates within systems and is exposed to unstable turbulence. Snowmelt, overt precipitation, and groundwater all influence the average flow rate in lotic water, as it may obstacle sinuosity gradient, as well as the variability of contact with the sides and bottom of the channel. By erosion and deposition, the lotic structures will change the morphology of the creek bed and generate a range of ecosystems such as riffles, glides, and pools by depositions and erosion (Parwal, 2015).

1.4 RIVERS

Rivers, which have existed for generations, are formed by well-defined channels draining all of the water that has been collected from the land in the form of rainfall and snowmelt. Rivers in their headwaters are often cold and oxygen-rich because most of the rivers originated from ice. As rivers move down the mountains, they get larger, colder, slower, and deeper, and their oxygen level decreases. As the land and surroundings from which it flows change, so do the characteristics of the river.

1.5 STREAMS

Rainwater seeps through the surface and enters the freshwater supply. There is increased hydrostatic pressure in the soil mass when the natural relief is such that the land surface drops below the upper most surface of the groundwater reservoir at some point, and the groundwater beneath exerts pressure and makes it way deep into the soil, creating a drain. Streams can be torrential or scattered in nature. Streams that

are torrent only contain surface runoff and only receive water during the rainy season and subsequent runoff season.

1.6 LENTIC WATER

Stagnant water consisting of closed structures is referred to as lentic water. This
generally emerges in small or wide depressions on the Earth surface that lack an
outlet for the water to drain out. Lakes and ponds are examples of lentic or standing water. Springs, lakes, and waterways are examples of lotic or flowing water.
Wetlands include marshes and swamps where water levels rise and fall seasonally
and annually (Sharma and Giri, 2018). Lentic waters typically rot, decompose, or
remain as such within the lentic water body, and the normal course of succession
eventually transforms such a lentic water body into a marsh, swamp, wetland, and
finally a dry land. Any lentic habitats may provide water with a higher salt content
(for example, the Pangong Lake in Jammu and Kashmir and the Great Salt Lake in
Utah) (Mushtaq *et al.*, 2020).

1.7 GROUNDWATER

Around 70 per cent of rainwater goes into lakes, rivers, drains, reservoirs, and other
bodies of water, with the other 10 per cent evaporating. The remaining 20 per cent
steeps into the soil and contributes towards the groundwater by moving a set of
distances beneath the ground into soil particles. As a result, runoff, also known as
plutonic water, becomes the principal supply of freshwater (Mushtaq *et al.*, 2020).

Groundwater provides a substantial amount of usable water in locations where
surface water is short in supply. The volume and kind of rainfall, as well as the form
and slope of the soil, determine its availability. In locations with heavy rainfall and
permeable rocks, simple percolation makes underground water readily accessible in
large volumes and at shallower depths.

1.8 TYPES OF WATER

The river systems have been classified according to Sioli (2012): white water, clear
water, blue water, grey water, and black water. Each form of water system has a specific visual behaviour that is linked to the drainage area's physical properties.

1.8.1 WHITE WATER

White water rivers are characterized by muddy, ochre-coloured water. Because of
the large amount of suspended clay particles, white water has a unique brownish
colour and a pH that is almost neutral (Junk, 1970). The white-water river floodplains
are referred as varzea (Junk *et al.*, 2015) (Fig. 1.2).

1.8.2 CLEAR WATER

Clear water is translucent and poor on suspended sediments, yielding a pH range
from acidic to neutral. Clear water drains through catchment areas with smooth,

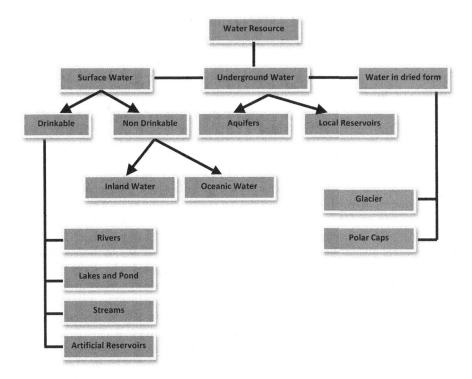

FIGURE 1.2 Classification of water resources (Mushtaq *et al.*, 2020).

level relief, particularly those shielded from surface erosion by dense forest (Sioli, 2012).

1.8.3 BLUE WATER

Blue water is the water available at the surface and groundwater reservoirs. Blue water is drawn to irrigate crops that would sustain transpiration process. It is critical to be used effectively.

1.8.4 GREY WATER

Grey water is the water source that has been contaminated throughout the production process, such as in agriculture due to the leaching of fertilizer and pesticide and waste water from the industries. The volume of grey water can be calculated by calculating the amount of blue water required to dilute the receiving water body to an acceptable quality standard (Clothier *et al.*, 2010).

1.8.5 BLACK WATER

Black water is clear and sediment-free, yet it's high in decomposing plant matter, which gives it its distinctive reddish and black hue and acidic in nature pH (Stefanelli-Silva *et al.*, 2019).

1.9 IMPACT OF CLIMATE CHANGE ON WATER RESOURCES

Water is present in every component of the climate system, including the atmosphere, cryosphere, land surface, hydrosphere, and biosphere. Changes in precipitation patterns, intensity, and extremes; extensive melting of snow and ice; increased atmospheric water vapour; increased evaporation; and changes in soil moisture and runoff are all effects of climate change on the hydrological cycle. Water acts as a critical resource for social and economic development as well as environmental protection, but due to the sudden increases in the extreme climatic conditions that may worsen its quality, causing floods and droughts in an area, resulting in the scarcity of water. Climate change has been linked to changes in the global hydrological cycle in recent decades, including rising atmospheric water content and shifting rainfall patterns (Allan, 2011). Human activities in Earth's environment have had an effect on ecosystems in many regions of the world, including freshwater, marine, and terrestrial ecosystems, and more changes are expected as a result of rapid global change (Mishra *et al.*, 2021). The climate and water systems are intricately linked and their impacts can be exacerbated when they occur in areas that already have low water resource levels and frequent droughts, resulting in imbalances between water demands and available resources. The water quantity and quality have been affected due to anthropogenic activities that have an impact on water use (WU) and the availability. Further warming is predicted to exacerbate the hydrological cycle, increasing the risk of floods and droughts around the world (Giorgi *et al.*, 2011). Temperature of the water, together with its flow, is one of the most important factors influencing ecosystem state in surface waters (Webb *et al.*, 2008). Water temperature has a direct influence on the metabolisms and lifecycles of aquatic species, as well as on water quality, with most chemical and bacterial processes functioning faster at higher temperatures, leading to a rise in the occurrence of algal blooms (Watts *et al.*, 2015). These algae cause the lack of oxygen in the water leading to a threatening impact on the aquatic ecosystem underwater. Increase in surface air temperature will vary seasonally and will be greater in some areas than others, affecting rivers more severely. Even though streams and rivers are generally well mixed and turbulent, they respond quickly to changes in atmospheric conditions, becoming warmer. Shifts in river flow are one of the prominent reasons that climate change or other stressors affect river ecosystems. The shifts in pattern may increase runoff in the Asian region that would be a disadvantage because they tend to occur during the wet season, and the excess water may be unavailable during dry season lacking to fulfil the need in required times (Arnell, 2004). Rivers vary extensively in terms of their natural flow regime, and this variation is important for stream and river ecological management. Along with temperature, the climate affects the precipitation patterns. Precipitation forecasts are more uncertain than temperature forecasts. Increased temperatures may cause a shift from winter snow to rain or rain plus snow in regions that receive the majority of their precipitation as snow (Palmer *et al.*, 2009). Additional resources of water are groundwater resources which have static and dynamic components. It's also necessary to be careful about the climate crisis on groundwater resources. Variations in recharge, which includes changes in precipitation and evapotranspiration, as well as changes in the nature of the interactions between groundwater and surface water systems and changes in irrigation use, are all expected to affect groundwater systems as part of the hydrologic cycle. The alarming concern for the water resources urges the requirement for better conservation technology and alternatives.

1.10 METHODOLOGY

1.10.1 ALTERNATIVE TO WATER

Alternative waters are prolonged sources of water that are not derived from groundwater or surface water that may help to meet the requirement for freshwater use. Alternative water can help federal agencies support water resilience by providing a variety of water sources. Grey water can be used as an alternative to water scarcity and is an alternative to the problem.

1.10.2 WATER SAVING TECHNIQUES

Several ways can be used to alleviate water scarcity in arid places without relying on irrigation water from permanent rivers, reservoirs, or raised groundwater. Various water conservation techniques need to be adopted for better management of available freshwater. Water acts as an ecosystem for various aquatic life forms and thus needs to be conserved to protect the natural habitat as well as to meet the demands of humans.

1.11 GREY WATER AS ALTERNATIVE WATER RESOURCES

Due to the increasing temperature causing global climate change and population growth, many municipal water sources are in jeopardy. As a result, many strategies for reducing the use of fresh water, such as the implementation of water demand control and the substitution of fresh water with the renewable water supplies, have been adopted (Marleni et al., 2015). Grey water is the portion of waste water collected from residential use that does not contain flush water. The most prevalent classifications are light and dark grey water. Light grey water is water with very low levels of contaminants that comes from swimming pools and hand-washing sinks (Alagirisamy, 2021).

1.11.1 COMPOSITION OF GREY WATER

Based on waste water sanitation levels, it can be categorized as follows:

- Grey water, or wash water or waste water drained from a household, inclusive of water from toilets, bathroom, drains, kitchens, washing machines, and dishwashers – can be classified based on its cleanliness. Black water is heavily contaminated with biological pollutants such as excrement and chemicals (Oron et al., 2014).
- Black water, which is deeply tainted with biological pollutants including faeces and chemicals (Oron et al., 2014).

1.12   COMPLEXITIES OF AGRICULTURAL WATER MANAGEMENT PROBLEMS

Many countries have successfully created grey water technologies and methods, as well as adapted them to agricultural and domestic applications, in response to the challenge of rapidly dwindling water sources. The type of action determines the chemical composition of grey water. The formulation is inconsistent, which could be due to the use of solvents or kitchen waste water (Alagirisamy, 2021). Grey water from kitchen outlets, degradable components and residues from detergents and soaps are found in significant concentrations. Pathogens can only be found in faeces-contaminated fractions of grey water. Grey water provides an ideal environment for germs to proliferate quickly, needing meticulous preparation prior to disposal. Kitchen waste water accounts for only 10 per cent of all grey water produced in a home. The volume of grey water produced by houses in the kitchen amounts for about 10 per cent of overall grey water volume (Pachkor and Parbat, 2017; Alagirisamy, 2021). Services for waste water collection and treatment are provided to approximately 4 million persons (62 per cent of the population), producing approximately 100 million m^3 of waste water per year, of which the Amman's effluent is combined with storm water in the King Talal Dam and used for agriculture in the Jordan Valley. The majority of Tunisia's 1941 million m^3/year of municipal effluent is treated and irrigated (Shevah, 2017).

The unpredictability of agricultural system elements and their interrelationships affects many outlooks of hydrology, technology, and economic. When there is an uncertainty, related to the optimization as process becomes more difficult. Pollutant discharges differ depending on the type of farming, location, and season. The reduction of pollutants in separate plantation regions is essentially accomplished by multiple water sources due to the breadth of the farming environment. Pollutants instigated by farming activity in a particular place might discharge to nearby and isolated water bodies due to the fluidity of water and human influences (Wang *et al.*, 2015).

The grey water footprint is calculated by the contaminant that needs the maximum dilution. Crop cultivation's grey water footprints are a source of uncertainty (Chukalla *et al.*, 2018). In South Korea, the grey water plants operate at a level of 265 per cent with 200 million tons of grey water supplies (Hoekstra and Mekonnen, 2012).

1.13   WATER RESOURCE MANAGEMENT IN AGRICULTURE

Rivers are under strain as expansion of water supplies increases competition and lowers water protection for urban and ecological environments throughout the world. In Europe, over the past decades, farming has seen the introduction of ICTs that have the capability to improve farmers' access to the private and public resources and increase their productivity in water resources, among other things (Aker *et al.*, 2016). Inadequate availability of information is viewed as a serious delinquent in the agricultural sector since wasting of water continues. Farming techniques and the demand for resources, particularly on supply of water (Nakasone and Torero, 2016).

The belief that the natural development variability not depends on the management of water resources prevents improved water supplies and irrigation technologies from being implemented due to the volatility of the results. However, the high potential of catastrophic repercussions in the actual world leads to the creation of novel irrigation and water management methods.

1.14 WATER SAVING TECHNIQUES

The amount of fresh, usable water on Earth is limited. The hydrologic cycle, fortunately, recycles (collects, cleanses, and distributes) water naturally. Water conservation is defined as the wise use of water and the prevention of waste. Scientists have made technology to accelerate this process. Water resources, on the other hand, may not be adequate to meet a community's needs due to a variety of circumstances (drought, flood, population increase, contamination, and so on). Water conservation can ensure that fresh water supplies are available for everyone now and future. Water management allows species such as water voles, otters, herons, and fish to flourish by conserving resources in the ecosystem.

1.15 ENHANCING WATER USE EFFICIENCY

The removal of water losses and improvement of WU efficiency (WUE) are a major problem for the management of water supply. The increase in population, the need for economic growth, and climate change are all factors to be considered for making it difficult to solve this challenge. The main targets for the improvement of the WU are the agricultural and home sectors that make up the great bulk of WU worldwide (>95 per cent). In some instances, significant improvements can be made. The proportion of irrigation water used by an irrigated land or project cultivations by transferring water from a river or other natural sources of water to a project or waterway has been characterized as the traditional irrigation efficiency (Lankford, 2012). The primary productivity difference between agronomists and physiologists is that the latter cannot account for agricultural processing for potential water losses. Irrigation efficiency is defined as the level of irrigation that maximises revenue production by taking into consideration all expenses, rates, and the reaction to crop irrigation. The multidisciplinary nature of this definition requires that multiple perspectives be included to address the global problems that researchers face during discussions concerning WU (Boutraa, 2010). When it comes to domestic water consumption, household-centred interventions are able to lead to a drinking water consumption is reduced by 30 per cent or more with little effort and comfort or more (50 per cent) (Schuetze and Santiago-Fandiño, 2013).

1.16 QUALITY OF WATER RESOURCES

The misuse of nitrogen and pesticides as well as the use of non-sustainable technologies for field management is a serious concern in the intensively regulated agricultural areas. The high rates of urbanization and industrialization have raised the problems associated with the ecological degradation of the rivers. A comparison of geographical availability and demand and little concern for consistency have been extensively explored for volumetric water scarcity (Liu *et al.*, 2017). According to Dalezios *et al.* (2018), a systematic examination defines water scarcity as the ratio of demand to water available for most commonly used methodologies.

1.17 IRRIGATION SCHEDULING

Irrigation scheduling is the method for deciding how much water it needs to be added and when it should be used WUE: Use more water to optimize the use of the plant decreases irrigation WUE (Koech and Langat, 2018). In order to handle irrigation water management problems, besides simulation modelling, optimization methodologies are another successful option. The optimization approaches describe and generalize the irrigation system through the development of a set of mathematical equations and the application of optimization solution technology. As shown above, the simulation optimization model's structure is divided into two portions for irrigation scheduling (the simulation and the optimization). In the initial portion of the soil water balance simulation model, experimental soil moistures were used to optimize and evaluate it. Then, to simulate the objects, the simulation model was combined with water production functions (Li *et al.*, 2020). Farmers readily adapt new technology, such as auto-control machines and irrigation pivot systems that do not need new preparation or experience. However, it is typically restricted to adopt soil or plant moisture sensors, as well as related applications, that are data-intensive innovations. These systems require more education and/or the usage of data collection sources from third parties. Management practises must frequently be adapted to fit data-driven decisions. This can lead farmers to shift their profiles from that of an agricultural management that grows crops based on their practical experience to that of a field technician, whose work should be based on their real-world experience. These conditions can transform farmer profiles from the profile of a manager who cultivates crops on the basis of his or her practical experience to that of a field technician, whose job must also be founded on professional skills (Fernández García *et al.*, 2020).

1.18 ROOFTOP RAINWATER HARVESTING (RRWH) TECHNOLOGIES

In different places of the world, many traditional methods of rainwater gathering were employed over the years. In dry rangelands, animal nutrition is lowered more by a paucity of dry farmlands, drinking water is more important than a scarcity of forage, and the abilities of animal nutrition are lowered. Rainwater harvesting (RWH) is a long-standing technique of water conservation used in areas under water stress to supplement drinking and non-drinking water sources. After the fall in RWH system development in the past 100 years, renewed attention has arisen as a result of increasing demands on water due to the increasing population, development, climatic variability, and food safety since the latter half of the twentieth century (König and Sperfeld, 2007). Collecting water from the roofs and other areas of the rainfall is used for rainwater (RRWH). Water is channelled into storage tanks or cisterns via gutters and pipes. Stone, bamboo, galvanised iron boards, and poly Vinyl chloride (PVC) are used to construct the pipes and gutters. Usually, iron sheets are constructed of tanks. Dust, leaves, sand, insects, or bird droppings might obstruct pipes and all conveying ducts, so that filters are mounted on the roof near the entrance to them. A spigot, manual bucket, or water pump should be used to empty the water tank (Lim and Jiang, 2013). If properly developed, water management systems will reduce organic matter and clay deposition on rocky and textured soils on flood plains and improve the physical and chemical qualities of soils and plant cover. Incorporating rainwater collection and grey water reuse technologies can lead to substantial water conservation. There have been studies done to analyse the possibility of conserving drinking water by using rainfall and grey water for housing and residential buildings with water savings in houses of 36 per cent and in multi-story structures of 42 per cent (Ghisi and Ferreira, 2007; Teh et al., 2015). If properly developed, water management systems will reduce clay and matter deposition on rocky and textured soils on flood plains and improve the physical and chemical qualities of soils and plant cover. On the other hand, flood management technologies would improve water infiltration into deep soil and retention of soil thus providing cover for plants (Abdollahi et al., 2015). Maintaining these structures attentively helps in erosion control and soil deterioration prevention methods (Qadir et al., 2007). Various experiences with various irrigation systems have led to advances in crop yields and water efficiency across the world, as well as their contribution to reducing climate change impacts on agriculture (Bafdal and Dwiratna, 2018).

1.19   DISCUSSION

In the economic progress of the country, water services play a significant role. Water supplies are interconnected with all types of companies during the construction phase (Ma, 2020). It is anticipated that approximately 70 per cent of fresh water abstraction will irrigate 25 per cent of the total global cultivated land (399 million hectares), which provides 45 per cent of global foodstuffs (Thenkabail et al., 2011). Today's water conservation issues differ in size, breadth, and design from those of the past. Activity standards and learning approaches therefore need to adapt in order to tackle these developing realities. The irreducible level of ambiguity and sophistication inside socially ecological systems are not sufficiently matched by modern activities and old planning methods. In short, planners, engineers and policy-makers are insufficient for solving problems to meet difficulties (Medema et al., 2014). A widely applicable, quantitative, repeatable tool for connecting hydrology and aquatic ecology characteristics defines and uses the lens-like activity of a system based on purposeful physical parameters and associated ecological indicators whilst integrating time-variable components of surface water sources. The main cause of water savings is the improvement in performance. The impact of land use improvement on surface water conditions should be understood (Liu *et al.*, 2020). Due to weakness of faulty irrigation and field equipment, crops cannot fully use all the water supply that flows into farming systems. In the future, a comprehensive combination of generic efficiency, irrigation quality, and water productivity can be used to calculate the quantitative connection among regional water use, conversion, usage and crop output (Cao *et al.*, 2018). Water Scarcity Pinch Analysis (WSPA) outlines water shortages as well as recommendations to reduce water shortages in terms of quality and volumes of waters. WSPA takes water quality and scarcity into account as well as using gradient water quality methodologies for calculating minimum water shortages. Adequate water supply would provide substantial growth in economy because water resources are crucial to everyday lives (Jia *et al.*, 2020). Climate alteration and the variation of RWH is a 100-year-old water delivery system needed to address ever-growing water demand (Rahman, 2017; Hanson and Vogel, 2014). The RWH system has several other advantages aside from saving electricity. In this respect, Melville-Shreeve *et al.* (2016) used a theoretical multi-criteria investigation to evaluate the RWH structure in the UK against a variety of new criteria. A series of RWH system architectures have been presented in terms of benefits and costs, which outperform conventional RWH systems. The research findings will lead to the economic introduction of a RWH plan in the UK. The usage of RWH in urban settings becomes more and more important. Liang and Van Dijk (2016) looked into the part of non-technical aspects of RWH for cultivation and irrigation in Beijing. The use of RWH method for irrigation in urban areas gains prominence. Liang and Van Dijk (2016) searched the role of farm irrigation in Beijing of non-technical variables at RWH. The study focuses on water resources and water conservation measures so that only minimum optimization is required for the management of water bodies.

1.20 CONCLUSION

Water is the most crucial and sustainable natural resource, yet it is greatly in jeopardy as demand rises while supply falls. With increase in demand due to rising industrialization, globalization, urbanization, and other similar trends, which have been aggravated by increased population water consumption, resulting in various sorts of wastes and sewerages water. Water is available for use from various sources, such as surface water and ground water, and can be broadly classified further into lentic, lotic, river system, and stream waters. With the jeopardy of natural available forms of water, the quality is greatly affected by depletion in the availability. To reduce the depletion of water and mismanagement, proper strategies need to be adopted for sustainable management of water resources. Alternatives need to be derived for replacement of freshwater in various area such as to create a balance with the availably of freshwater. One such alternative is grey water. Household water waste from basins, toilet, dishwashing can be used as an alternative to freshwater. The water shortage due to water table depleted in various areas globally, some conservation techniques practices will also help in fulfil the growing demand for water in water scarcity areas. To improve the WUE in agricultural farms, adaptation of rooftop RWH and better irrigation scheduling are some of the techniques which would help to achieve sustainable water management. Jordan valley has been using waste water along with storm water. Approximately 100 million m^3 of waste water has been used as alternative in the valley. Through this, the use of grey water and scarcity of water can be reduced to some extinct and be managed for sustainable management of water globally.

ACKNOWLEDGEMENT

We wish to express our heartfelt gratitude to the division of Soil and Water Engineering, and Division of Irrigation and Drainage, College of Agricultural Engineering and Technology, Sher-e-Kashmir University of Agricultural Sciences and Technology.

REFERENCES

Abdollahi, V., Zolfaghari, F., Jabbari, M., Dehghan, M. and Rafie, M. 2015. Investigating the effect of crescent construction on some parameters of vegetation and soil in Saravan Rangelands (Sistan and Baluchestan Province). *Iranian Range and Desert Research* 22(4): 675–667.

Aker, J.C., Ghosh, 1 and Burrell, J. 2016. The promise (and pitfalls) of ICT for agriculture initiatives. *Agricultural Economics* 47(S1): 35–48.

Alagirisamy, B. 2021. Greywater recycling and utilization. *AgroScience Today* 2(2): 66–72.

Allan, R.P. 2011. Human influence on rainfall. *Nature* 470: 344–345.

Arnell, N. W. 2004. Climate change and global water resources: SRES emissions and socio-economic scenarios. *Global Environmental Change* 14(1): 31–35.

Bafdal, N. and Dwiratna, S. 2018. Water harvesting system as an alternative appropriate technology to supply irrigation on red oval cherry tomato production. *International Journal on Advanced Science, Engineering and Information Technology* 8(2): 561–566.

Beal, C.D., Stewart, R.A. and Fielding, K. 2013. A novel mixed method smart metering approach to reconciling differences between perceived and actual residential end use water consumption. *Journal of Cleaner Production* 60: 116–128.

Boutraa, T. 2010. Improvement of water use efficiency in irrigated agriculture: A review. *Journal of Agronomy* 9(1): 1–8.

Cao, X., Ren, J., Wu, M., Guo, X., Wang, Z. and Wang, W. 2018. Effective use rate of generalized water resources assessment and to improve agricultural water use efficiency evaluation index system. *Ecological Indicators* 86: 58–66.

Chukalla, A.D., Krol, M.S. and Hoekstra, A.Y. 2018. Grey water footprint reduction in irrigated crop production: Effect of nitrogen application rate, nitrogen form, tillage practice and irrigation strategy. *Hydrology and Earth System Sciences* 22(6): 3245–3259.

Clothier, B., Green, S. and Deurer, M. 2010, August. Green, blue and grey waters: Minimising the footprint using soil physics. In *19th World Congress of Soil Science, Soil Solutions for a Changing World*: 1–6.

Dalezios, N.R., Angelakis, A.N. and Eslamian, S. 2018. Water scarcity management: Part 1: Methodological framework. *International Journal of Global Environmental Issues* 17(1): 1–40.

Fernández García, I., Lecina, S., Ruiz-Sánchez, M.C., Vera, J., Conejero, W., Conesa, M.R., Domínguez, A., Pardo, J.J., Léllis, B.C. and Montesinos, P. 2020. Trends and challenges in irrigation scheduling in the semi-arid area of Spain. *Water* 12(3): 785.

Ghisi, E. and Ferreira, D.F. 2007. Potential for potable water savings by using rainwater and greywater in a multi-storey residential building in Southern Brazil. *Building and Environment* 42(7): 2512–2522.

Giorgi, F., Im, E.S., Coppola, E., Diffenbaugh, N.S., Gao, X.J., Mariotti, L. and Shi, Y. 2011. Higher hydroclimatic intensity with global warming. *Journal of Climate* 24(20): 5309–5324.

Hanson, L. S. and Vogel, R. M. 2014. Generalized storage–reliability–yield relationships for rainwater harvesting systems. *Environmental Research Letters* 9: 075007. DOI: 10.1088/1748-9326/9/7/075007

Hoekstra, A.Y. and Mekonnen, M.M. 2012. The water footprint of humanity. *Proceedings of the National Academy of Sciences* 109(9): 3232–3237.

Jafari, M., Tavili, A., Panahi, F., Zandi Esfahan, E. and Ghorbani, M. 2018a. Management of water resources. In *Reclamation of arid lands (Environmental science and engineering)*. Cham: Springer.

Jafari, M., Tavili, A., Panahi, F., Esfahan, E.Z. and Ghorbani, M. 2018b. *Reclamation of arid lands*. Springer Nature Switzerland AG: Springer International Publishing.

Jia, X., Klemes, J.J., Alwi, S. and Varbanov, R.W. 2020. Regional water resources assessment using water scarcity pinch analysis. *Resources, Conservation and Recycling* 157: 104749.

Junk, W. 1970. Investigations on the ecology and production biology of the 'Floating Meadows' (Paspalo-Echinochloetum) on the Middle Amazon. I: The floating vegetation and its ecology. *Amazoniana* 2: 449–495.

Junk, W.J., Wittmann, F., Schöngart, J. and Piedade, M.T. 2015. A classification of the major habitats of Amazonian black-water river floodplains and a comparison with their white-water counterparts. *Wetlands Ecology and Management* 23(4): 677–693.

Koech, R. and Langat, P. 2018. Improving irrigation water use efficiency: A review of advances, challenges and opportunities in the Australian context. *Water* 10(12): 1771.

König, K.W. and Sperfeld, D. 2007. Rainwater harvesting – A global issue matures. *Fachvereinigung Betriebsund Regenwassernutzung eV Disponible el* 25(09): 2015.

Lankford, B. 2012. Fictions, fractions, factorials and fractures; on the framing of irrigation efficiency. *Agricultural Water Management* 108: 27–38.

Li, J., Jiao, X., Jiang, H., Song, J. and Chen, L. 2020. Optimization of irrigation scheduling for maize in an arid oasis based on simulation-optimization model. *Agronomy* 10(7): 935.

Liang, X. and Van Dijk, M.P. 2016. Identification of decisive factors determining the continued use of rainwater harvesting systems for agriculture irrigation in Beijing. *Water* 8(1): 7.

Lim, K.Y. and Jiang, S.C. 2013. Reevaluation of health risk benchmark for sustainable water practice through risk analysis of rooftop-harvested rainwater. *Water Research* 47(20): 7273–7286.

Liu, J., Yang, H., Gosling, S.N., Kummu, M., Flörke, M., Pfister, S., Hanasaki, N., Wala, Y., Zhang, X., Zheng, C. and Alcamo, J. 2017. Water scarcity assessments in the past, present, and future. *Earth's Future* 5(6): 545–559.

Liu, X., Zhang, G., Xu, Y.J., Wu, Y., Liu, Y. and Zhang, H. 2020. Assessment of water quality of best water management practices in lake adjacent to the high-latitude agricultural areas, China. *Environmental Science and Pollution Research* 27(3): 3338–3349.

Ma, L., 2020. Explore the role of sustainable utilization of water resources and water resources management. *IOP Conference Series: Earth and Environmental Science* 560(1): 012057. DOI: 10.1088/1755-1315/560/1/012057

Marleni, N., Gray, S., Sharma, A., Burn, S. and Muttil, N. 2015. Impact of water management practice scenarios on wastewater flow and contaminant concentration. *Journal of Environmental Management* 151: 461–471.

Medema, W., Light, S. and Adamowski, J. 2014. Integrating adaptive learning into adaptive water resources management. *Environmental Engineering & Management Journal (EEMJ)* 13(7):1819–1834, DOI: 10.30638/eemj.2014.201

Melville-Shreeve, P., Ward, S. and Butler, D. 2016. Rainwater harvesting typologies for UK Houses: A multi criteria analysis of system configurations. *Water* 8(4): 129.

Mishra, A., Kumar, R. and Richa, R. 2021. Biodiversity conservation to mitigate the impact of climate change on agro-ecosystems. In *Biological diversity: current status and conservation policies* (pp. 89–107), Volume 1, Eds. Kumar, V., Kumar, S., Kamboj, N., Payum, T., Kumar, P. and Kumari, S. Haridwar: Agro Environ Media, Publication Cell, India Agriculture and Environmental Science Academy.

Mushtaq, B., Bandh, S.A. and Shafi, S. 2020. *Environmental management: environmental issues, awareness and abatement*. Springer Nature.

Nakasone, E. and Torero, M. 2016. A text message away: ICTs as a tool to improve food security. *Agricultural Economics* 47(S1): 49–59.

Oron, G., Adel, M., Agmon, V., Friedler, E., Halperin, R., Leshem, E. and Weinberg, D. 2014. Greywater use in Israel and worldwide: Standards and prospects. *Water Research* 58: 92–101.

Pachkor, R.T. and Parbat, D.K. 2017. Assessment of works under Jalyukta Shivar Campaign–A case study of Pusad region. *International Journal for Research in Applied Science & Engineering Technology* 5(4): 1614–1619.

Palmer, M.A., Lettenmaier, D.P., Poff, N.L., Postel, S.L., Richter, B. and Warner, R. 2009. Climate change and river ecosystems: Protection and adaptation options. *Environmental Management* 44(6): 1053–1068.

Parwal, M. 2015. A review paper on water resource management. *International Journal of New Technology and Research* 1(2): 263701.

Qadir, M., Sharma, B.R., Bruggeman, A., Choukr-Allah, R. and Karajeh, F. 2007. Non-Conventional water resources and opportunities for water augmentation to achieve food security in water scarce countries. *Agricultural Water Management* 87(1): 2–22.

Rahman, A. 2017. Recent advances in modelling and implementation of rainwater harvesting systems towards sustainable development. *Water* 9(12): 959.

Schuetze, T. and Santiago-Fandiño, V. 2013. Quantitative assessment of water use efficiency in urban and domestic buildings. *Water* 5(3): 1172–1193.

Sharma, P. and Giri, A. 2018. Productivity evaluation of lotic and lentic water body in Himachal Pradesh, India. *MOJ Ecology & Environmental Sciences* 3(5): 311–317.

Shevah, Y. 2017. Challenges and solutions to water problems in the Middle East. *Chemistry and Water* 207–258. https://doi.org/10.1016/B978-0-12-809330-6.00006-4

Sioli, H. ed. 2012. *The Amazon: limnology and landscape ecology of a mighty tropical river and its basin*, Volume 56, Page 519, Springer Science & Business Media.

Stefanelli-Silva, G., Zuanon, J. and Pires, T. 2019. Revisiting Amazonian water types: Experimental evidence highlights the importance of forest stream hydrochemistry in shaping adaptation in a fish species. *Hydrobiologia* 830(1): 151–160.

Teh, X.Y., Poh, P.E., Gouwanda, D. and Chong, M.N. 2015. Decentralized light greywater treatment using aerobic digestion and hydrogen peroxide disinfection for non-potable reuse. *Journal of Cleaner Production* 99: 305–311.

Thenkabail, P.S., Hanjra, M.A., Dheeravath, V. and Gumma, M. 2011. *Global croplands and their water use from remote sensing and nonremote sensing perspectives* (pp. 383–419). FL: CRC Press, Taylor and Francis Group.

Wang, D.Y., Li, J.B., Ye, Y.Y. and Tan, F.F. 2015. An improved calculation method of grey water footprint. *Journal of Natural Resources* 30: 2120–2130.

Watts, G., Battarbee, R.W., Bloomfield, J.P., Crossman, J., Daccache, A., Durance, I., Elliott, J.A., Garner, G., Hannaford, J., Hannah, D.M. and Hess, T. 2015. Climate change and water in the UK – Past changes and future prospects. *Progress in Physical Geography* 39(1): 6–28.

Webb, B.W., Hannah, D.M., Moore, R.D., Brown, L.E. and Nobilis, F. 2008. Recent advances in stream and river temperature research. *Hydrological Processes: An International Journal* 22(7): 902–918.

Yazdanpanah, M., Feyzabad, F.R., Forouzani, M., Mohammadzadeh, S. and Burton, R.J. 2015. Predicting farmers' water conservation goals and behavior in Iran: A test of social cognitive theory. *Land Use Policy* 47: 401–407.

2 Water Management Issues and Challenges

Mukesh Kumar and Rohitashw Kumar

2.1 INTRODUCTION

Life cannot be sustained in this universe without water as it is the most important natural resource, which is essential for the survival of humans, animals, plants and other living organisms. The role of water in human development cannot be underestimated as it is required for each and every aspect of people's lives and livelihoods. The ever-increasing population and urbanization create pressure on water resources resulting in threat to climate change affecting the complex relationship between water and human beings. Water is the most widely distributed key resource to meet the basic needs of a growing population, social and economic ambitions, demanding agriculture, expanding urbanization, increasing industrialization and many other causes (United Nation's Brundtland Report, 1987). Drinking water is one of the basic needs for all. In India, about 5.76 lakh villages have been suffering with the problem of water scarcity. In many places, particularly in Rajasthan, Gujarat and Madhya Pradesh, women have to travel long distance to collect water. In certain pockets of Rajasthan, people have to wait for hours to collect one bucket of water from water tankers brought by the trains, trucks and tractors. Even in hilly areas, where rainfall is abundant, the situation is almost the same in these places. In many tribal and backward areas, during rainy season, un-hygienic and unsafe water is collected from ponds, tanks etc., which is the only source of drinking water.

Water has social significance as it plays an important role in social, economic and overall development. The demand for water in all sectors is continuously increasing and is becoming a big challenge to sustain this important natural resource in the coming times. Hence, water in all its forms (solid, liquid and gas) should be harnessed appropriately.

In present-day policy debates, water occupies a central position and it is increasingly argued that the growth and development of a society in particular and nation as a whole is determined by the sensitization of each and everyone about water. This chapter discusses the significance of water harvesting conservation and effective ways, policy and government initiatives to deal with the question of sustainable management of water resources in the 21st century.

2.2 WATER MANAGEMENT: AN OVERVIEW

Water is a valuable input to sustain human, animal as well as plant life on this planet. Water is the primary natural resource and plays a significant role in nation development such as socio-economic development as well as maintaining healthy ecosystems.

DOI: 10.1201/9781003351672-2

In this universe, water covers about two third earth's surface and out of the available water, maximum (around 97%) water is saline which is not potable. Only 3% of water is fresh water, of which just 1% is accessible for different purposes of human use, i.e. domestic, industry, agriculture etc.

The importance of this natural resource is mentioned in different literatures, including religious books. Water is a phenomenal element of life and is one of the five *panchmahabhutas*. The need and significance of water and its conservation have been mentioned in Indian literature belonging to different religions, including Hinduism, Buddhism, Jainism and other traditions. Water is defined in ayurvedic literature as *jiva* (life) that has countless treatments for life. However, this tonic of life is under serious stress and is becoming scarce due to various reasons, including ever-growing population, urbanization, changing lifestyle etc.

We all are aware of the fact that all the civilizations in this universe have moved around water, i.e. appeared on the banks of the rivers.

Water is the core required for human development. The water is required for different purposes in our day-to-day life for the sustainable development. Fig. 2.1 describes the purposes for which water is required under different levels.

Water resources' development, optimal use and management of water and maintenance are necessary for the sustainability of services from both existing and new developmental activities of all sectors. The ever-increasing population growth, change in food patterns and lifestyle, industrialization, urbanization, globalization and improper water use have resulted in serious problems of usable water in different parts of the globe, particularly in developing countries like India. The declining trend in the per capita water availability is continuous. Fig. 2.2 shows the trending annual per capita water availability in India.

It is clear from Fig. 2.2 that in the year 1951, the per capita water availability was 5200 m^3, which has drastically reduced to 1545 m^3 in 2011 and may further reduce to 1140 m^3 in the year 2050. It shows that there is a considerable gap between the

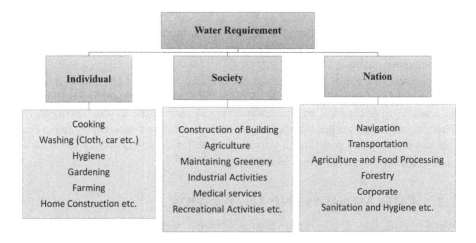

FIGURE 2.1   Water required for various purposes at different levels.

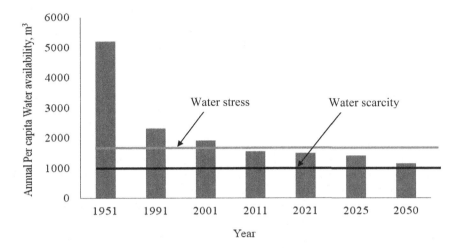

FIGURE 2.2   Trend in annual per capita water availability in India.

water demand-supply since the water requirement of different sectors is continuously increasing.

According to a NITI Aayog Report (2018), almost two lakh people die every year due to inadequate access to safe, good quality potable water and presently about 60 crore citizens of the country face high to extreme water stress conditions.

Keeping above in view, it is clear that there is an urgent need for adopting advanced and efficient methods for water harvesting, storage, conservation and management, including measures like rainwater harvesting, preventing pollution and management of surface and groundwater resources. It is envisaged that active participation of local people plays a pivotal role in management of this important natural resource.

2.3   SUSTAINABLE DEVELOPMENT GOAL ON WATER AND SANITATION

The growing urban and industrial water demand can be fulfilled by optimizing water use for agriculture by different techniques like micro-irrigation, on-farm water management, etc. The growing population needs more water for food production. This has resulted in increasing water requirements in many countries to fulfill the requirement for food production and processing. The problem is augmented by the alarming levels of water pollution from untreated urban and manufacturing waste and runoff from agricultural fields. The costs of treating urban and industrial wastes produced currently in developing countries far exceed the available investment funds. Solving this problem will require new technologies and business models (Molden, 2007; Amarasinghe & Smakhtin, 2014).

Water and sanitation have great impact on sustainability in agriculture. It is a multidimensional challenge with complex undertones, as water security and sanitation in an increasing concern both at national and international levels for sustainable development.

The Sustainable Development Goals (SDGs) are mainly focused to provide an improved water resource that involves building overhead tanks, tube wells or toilets for drinking and domestic water supply at micro level. The development of mega structures like dams and embankments at the macro level is required and maintained to supply water for irrigation to support agriculture/food production.

Most of the SDGs have positive interlinkages with SDG-6, which has eight targets and suggests not only supply of water to everyone but also implies that safe and good quality drinking water is available continuously. The main emphasis of SDG-6 is to increase water and sanitation access, which reduces poverty and ultimately have good relationship with health and education outcomes. There is a negative impact on ambient water quality and ecosystems due to agricultural activities (SGD-2) and energy services (SDG-7). The SDG-6 primarily focuses on Target 6.1, which states that "By 2030, achieve universal and equitable access to safe and affordable drinking water for all" and Target 6.2 i.e. "By 2030, achieve access to adequate and equitable sanitation and hygiene for all and end open defecation, paying special attention to the needs of women and girls and those in vulnerable situation".

The government of India had already taken several initiatives to accomplish the requirements of SDG-6. The current national policy and program in water and sanitation sector has been aligned with SGD-6.

2.4   WATER: ISSUES AND CHALLENGES

Water safety and security is a major concern and has direct effect on the sustenance of flora and fauna. Water occupies a central position in all policy debates now-a-days, both at national and international levels. It is more frequently argued that the development and progress of a society and nation is directly related to the sensitization of masses about different aspects of water use and management. There are many issues and challenges w.r.t. water at state, national and international levels. There is a need for effective ways, policy and government initiatives to deal with the question of sustainable management of water resources in the 21st century. In this section, the issue of water management at state, national and international levels is discussed in detail.

Indian governance model is based on a federal system and subjects are identified for Centre and State to take policy decisions. Water is listed under state list and different state governments take a call on water issues from time to time. When we are discussing issues and challenges in our country with respect to water from the development point of view, it is important to take note that almost all states are stuck in legal battles for water.

2.4.1   Interstate River Water Dispute

It is the most aggressive issue on water with different state governments in the Indian federalism today. This type of dispute arises when two or more states argue on the use, distribution, and control of rivers for sharing water from rivers common to them. Interstate river water disputes are governed by the Interstate Water Disputes Act, 1956. A list of few Interstate Water Disputes and different states are presented in Table 2.1.

TABLE 2.1

Few Interstate Water Disputes and Different States

Sl. No.	Interstate Water Disputes	States Involved
1.	Narmada Water Dispute	Gujarat, Maharashtra, Madhya Pradesh and Rajasthan
2.	Mahi River Dispute	Gujarat, Rajasthan and Madhya Pradesh
3.	Ravi and Beas Water Dispute	Punjab, Haryana, Himachal Pradesh, Rajasthan, Jammu and Kashmir and Delhi
4.	Satluj-Yamuna Link Canal Dispute	Punjab, Haryana and Rajasthan
5.	Yamuna River Water Dispute	Uttar Pradesh, Haryana, Himachal Pradesh, Punjab, Rajasthan, Madhya Pradesh and Delhi
6.	Karmanasa River Water Dispute	Uttar Pradesh and Bihar
7.	Barak River Water Dispute	Assam and Manipur
8.	Cauvery Water Dispute	Tamil Nadu, Kerala and Karnataka
9.	Krishna Water Dispute	Maharashtra, Karnataka and Andhra Pradesh
10.	Tungabhadra Water Dispute	Andhra Pradesh and Karnataka
11.	Aliyar and Bhavani River Water Dispute	Tamil Nadu and Kerala
12.	Godavari River Water Dispute	Andhra Pradesh, Odisha, Chhattisgarh, Karnataka, Madhya Pradesh

First, the concerned states try to resolve the issues with discussion or negotiation, but if they are unable to resolve, then the Central Government constitutes a water dispute tribunal for resolving the water dispute as per the Inter-State River Water Disputes Act, 1956. There are many tribunals in the country to resolve water disputes between different states. Few active tribunals are listed in Table 2.2.

It is clear from the above that there are long pending issues between different states like Karnataka, Tamil Nadu, Maharashtra, Telangana and Andhra Pradesh such as Krishna water dispute. Delhi, Haryana and Uttar Pradesh have to target each other every year for the sake of water. The situation in Gujarat and Maharashtra and most other states is not great either. There is an urgent need to sort out this policy paralysis to overcome the water crisis in India. Some experts have argued that water should be included in the concurrent list so that both Centre and State can take decisions on water issues in a harmonious and progressive way.

TABLE 2.2

Tribunals in the Country to Resolve Water Disputes between Different States

Tribunals	States Involved	Year of Formation
Mahanadi Water Disputes Tribunal	Chhattisgarh and Odisha	2018
Vansadhara Water Disputes Tribunal	Odisha and Andhra Pradesh	2010
Mahadayi Water Disputes Tribunal	Karnataka, Goa and Maharashtra	2010
Krishna Water Disputes Tribunal II	Andhra Pradesh, Maharashtra, Telangana and Karnataka	2004
Ravi & Beas Water Tribunal	Rajasthan, Haryana and Punjab	1986

The above-mentioned tribunals exist to resolve state-level issues/disputes related to water particularly sharing water flowing in the rivers in different states. The Union Government has always tried to resolve the disputes at the earliest. However, large establishment and permanent office space and infrastructure are needed to set up a separate Tribunal for each water dispute and it consumes a good time to give decisions. Inter-State River Water Disputes (Amendment) Bill, 2019 was introduced by the Union Government to overcome this issue. The Bill 2019 proposes a Dispute Resolution Committee set up by the Central Government for amicably resolving interstate water disputes within 18 months. Any dispute that cannot be settled by negotiations would be referred to the tribunal for its adjudication.

In view of major issues and challenges, it is relevant to point out that per capita water availability has shown an unswerving declining trend over the years in the country. We have already discussed the trends of water availability in Section 2.2. According to NITI Aayog report (2018), India is ranked 120 among 122 countries on the Water Quality Index.

The problem of water does not end at state or national level; it also exists at international level. The country is facing the water security challenges at international level and potentially affected by the water-related actions with neighboring countries, including China, Nepal, Pakistan and Bangladesh on the Indus and the Ganges, respectively.

Water is also a politically contested issue in South Asian countries. These countries are facing acute water shortage and difficulties of good & safe water. The region is facing issues of increasing demands on energy and water with rapid industrialization.

Over-extraction of groundwater is a very important issue because around 23 million pumps are installed and used in different South Asian countries, including Bangladesh, India, Nepal and Pakistan.

2.4.2 GROWING WATER DEMAND

The geographical area of country is about 329 m ha, which is 2.4% of the world's land area and supports about 16% of the world population with only 4% of the world's fresh water available in the country. In India, main source of fresh water is precipitation mainly in the form of rainfall or snowfall. Water cycle or hydrologic cycle is the entire unending process of circulation and redistribution of water by the atmosphere and the earth.

As discussed in Section 2.2, the country's per capita water availability has touched the water-stressed benchmarks, i.e. annual per capita renewable freshwater availability is less than 1700 m^3 and expected to decline further toward water-scarce conditions (annual per capita water availability is less than 1000 m^3) by 2050. As reported by the Water Resource Institute (WRI, 2019), India ranked 13th among the world's 17 "extremely water-stressed" countries and was under the category of "extremely high" levels of baseline water stress due to high population growth and rapid urbanization, which has exacerbated pressure on freshwater sources. According to the report published by the NITI Aayog anonymous (2018), country's' water demand is predicted to be twice the available supply by 2030.

It is clear from the perusal of various reports published recently that surface and groundwater resources have been exploited to the maximum and are fast depleting in many parts of the country. The groundwater level in most parts of the country has been depleting due to over pumping of groundwater for meeting requirement of high-water-requiring crops. In present-day scenario, demand of groundwater is increasing to fulfill the requirement of different sectors, particularly agriculture, industry and domestic use, resulting in depletion of the groundwater level at an alarming rate in many states like Rajasthan, Gujarat, Tamil Nadu, Punjab, Delhi and Haryana. Almost 60 and 40% of blocks in Punjab and Haryana, respectively, have turned dark due to overexploitation of groundwater. This is one of the biggest challenges in the country. It is, therefore, absolutely essential to optimally use available surface and subsurface water resources, particularly rainwater management.

2.4.3 WATER LOGGING

It can be defined as when soil is completely saturated with water and water stagnation takes place on flat land and low-lying areas. The major reasons for water logging are excess rainfall, floods, seepage high water table, obstruction to natural drainage, over irrigation etc. The major effects of water logging in agricultural fields are delayed sowing or less crop production, accumulation of salts on the soil surface resulting in salinity. Water logging is the second major challenge next to soil erosion.

2.4.4 WATER MANAGEMENT IN AGRICULTURE

Agriculture is the highest water-demanding sector as it consumes over 70% of fresh water followed by the industrial and domestic sectors. The agricultural sector is the main stay of economy. Agriculture consumes an enormous amount of water to produce crops: one to three cubic meters to yield just one kilo of rice, and 1000 tons of water to produce just one ton of grain. It is, therefore, of paramount importance to alter the existing cropping pattern from high to low water requiring crops and use the water resources, including rainwater, in the most optimal way by adopting efficient on-farm water management practices.

Therefore, it is extremely important to use water efficiently and conserve it in order to increase command area with the limited water resource. Efforts should be directed in order to use every drop of water efficiently to maximize crop production. Farmers should be forced to grow more water-efficient crops and adopt less wasteful irrigation techniques. The water use efficiency in agriculture is still not reached the significant level, even though we can say that it is low due to poor management and improper use of water resources resulting in severe water shortages, floods, droughts and associated economic losses.

Water use efficiency needs to be enhanced by adopting on-farm water management practices and using water optimally resulting in maximum productivity with minimum water. In irrigated agriculture, water needs to be used as efficiently as possible along with adoption of recommended packages of practices so that it results in maximum production. Presently, in rainfed agriculture, crop productivity is very low, which can be increased by maintaining or storing soil moisture in the root zone to maximize crop production. In India, almost two thirds of cultivated areas are rainfed. The crop productivity in such areas needs to be enhanced and sustained in order to meet the food requirement of ever-growing population of the country. In both irrigated and rainfed areas, highest efforts should be made to use existing water resources, including rainwater efficiently and optimally and conserve soil moisture for long-term agriculture use.

2.4.5   WATER SECURITY

In the present-day scenario, water security is a serious concern and is discussed in all the debates at state, national and international levels. Achieving "water security" is considered a prerequisite to long-term sustainable development. However, most definitions of "water security" are imprecise and qualitative. At household level, water security is generally defined as "access by all individuals at all times to sufficient safe water for a healthy and productive life". This includes three key dimensions: water availability, access and usage, similar to the food security concept (Ringler et al., 2016). Water security at the national level has been defined as "the availability of an acceptable quantity and quality of water for health, livelihoods, ecosystems and production, coupled with an acceptable level of water-related risks to people, environments and economies" (Grey & Sadoff, 2007). The various regions within countries that have not achieved water security are usually the poorest and most vulnerable. Failure to minimize risks posed by water-related disasters (floods and droughts) leads to cycles of serious economic loss compounded over time. Reducing those risks and ensuring a basic supply of water for productive and other purposes creates a firmer foundation for the development. "Har Ghar Nal se Jal" is one of the most important initiatives of the Government of India to fulfill the objective of water security and safety.

2.4.6   WATER SAFETY

The availability of good quality water is a serious problem not only in rural areas but it is equally critical in urban areas particularly big cities in India. Almost 50% of population does not have access to good quality drinking water even after almost eight decades of independence. All civilizations not only lived along the rivers but thrived and prospered as well when the water bodies including groundwater were far cleaner and uncontaminated, but the modern civilization is reversing the trends.

Surface and subsurface water quality is deteriorating due to water pollution, which is posing a serious concern to the environment, human & animal health all over the world, particularly in the developing countries. Pollution is the by-product of misuses of resources and developmental activities and has not only become a major concern all over the world, but it has become the critical issue in developing countries. Deterioration of water quality (surface and groundwater) due to unplanned urbanization, industrialization, modernization, changing lifestyles etc., untreated or partially treated industrial effluent discharged, improperly managed landfills, poor sanitation and other pollution sources like fertilizers and pesticides from the agricultural sector is an area of grave concern (Anonymous, 2019). Water pollution is posing a serious threat to the environment, which has direct impact on human and animal health. The water

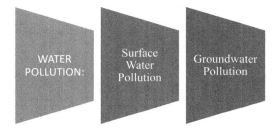

FIGURE 2.3 Type of water pollution.

bodies are getting polluted and contaminated by untreated sewage water, leaching of agrochemicals from the agricultural fields, industrial effluents etc. The situation has worsened to the extent that almost 70% of rivers and streams not only in our country but also all over the world contain polluted water. Salinity and arsenic contamination in groundwater are big concerns in South Asian countries. It is reported that salinity and arsenic contamination affect over 60% of groundwater in the Indo-Gangetic plain. Only 37% of the municipal sewage and 62% of the industrial effluents are treated in India (MoEFCC, 2019). Many districts in the country have contaminated water sources, thereby affecting human health at large (Anonymous, 2019).

Rapid urbanization, industrialization and change in lifestyle have further worsened the freshwater sources leading to reduced water availability. Water pollution is posing a severe threat to the environment and consequently human and animal health. Water pollution in most rivers and groundwater countries needs to be controlled urgently in order to keep our environment healthy and clean. Water pollution can be classified as surface and groundwater pollution, discussed in the following sections (Fig. 2.3).

2.4.6.1 Surface Water Pollution

Rivers and small water bodies in the local areas are the lifeline of the people and need to be kept pollution free. Most of the rivers are polluted due to regular release of large quantities of human, animal and untreated industrial wastes. Surface water pollution can be defined as water on the surface bodies polluted by physical, chemical and biological pollutants. A suspended material is one of the physical pollutants causing pollution.

The sources of water pollution are discussed in the following sections (Fig. 2.4).

2.4.6.1.1 Natural Sources

- **Siltation:** It is due to a natural phenomenon in most water bodies by which silt, sand and mineral particles deposit resulting in considerable reduction of capacity of water body like rivers, lakes and reservoir.
- **Fluorides Content:** High concentration of fluoride ions (F) in natural water is a serious health risk that affects bones, joints and teeth. Fluorosis disease in human beings can occur when fluoride concentration is beyond 0.7 ppm.
- **Arsenic Content:** High concentration of arsenic content in water is critical for animal and human health. Arsenicosis (arsenic poisoning), caused by using water with high arsenic content, is a type of chronic disease that affects different parts of the human body.

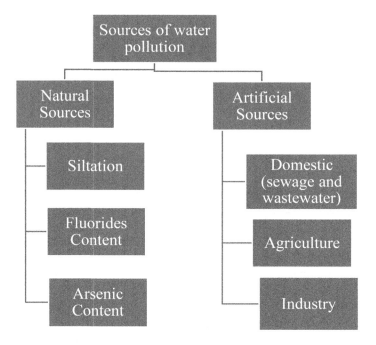

FIGURE 2.4   Sources of water pollution.

2.4.6.1.2 Artificial Sources

Domestic (sewage and wastewater), agricultural and industrial wastes are the major sources of polluting the surface water bodies, *viz.* rivers, lakes and streams.

The following are the main sources of water pollution:

- Application of agrochemicals, *viz.* fertilizers, pesticides and weedicides, are the major groundwater pollutants. These chemicals applied to the soil are leached through rain or irrigation water.
- Indiscriminate disposal of effluents from industries and urban sewerage further aggravates the problem due to reduction in oxygen content.
- Similarly, presence of excess quantity of arsenic, fluoride and iron in groundwater more than the permissible limits make water unsuitable for drinking purposes.
- Excessive use of synthetic detergent for washing purposes produces a lot of foam and pollutes water.
- Wastes containing organic materials may produce excess nitrate or growth of algae and bacteria, which leads to clogging of infiltration surfaces during recharge.
- Sharp increase in population and industrialization are causing severe stress on groundwater availability.

The water and sanitation sectors are highly interlinked. The sanitation sector deals with safe disposal of human waste, wastewater management, solid waste management, water supply, control of vector-borne diseases and domestic and personal hygiene. Poor sanitation and unsafe hygiene practices have impacted health by the manifestation of disease and infections, notably diarrhea and Respiratory Tract Infections (RTIs) resulting in higher morbidity and mortality, particularly among children (Azupogo et al., 2019).

Thus, it is vital to control water pollution for sustaining water resources not only to provide safe water but also a healthy and good life for every citizen of the country.

2.5 EFFECT OF CLIMATE CHANGE ON WATER RESOURCES

Climate change leading to rise in water temperature, increase in evaporation rates, high intensity and frequent rainfall, advanced and shorter runoff seasons and decrease in water quality (both inland and coastal areas) is significantly affecting water resources. The climate change has caused significant variations in the inter-annual and inter-seasonal variability of monsoon rainfall resulting in stress on water resources in the country. Overall, it may cause more severe droughts and floods and limited water access for people. Due to the impact of climate change, many factors lead to reduction in the amount of safe water availability in the river basin particularly due to changing patterns of water flow in different river basins.

Climate change may cause a serious impact on water cycle by influencing when, where and how much precipitation occurs in terms of rainfall and snowfall. More evaporation from the large water bodies due to increasing global temperatures resulted in higher levels of atmospheric water vapor causing more frequent, heavy and intense rains in the coming times. These changes in the form and timing of precipitation and runoff, precisely in snow-fed basins, may cause more frequent droughts, particularly during summer. Recent research reports have revealed that these changes are already taking place in the western United States. Changes in precipitation and runoff are the serious concerns for the water managers to manage a good number of situations, such as rural and urban water supply, water supply for irrigation, hydropower generation and drought and flood protection.

Climate change may also affect water quality in both inland and coastal areas because of more frequent high-intensity rainfall events that may cause more runoff and erosion resulting in additional sediments and chemical runoff transported into streams and groundwater. Water quality may further deteriorate as a result of decrease in water supply leading to more concentrations of nutrients and contaminants.

Rising air and water temperatures will also impact water quality by increasing primary production, organic matter decomposition and nutrient cycling rates in lakes and streams, resulting in lower dissolved oxygen levels. Lakes and wetlands associated with return flows from irrigated agriculture are of particular concern as they may increase the number of water bodies in violation of today's water quality standards, worsen the quality of water bodies that are currently in violation and ultimately increase the cost of meeting current water quality goals for both consumptive and environmental purposes.

2.6   RAINWATER HARVESTING AND CONSERVATION

Rainwater harvesting is not only important to fulfill the demand of water in different sectors but also to sustain water resources for future requirements. Rainwater harvesting is the process by which excess rainwater is harvested, collected, conserved and stored to use in the best possible manner to control runoff, evaporation and seepage.

2.6.1   Objectives of Rainwater Harvesting

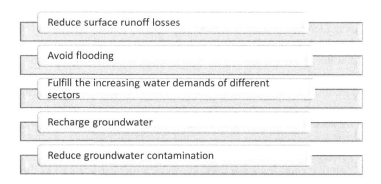

Reduce surface runoff losses

Avoid flooding

Fulfill the increasing water demands of different sectors

Recharge groundwater

Reduce groundwater contamination

The rainwater harvesting is not only done for groundwater recharge but also can be stored on surface, which can be used for different domestic agriculture and many day-to-day purposes. The basic purpose of artificial groundwater recharge is to replenish depleted groundwater.

2.6.2   Is Rainwater Harvesting Practiced in Ancient Times?

Yes, in India, the usable surface and subsurface water availability are not adequate in terms of quality and quantity to meet water requirements of different sectors like community, health, industry and agriculture. Rainwater harvesting for direct use and groundwater recharge is a common practice since ancient times. A good number of traditional methods in various parts of the country, particularly in arid regions, are still used for rainwater harvesting. At present, different cities are facing severe water crises particularly lack of availability of safe drinking water and the scenario is even more dismal in rural areas. Water requirement is likely to further increase in future due to constant increase in population, urbanization, industrial growth and change in lifestyle.

In order to fulfill the demand of increased food production in the country, the water availability will have to be stepped up. The water scarcity is going to pose a serious threat to the survival of humankind if remedial steps are not taken to ensure reliable water supply. This clearly illustrates the relevance and importance of rainwater harvesting to enhance water availability and recharge the depleting groundwaters in the country to meet water demand of different sectors.

2.6.2.1   Main Advantages of Rainwater Harvesting

- Free storage space in the ground for artificial groundwater recharging;
- controls soil & water erosion because of minimum runoff;
- negligible evaporation losses;
- improve groundwater quality by dilution of salts;
- no adverse social impacts for construction of storage tanks such as displacement of population;
- controls flood and mitigates effects of drought;
- increases productivity of aquifers; and
- environmental friendly process.

2.6.3   MODERN RAINWATER HARVESTING

Safe and secure water availability is a major concern in the urban areas due to searching for livelihood and rehabilitation of population from the rural to urban areas. Groundwater is the main source of water in most of the cities, which is supplied in most of the cities only for few hours. Groundwater level is falling day by day in most of the cities in the country as these areas are mainly dependent on groundwater. The problem is further compounded by the fact that most of the open land surfaces in urban are covered by concrete structures that do not allow rainwater to infiltrate (entry of water from the atmosphere to the soil) resulting in hindrance to groundwater recharge. The rooftop rainwater harvesting is considered to be an effective method and a good solution for this problem. The rooftop rainwater harvesting is the method in which rainwater from the rooftops is collected and stored in the underground tank or surface tanks for future use or diverted through dug wells or pits for groundwater recharge, which otherwise flows down in the drain. In many cities, this practice is followed.

2.7   MAJOR SCHEMES AND INITIATIVES BY THE GOI

Water is a state subject. Water resources development & management are planned and maintained by the State Governments themselves as per their own resources and urgencies. In India, Union Government also provides technical and financial assistance to State Governments in order to supplement their efforts, to encourage sustainable development and efficient management of water resources through various schemes and programs. In order to overcome and mitigate the water scarcity stress, governments are taking remedial measures at the national level and also in collaboration with inter-governmental organizations such as SAARC and UN. We are going to discuss briefly the important initiatives of Government of India that have made significant impact from the Indian Perspective.

It is relevant to discuss at this point of time that Indian water management and conservation started during the colonial phase and several schemes were initiated, which have strengthened the water security in the country. Prime Minister Sh. Narendra Modi in 2016 highlighted the role and contribution of Dr. Ambedkar in creation of two powerful technical organizations, *viz.* the Central Waterways, Irrigation and Navigation Commission (CWINC) and the Central Technical Power Board, which were the forerunners of the present-day Central Water Commission and Central Electricity Authority. He emphasized that Dr. Ambedkar's pioneering role in securing the water resources in our country cannot be understated.

In independent India, the Central Water and Power Commission was put in place (1952–56), which initiated various plans and schemes for groundwater conservation, irrigation development and drinking water. Indian government has adopted all major international Treaties and Conventions announced by UN agencies from time to time. National Rural Drinking Water Programme (NRDWP) was launched in 2009 with an aim to provide safe and adequate water for drinking, cooking and other domestic needs to every individual, particularly in rural areas on a sustainable basis. Its major objective was to provide all rural habitations, government schools and anganwadis access to safe drinking water.

As you must have noticed, India is microcosm of the whole world, which has coastal areas on the one hand and deserts on the other. Therefore, the nation requires a comprehensive plan and strategy to ensure seamless water supply for its large-scale population and agricultural needs. To address this need, the Government of India has initiated Namami Gange Programme which is an Integrated Conservation Mission, approved as "Flagship Programme" by the Union Government in June 2014 with budget outlay of Rs.20,000 Crore to accomplish the twin objectives of effective abatement of pollution, conservation and rejuvenation of National River Ganga. The program seeks to cater to the medium and long-term needs of the country with a focused approach on sewerage Treatment Infrastructure, River-Front Development, River Surface Cleaning, Bio-diversity, Afforestation, Public Awareness, Industrial Effluent Monitoring etc.

To fulfill the objectives of sustainable development and efficient management of water resources, the Government of India has started the Pradhan Mantri Krishi Sinchayee Yojana (PMKSY) during 2015–16. The PMKSY mainly concentrate on water sources, distribution network, efficient farm level applications, extension services on new technologies & information, etc., to give end-to-end solution in irrigation supply chain. Promotion of underground piping system, micro-irrigation (Drip & Sprinklers, pivots, rain-guns) and other application devices etc. within the farm. The main attentions of PMKSY on irrigation techniques are to:

• reduces the evaporation/percolation losses;
• improve the conveyance and field application water efficiency;
• create opportunities for recharging groundwater (shallow tube/dug wells) through scientific moisture conservation and runoff control measures;
• help farmer to have easy access to water resources; and
• create new water sources; repair, restore and renovate defunct water sources; construct water harvesting structures, secondary & micro storage, groundwater development, enhance potentials of traditional water bodies at village level.

During 2016–17, major/medium irrigation projects in the country have been given priority under PMKSY-Accelerated Irrigation Benefits Programme (PMKSY-AIBP) in consultation with States for completion in phases. The Government has provided funding mechanism that includes the Central and State share under Long Term Irrigation Fund (LTIF) through National Bank for Agriculture and Rural Development (NABARD).

The major objectives of Har Khet Ko Pani (HKKP)-Command Area Development and Water Management (CADWM) program are to:

- use Irrigation Potential Created (IPC) under the project soon after its creation;
- improve water use efficiency;
- increase agricultural productivity and production; and
- bring sustainability in the irrigated agriculture in a participatory environment.

The States are also encouraged to use the underground pipeline network instead of conventional field channels under Command Area Development and Water Management (CADWM) scheme which also promote the Participatory Irrigation Management (PIM).

The Surface Minor Irrigation (SMI) and Repair, Renovation and Restoration (RRR) of Water Bodies schemes have multiple objectives that include the following:

- expanding cultivable area under assured irrigation;
- improving water use efficiency;
- groundwater recharge;
- improvement and restoration of water bodies thereby increasing the tank storage capacity and revival of lost irrigation potential;
- increased availability of drinking water; and
- improvement of catchment of tank commands.

Central Ground Water Board is implementing innovative schemes for selected overexploited blocks of the Aspirational districts on pilot basis to Aquifer Rejuvenation under "Ground Water Management and Regulation" scheme. Water Harvesting and Recharge Augmentation are completed as a pilot project through construction of Bridge cum Bhandaras in districts of Wardha and Amravathi in Maharashtra. Further, a joint Action Plan has been prepared with the Ministry of Rural Development for effective implementation of water conservation and artificial recharge structures in convergence with Mahatma Gandhi National Rural Employment Guarantee Scheme (MGNREGS).

Groundwater Irrigation component of PMKSY-HKKP aims to provide financial assistance to States to provide assured groundwater irrigation for small and marginal farmers, in rainfed areas. This scheme is applicable only in areas having stage of groundwater development of less than 60%, average rainfall of more than 750 mm rainfall and with shallow groundwater levels (less than 15 m bgl). The scheme is effectively launched in 2019–20, after revision of operational guidelines of the scheme.

Department of Agriculture, Cooperation and Farmers' Welfare is implementing "Per Drop More Crop" component of PMKSY. It mainly focuses on increasing the water use efficiency at farm level through precision/micro-irrigation. This component of PMKSY also supports micro level water storage or water conservation/ management activities to supplement micro-irrigation besides promoting precision irrigation (drip and sprinkler irrigation system) and better on-farm water management practices to optimize the use of available water resources.

National Water Mission (NWM) has taken up Jal Shakti Abhiyan: Catch the Rain and Sahi Fasal campaigns for water conservation.

- "Jal Shakti Abhiyan: Catch the Rain" is under implementation in all districts, rural as well as urban areas, of the country with the main theme "Catch the Rain, where it falls, when it falls". The period of the campaign was from 22 March 2021 to 30 November 2021 during the pre-monsoon and monsoon periods in the country. The "Jal Shakti Abhiyan: Catch the Rain"-2021 campaign was launched by Hon'ble Prime Minister Shri. Narendra Modi with a view to take water conservation at grass-root level through people's participation to accelerate water conservation across the country.
- "Sahi Fasal" campaign was launched by NWM to nudge farmers in the water-stressed areas to grow crops that are not water intensive but use water very efficiently; and are economically remunerative; are healthy and nutritious; suited to the agro-climatic-hydro characteristics of the area; and are environmentally friendly.

When you are reading and analyzing these campaigns, it is important to add your own individual set of ideas into these campaigns and create local success stories. When we are discussing policies and initiatives, it is important to understand that the spirit of these policies has to be taken to the masses at large. If you take a stock of water-related program and their success, several experts have argued that we have failed to generate a massive public understanding and sensitization about water issues among common people at large. Therefore, for the welfare of the society in particular and nation as whole, we should join together and support the initiatives and Schemes of the Government for their effective implementation.

It is all the more important now that we are facing such a severe water crisis that we reach out to each and every citizen of the country and impress upon the need of water conservation. Some experts have pointed out that water conservation has to be made a part of Indian educational system right from the school curriculum so that our next generation is trained in the mission to save and conserve water on a day-to-day basis. We need to disseminate and propagate the idea of rainwater harvesting, water conservation and its optimal use at all levels through the mode of documentaries, movies and all other available media sources.

2.8 WATER AND EDUCATION

It is predicted that the next world war will be due to the issue of water. The water resources are depleting day by day and if we will not act immediately, we may fight with each other for potable water. Hence, every citizen in the country as well as in the world should be sensitized about the worst situation which we or our future generation may face. Hence, amalgamating the water management component in the education system right from the school education to the research degree program is essential to sustain this important natural resource for long duration.

Some of the institutions are already working in this area. For example, Indira Gandhi National Open University (IGNOU) is offering a six-month Certificate in Water Harvesting and Management program as well as one-year Diploma in Watershed Management program through Distance learning mode to create awareness and sensitize the masses regarding the water harvesting, conservation, storage, use and management.

The school education in India also focuses on the importance of water through value-education. The school children are taught about the water – sources, use, conservation, management as well as pollution. This not only sensitizes the children at early age about the importance of water in our lives but also equip them to follow measures to prevent misuse and conserve water.

Water literacy is an important aspect to respond to water-related sustainability challenges. For ensuring safer water availability and use, appropriate knowledge about various aspects of water use and management is vital. Therefore, it is essential to promote water literacy for acquiring knowledge, skill, and competencies at all levels of education.

2.9   CONCLUSION

Water plays a critical role in the Human Development and associated issues. In the course of discussion, we have taken up the various issues and challenges and suggested the solutions. Since the economy of the country depends on agriculture which consumes a big share of water, the on-farm water management including water modern ways of agriculture practices can help in minimize input and maximize output. The government of India has started various schemes in order to overcome and mitigate the water scarcity stress problem at the national level. The government has collaboration with inter-governmental organizations such as SAARC and UN particularly to solve the water scarcity and safety issues. In a nutshell, the authors have tried to underline the importance of water for the People's well-being besides its role in economic development.

REFERENCES

Amarasinghe, U.A., and Smakhtin, V. 2014. *Global water demand projections: Past, present and future.* Colombo: International Water Management Institute (IWMI). 32 p. (IWMI Research Report 156). doi: 10.5337/2014.212

Anonymous (2018). Composite Water Management Index: A Tool for Water Management. Niti Aayog Report. p. 180.

Anonymous (2019). National Compilation on Dynamic Ground Water Resources of India, 2017. Ministry of Ministry of Jal Shakti, Department of Water Resources, RD & GR Central Ground Water Board Faridabad, Govt of India. p. 306.

Azupogo, F., Abdul-Rahaman, F., Gyanteh, B., and Atosona, A.. (2019). Hygiene and sanitation practices and the risk of morbidity among children 6–23 months of age in Kumbungu district, Ghana. *Advances in Public Health* 2019: 1–12. https://doi.org/10.1155/2019/4313759

Grey, D., and Sadoff, C.W. (2007). Sink or swim? Water security for growth and development. *Water Policy* 9: 545–71.

Molden, D. (2007) *Water for food, water for life: A comprehensive assessment of water management in agriculture.* Earthscan and Columbo: International Water Management Institute, London, 40 p.

MoEFCC. (2019). National Status of Wastewater Generation and Treatment. 8 May 2019.

Ringler, C., Zhu, T., Gruber, S., Treguer, R., Auguste, L., Addams, L., and Cenacchi, N. (2016). Role of water security for agricultural and economic development – Concepts and global scenarios. In *Handbook on water security*, C. Pahl-Wostl, A. Bhaduri, and J. Gupta (Eds.). Aldershot: Edward Elgar Publishing Ltd,.

United Nation's Brundtland Report. (1987). Report of the World Commission on Environment and Development: Our Common Future. United Nations UN Documents: Gathering a Body of Global Agreements has been compiled by the NGO Committee on Education of the Conference of NGOs from United Nations web sites with the invaluable help of information & communications technology. p. 247.

World Resources Institute (WRI) (2019) Aqueduct tools, WRI Aqueduct Web: https://www.wri.org/aqueduct.

3 Soil and Water Conservation Concept

A Forward Step to Manage Groundwater Depletion and Greenhouse Gas Emission

Pradeep Kumar Verma, Ashish Kumar Yadav, Naushad Khan, and Sanjeev Kumar

3.1 INTRODUCTION

The fast pace of development and industrialization and intensive agriculture system, indiscriminate destruction of natural environment, more so in the last century, have altered the concentration of atmospheric gases that result in global warming. The gaseous composition of the atmosphere has experienced a notable change due to an increase in industrial discharges, fossil fuel burning, extensive deforestation and combustion of biomass as well as deviations in land use patterns and land management practices. These anthropogenic activities have resulted in an increased emission of gases, e.g. carbon dioxide (CO_2), methane (CH_4) and nitrous oxide (NO), popularly known as "greenhouse gases". Agriculture, Forestry and Other Land Use (AFOLU) activities accounted for around 13% of CO_2, 44% of methane (CH_4) and 81% of nitrous oxide (N_2O) emissions by human activities globally during 2007–2016, representing 23% (12.0 ± 2.9 GtCO$_2$eq year^{-1}) of total net anthropogenic emissions of greenhouse gases (GHGs). The natural response of land to human-induced environmental change caused a net sink of around (11.2 GtCO$_2$ year^{-1}) during 2007–2016 (equivalent to 29% of total CO_2 emissions). If emissions related to pre- and post-production activities in the global food system are included, the emissions are estimated to be 21–37% of total net anthropogenic GHG emissions. The process, generally referred to as the greenhouse effect, adds net energy to the lower atmosphere and, therefore, results in atmospheric warming. Greenhouse effect is one of chief causes of climate change on Earth. Global warming, in turn, leads to regional changes in climate-related parameters, such as temperature, rainfall, soil moisture and sea level. The extensive and frequent occurrence of climatic extremes, such as droughts, heat waves and floods, in the last decade in many parts of the world may be due to global warming. Soil moisture is one of the key factors affecting GHG production in agricultural soil. Optimal irrigation can reduce GHG emissions by regulating the N and carbon turnover process in soil via manipulating the level of soil moisture (Li et al., 2003).

DOI: 10.1201/9781003351672-3

Nitrification process in the soils easily takes place with moderate water contents, while more soil water contents restrict the amount of oxygen in the soil. However, oxygen levels continuously decrease in the soil with the amount of N_2O produced per unit of ammonia processed. It specifies that substantial amounts of N_2O can be formed by nitrification with low levels of oxygen, but more than zero (Kallenbach et al., 2010). Due to climate change, the more intense and erratic and scattered precipitation may be significant effects on hydrology and regional water resources availability, increased precipitation may lead to higher runoff. The rising CO_2 and climate change due to global warming directly affect both precipitation and evapotranspiration, accordingly, increasing the irrigation water demands. Moreover, the irrigation water requirements of the crops change as a result of climate change. Elgaali et al. (2007) studied the regional impact of climate change on irrigation water demand by considering rainfall and evapotranspiration in the Arkansas River Basin in southeastern Colorado. They assumed that the no change in crop phenology and found an overall increase in irrigation water demands due to climate change (Elgaali et al., 2007). De Silva et al. (2007) studied the impacts of climate change on irrigation water requirements in the paddy field of Sri Lanka and predicted an increase of 13%–23% in irrigation water demand depending on climate change scenarios (De Silva et al., 2007). The most common way to decrease water consumption and its contribution to GHG emission and climate change is to implement technical changes in the water-providing systems. Irrigation practices significantly influence GHG emissions due to their control of soil microbial activity and substrate supply. Irrigation increases crop productivity, but its accomplishment often increases operational energy demand and potentially GHG emissions (Smith et al., 2007). Understanding of GHG emissions is still poor, while new and efficient methods of irrigation, including drip and sprinkler irrigation, have been practiced to save a significant amount of water in irrigation. The timely and sufficient irrigation is a viable solution to boost agricultural production however, it not only varies soil biogeochemical characteristics but also alters the soil structure, which can unpleasantly influence soil carbon sequestration potential (Lal, 2004). Soil water plays a significant role in the release and consumption of GHGs that can be managed by methods of irrigation in most agricultural soils (Butterbach-Bahl et al., 2013; Stres et al., 2008). The more amount of water with frequent irrigations can increase the quantity of plant biomass and soil microbial activity resulting in more CO_2 and N_2O emissions compared to rainfed or non-irrigated soils (Fares et al., 2017). Higher water content in the soil accelerates microbial respiration of soil organic matter resulting in more CO_2 emissions (Trost et al., 2013). Table 3.1 presents the source of total GHG emission and their share.

3.2   WATER RESOURCES' STATUS IN INDIA

The geographical area of India is 3.29 million km^2, which contributes more than 18% of the world's population but has only own 4.2% of freshwater resources. The easily available freshwater for human being uses comprises water in lakes, river and groundwater. The total annual precipitation (including snowfall) is almost 4,000 billion cubic meters (BCM). However, physiographical, technological, socio-political,

TABLE 3.1

Source and Their Share towards Total GHG Emission

Source of GHG	Share of Total GHG Emission (%)
Direct emission from agriculture	10–12
N$_2$O from soil	4.2
CH$_4$	1.3
Biomass burning	1.2
Paddy	1.2
Manure handling	0.08
Direct emission from forest	12
Mineral fertilizers	1
Carbon sequestration	1
Arable land	

Source: Scialabba and Müller-Lindenlauf (2010).

legal and constitutional considerations limit now usable water resources to 1,121 BMC only, which includes 690 BMC of surface water and 431 BMC of groundwater.

3.2.1 GROUNDWATER RESOURCES

The total replenishable groundwater resources in the country are about 432 cubic km, out of which about 46% are in the Ganga and the Brahamaputra basins. The variation of groundwater use is very high in different parts of the country; some states are mainly dependent upon groundwater for different sectors. The use of groundwater is moderately high in the river basins lying in the north-western region and parts of south India. The states of Punjab, Haryana, Rajasthan and Tamil Nadu use more groundwater compared to Chhattisgarh, Odisha, Kerala, states *etc.*, which use only a small proportion of their groundwater potential. States like Gujarat, Uttar Pradesh, Bihar, Tripura and Maharashtra are using their groundwater resources at a moderate rate. If the present trend continues, the demand for water would need supplies. And such a situation will be detrimental to development and can cause social upheaval and disruptions.

3.3 IMPACT OF CLIMATE CHANGE ON IRRIGATION DEMAND AND GROUNDWATER DEPLETION

The irrigation water requirement of a crop is the function of edaphic and climatic factors besides agronomic practices adopted it. In a given land use pattern adopted in the field, there are many climatic factors, mainly temperature and rainfall regulate the irrigation water demand.

The semi-arid and arid regions in the country have long crop-growing seasons, which require more water withdrawals for providing water to sustain agricultural sector. Normally, the rates of groundwater recharge are low in the semi-arid and arid regions such that higher groundwater withdrawals for supplement irrigation, which is more than aquifer recharge resulted in groundwater depletion. Similar situations have been reported for many semi-arid and arid regions of the world in which aquifer depletion can be attributed due to agricultural withdrawals (Ahmed and Umar, 2009; Anonymous, 2004; Foster and Loucks, 2006). Upcoming scenario of change in rainfall, either amount or pattern and temperature, is likely to affect evapotranspiration and irrigation water demand in most parts of the country. Increasing the global temperature generally increased evaporative demand, leading to higher crop evapotranspiration and which hasten the crop water demand. Climate parameters' (precipitation, temperature and carbon dioxide levels) changes can affect the demand for water as well as supply. Increased water use efficiency is attributable to higher GHG emissions such as carbon dioxide levels (CO_2); this may tend to increase frequency of water application as temperature rises (Frederic, 1997). Changes in water demand for crops and irrigation practices will enhance the groundwater exploitation; this claims that climate change is likely to have dramatic impact on groundwater resources and lead to decrease groundwater table. In India, precipitation has always been extremely variable, with the number of annual rainy days varying from 12 to 100 and the rain events that have poured about 60% of the total annual precipitation occur within a few hours. It is projected that due to climate change, the inter-annual variability of the monsoon is expected to increase. Also, the rainy days will be less with concentrated rain within a few hours and an increase in dry spells (Mukerrji, 2009). The above said likely impact may create excessive runoff within a short period, thereby reducing the groundwater recharge potentials. Mall *et al.* (2006) reported that small-relative changes (about 3–5% until the 2020s and 5–8% until the 2070s) in net irrigation demand worldwide, but a larger increase (about +15%) in Southeast Asia and the Indian subcontinent due to climate change (Doll, 2002). The annual irrigation water requirements for paddy, sugarcane, permanent garden and semi-dry region crops were projected to increase in the Bhadra reservoir command area, suggesting that the effect of projected increase in rainfall on the irrigation demands is offset by the effect due to projected changes in other meteorological factors (Rehana and Mujumdar, 2012). However, due to the moderating effect of rising CO_2 concentration on evapotranspiration demand and shortening of crop growth period under raised CO_2 concentration, seasonal crop water demand may not increase significantly under the climate change scenarios (Islam et al., 2012). For India, increasing temperature does not always result in enhanced irrigation water demand because of relative increase in precipitation amount is projected to be larger in India than in other major irrigated countries, due to this outweighing the impact of warming. If rising temperature (>4°C) cancels out the effect of increasing precipitation, thus the magnitude of the increase in irrigation water demand widely depends on degree of continued global warming and associated change in precipitation pattern (Wada *et al.*, 2013). Table 3.2 describes the projected irrigation water demand (BMC Year^{-1}) under climate change for different RCPs.

Figure in the parenthesis indicates percentage change relative to the 2000s. Irrigation water demand during the 2000s for India and globe is 2,852 and 751 BCM Year^{-1}, respectively. RCP: Representative concentration pathway.

TABLE 3.2

Irrigation Water Demand (BMC Year^{-1}) under Climate Change for Different RCP$_S$

	Globe		India	
	2050	2080	2050	2080
RCP2.6	3,096	3,130	738	742
	(8.6)	(9.7)	(−1.7)	(−1.2)
RCP4.5	3,120	3,210	740	748
	(9.4)	(12.6)	(−1.5)	(−0.04)
RCP6.0	3,140	3,275	756	780
	(10.1)	(14.8)	(0.7)	(3.9)
RCP8.5	3,200	3,450	768	820
	(10.2)	(21.0)	(2.3)	(9.2)
Mean	3,140	3,270	750	773
	(10.1)	(14.7)	(−0.1)	(2.9)

Source: Wada *et al.* (2013).

3.4 IMPACT OF IRRIGATION ON GREENHOUSE GAS EMISSIONS

The population is continuously increasing and it is projected that global population will rise to 9 billion by 2050; to meet food demands of future generation, the food production will have to double (Ray *et al.*, 2013). As per FAO (2014), improved high-yielding varieties along with application of fertilizers and pesticides together with application of various irrigation practices were proved in increasing agriculture production through the green revolution. Nesheim *et al.* (2015) reported that intensified agriculture has also harmful impact on environment with increasing GHG emissions, namely nitrous oxide (N_2O), carbon dioxide (CO_2) and methane (CH_4), which account for 10%–13% of total global anthropogenic GHG emissions (Smith *et al.*, 2007). Timely application of irrigation increases agriculture productivity, but its execution often increases operational energy demand leading to the possible GHG emissions in the environment(Smith *et al.*, 2007). Besides it, the irrigation is a good solution for boosting agriculture production, but it can change soil biogeochemical feature and soil structure, which have harmful impact on soil carbon sequestration (Lal, 2004; Trost *et al.*, 2013). Hence, it is important to have good understanding of the different irrigation methods and their subsequent impact on GHG emissions.

The soil water content and microbial respiration significantly impact various biogeochemical processes that regulate the rate of GHG emissions from soil. Aerobic and anaerobic organic carbon and nitrogen respiration contribute to the processes of GHG emissions, namely CO_2 and N_2O emissions from soils (Bond-Lamberty and Thomson, 2010), which are mainly driven by three biological processes, namely microbial respiration, root respiration and faunal respiration (Hanson *et al.*, 2000; Rastogi *et al.*, 2002; Oertel et al., 2016); all of these are importantly influenced by water availability in the vicinity of the crop root zone (Bowden *et al.*, 1993; Maier and Kress, 2000; Orchard and Cook 1983; Skopp *et al.*,1990; Sulkava *et al.*, 1994). Several research studies published in decades have shown that soil microbial production of CO_2 is related to soil moisture potential, which can be described as a log-linear relationship when substrates and soil water are not limiting (Bowden *et al.*, 1993; Maier and Kress, 2000; Oertel *et al.*, 2016; Skopp *et al.*, 1990; Sulkava *et al.*, 1994).

Generally, the carbon mineralization rate in arid climates has a significant impact on wetting events than in humid climates (Borken and Matzner, 2009). In the agricultural soils, if the less frequent application of irrigation to the soil or less precipitation events take place, the CO_2 pulse is expected to be increased more with the irrigation event. Similarly, biologically driven autotrophic nitrification and heterotrophic denitrification processes help in production of N_2O in the soil under favorable soil moisture conditions based on temperature and soil texture (Maag and Vinther, 1996; Masscheleyn *et al.*, 1993).

Normally, in wet soils, biological denitrification, the reduction of nitrate (NO_3) or nitrite (NO_2), is performed by phylogenetically diverse bacteria for energy production, which are heterotrophic linking NO_3 or NO_2 reduction to the oxidation of organic compounds.

Ratering and Schnell (2001) reported based on their study that N_2O emissions were because of nitrification in the presence of O_2, although emissions of N_2O were more under anoxic conditions when denitrification dominated. The reduction of nitrite by Fe^{2+} and Mn^{2+} also contributes to soil GHG emissions under abiotic reactions. These reactions produce reduced redox active metals that are dominated by anaerobic microbial respiration predominantly in soils with high moisture content (Nealson and Saarini 1994; Ratering and Schnell 2001). Butterbach-Bahl *et al.* (2013) reported that application of water to the agriculture fields in the form of irrigation is the source of soil moisture content that plays a significant role in modulating the release and consumption of GHGs (Stres *et al*, 2008). The higher volume or more frequent irrigation helps in the production of more plant biomass and good microbial activity in the soil, which ultimately results in more CO_2 and N_2O emissions compared to rainfed or non-irrigated soils (Fares *et al.*, 2017). This is because more soil moisture contents accelerate microbial respiration of soil organic matter, which enhances CO_2 flux (Masscheleyn *et al.*, 1993). Islam *et al.* (2018) found that nitrification and denitrification are responsible for the release of N_2O, which is mainly influenced by the microbial metabolic processes under irrigation. The frequencies of irrigation affect the bacterial activity under anaerobic conditions, which resulted in elevated CH_4 emissions. Hence, it is clear that irrigation directly influences the GHG emissions; less irrigation can reduce GHG emissions by optimizing the nitrogen and carbon turnover processes in soil (Wu *et al.*, 2014). The less irrigation strategies can reduce GHG emission from the well-managed lands, particularly in arid regions. The results of several studies showed a significant reduction in CO_2 emissions with less quantity of water application through irrigation or with a change in irrigation strategies. Edwards *et al.* (2018) reported that comparative studies of surface and subsurface drip irrigation systems in Canada found negligible effects on CO_2 emissions.

3.5   NEED FOR MITIGATION AND ADOPTION OF STRATEGIES FOR GREENHOUSE GAS EMISSION

The increasing demand for food commodities due to growing population also causes increasing pressure on agriculture and consequently on the climate system. On the other hand, climate change is to be expected to intensify this pressure on agriculture. Therefore, continuous efforts are required for mitigation of GHG emissions to trim down the vulnerability of agriculture to impact of climate change; therefore, it is a need to achieve greater energy efficiency for controlling emission of GHG emission.

3.5.1   MITIGATION POTENTIAL OF SOIL AND WATER CONSERVATION PRACTICES

3.5.1.1   Cover Crops

Cover crops' effect on N_2O emissions is complex, which is dependent upon many factors, namely availability of soil water, carbon and nitrogen. Cover crops can be incorporated into any cropping system as it has potential benefits but requires careful management. The nitrogen in the atmosphere is mainly available in the form of nitrogen gas (N_2), which cannot be used directly by the plants. Plants can use this in the form of ammonia (NH_3). A group of microbes known as N-fixers, which convert N_2 into ammonia (NH_3), are present in the nodules of the leguminous plant roots. Nitrogen can be "fixed" in the soil by the microbes in the plant roots before it can be used. N-fixing microbes are housed in different leguminous plants' nodules in their roots like beans or alfalfa. The microbes use carbon compounds as food and fix nitrogen for plants. This Nitrogen is stored in the biomass of the plant, which can be accessible for different crops after the decomposition of biomass of the plants in the soil resulted in the less requirement of nitrogenous fertilization, which also resulted in increased N_2O emissions (Garland *et al.*, 2011). However, grasses such as ryegrass (Fig. 3.1) or millets and other grasses consume the required nitrogen from the soil that ultimately reduces the potential losses of nitrate leaching and nitrous oxide production (Feyereisen *et al.*, 2006; Shipley *et al.*, 1992).

The main driving factor is the type of cover crop that affects N_2O production rates. Basche *et al.* (2014) reported that the leguminous cover crops introduce additional nitrogen to the soil simultaneously enhancing N_2O emissions; however, the use of non-leguminous grasses helps in significantly less production of N_2O. The timing of sowing and variety of leguminous cover crops are the main factors that influence N_2O production. N_2O emissions by incorporation of the legume cover crops are affected by the irrigation and tillage practices on the one hand, which can be controlled by the farmers and precipitation and soil texture are the other factors that can be controlled by the growers. N_2O emissions can be reduced with the use of grasses as cover crops and improper management of non-leguminous plant cover crops may reduce the soil nitrogen availability because of immobilization.

3.5.2   MICRO-IRRIGATION SYSTEMS

The different irrigation practices, namely subsurface drip irrigation and micro-sprinkler irrigation systems, may reduce N_2O emissions when compared with flood

FIGURE 3.1   A ryegrass cover crop planted in a vineyard.

or furrow irrigation systems. The two important factors that play major role explain why these irrigation practices support less N_2O emissions. First, high levels of soil water stimulate N_2O production by nitrification and denitrification processes. Second, extreme variations in soil moisture content cause peak emissions events. A good quantity of N_2O gas can be emitted when dry soils get wet suddenly due to heavy irrigation during the first winter season or rains in a few days. Advanced and efficient irrigation systems help in prevention of conditions ripe for N_2O production due to the soil being consistently moist. Further, there are a number of prospects for denitrifying organisms to "catch" N_2O and convert it into N_2 when N_2O is produced in the different soil depths near the drip line (Neftel *et al.*, 2000; Yang *et al.*, 2016). Furthermore, water uptake by plant roots can help in limiting high levels of soil moisture in shallow soil layers, helping prevent large N_2O emissions (Wada *et al.*, 2013).

Drip irrigation system (Fig. 3.2) is an efficient method than furrow irrigation in the applying water and nutrients (through fertigation) to plant roots and saves both the time and area over which it is used simultaneously to promote N_2O production. In annual cropping systems, subsurface drip irrigation reduces N_2O emissions when compared to flood or furrow irrigation. Drip irrigation has proven technology to use less water with less nitrate leaching in soil (Schellenberg *et al.*, 2012; Sharma *et al.* 2012). N_2O emission is also less in subsurface drip system.

Mainly cotton is produced in the areas situated in semi-arid or arid regions. In these areas, for cotton cultivation, irrigation methods using less water are a key issue. Drip irrigation can be one of the suitable methods of irrigation for the semi-arid or arid regions as it saves significant amount of water since it reduces evaporation,

FIGURE 3.2   Irrigation systems.

surface runoff and deep percolation (Li *et al.*, 2003). As per the crop requirement, water and nutrients are directly applied to the crop root zone through drip irrigation resulting in higher water and nutrient use efficiency. The drip irrigation may have good impact on the nitrogen and carbon turnover in soil and reduce the N_2O or carbon-related GHG (e.g., CH_4) production compared to conventional furrow irrigation. It is reported in several reports based on previous studies, which showed that drip irrigation significantly decreased the N_2O emission from tomato and melon fields, as compared with furrow irrigation (Arce *et al.*, 2008; Kallenbach *et al.*, 2010; Sanchez-Martín *et al.*, 2010).

3.5.3 ALTERNATE WETTING AND DRYING (AWD) IRRIGATION APPROACH

Alternate wetting and drying (AWD) irrigation system approach has water-saving technology that has been developed by the International Rice Research Institute for rice cultivation in Asia. In this technique of irrigation, water is applied to the field several days after ponded water disappears. This technique is different from the traditional irrigation practices of continuous flooding (CF). In this technique, the field is allowed to be in "non-flooded" condition before irrigation is applied, which can vary from 1 day to more than 10 days. The underlying argument behind this irrigation approach is that the crop roots of the paddy plant are getting adequate water supplied for some period (due to the initial flooding) even if there is no observable ponded water in the field. Singh *et al.* (1996) reported that the AWD irrigation approach can save a good quantity of water and reduce water use by about 40–70% when compared with the traditional practice of continuous submergence conditions, without compromising yield loss.

Rice cultivation is a major CH_4 gas source that accounts for 9.0–13.2% of the total anthropogenic emission (Ciais *et al.*, 2013). Due to the development of soil reductive conditions after field flooding in rice, CH_4 is produced by methanogenic bacteria in the soil. In Asia, traditional rice cultivation uses continuous flooding of water as water regime. However, CF enhances CH_4 emissions from the soil. In contrast, draining the flooded field can effectively minimize the CH_4 gas emission (Smith and Conen, 2004). The production of paddy in flooded conditions generates more methane because the water blocks penetrating oxygen in the soil, creating conditions conducive for methane-producing bacteria. Frequent flooding irrigation and interrupted flooding water lower bacterial methane production and thus methane emissions. Techniques for reduced or interrupted flooding include (a) a single drawdown of water during the mid-season and (b) AWD, which frequently interrupts irrigation, so that water levels modestly decline below the soil level before reflooding. Other techniques include dry seeding instead of transplanting rice into flooded conditions in the fields, and various "aerobic rice" systems, in which rice is grown in well-drained soil. Evidence indicates that all of these modified techniques substantially reduce GHG emissions. Perfect water management strategies can theoretically reduce emissions by up to 90% compared to full flooding.

3.6 SYSTEMS OF REDUCING FLOODING AND EMISSIONS DURING THE CROP-GROWING SEASON

3.6.1 DRY SEEDING

The traditional method of seed transplanting, which is grown in nurseries in flowed fields, is followed in rice producing areas of Asia nurseries. But direct seeding of rice is growing in Asia and probably now accounts for a quarter of all rice production. Direct seeding can carry out in flooded fields or through drilling seeds into dry fields.

Most rice production area in Asia follows the traditional pattern of transplanting seedlings grown in nursery areas into already flooded fields. Now in about a quarter of the rice-producing fields, direct seeding of rice is growing in Asia. Direct seeding practice can change flooded fields into dry fields using drilling seeds. The dry field approach (dry seeding) reduces emissions because it shortens the flooding period roughly by a month period.

3.6.2 SINGLE MID-SEASON DRAWDOWN

Single drawdown during the crop production season sufficiently allows oxygen to penetrate the soils and considerably lowers GHG emissions. Itoh *et al.* (2011) reported that this kind of drawdown occurs mainly for 5–10 days to generate methane benefits. In China, Japan and South Korea, most of the farmers practice this draw-down to increase paddy yields.

3.6.3 ALTERNATE WETTING DRYING (AWD)

Alternate wetting and drying (AWD) practice involves repeatedly flooding with water up to 5 cm depth to the rice field and allows drying to field until the upper soil layer starts to dry out (typically when the water level drops down to around 15 cm below the soil surface) and again reflooding the field after drying of soil. This cycle can prolong up to 20 days after sowing till 2 weeks before flowering. Based on the research context and the country, this approach also refers to "controlled irrigation system" or multiple number of irrigations. Each drying cycle sets back the production of methanogenic bacteria; AWD accomplishes even larger reductions in methane compared to only one drawdown. It can be practiced along a continuum, with the frequency of drawdowns varying from more to less regular, although the level of methane reductions will depend on how rigorously it is practiced.

3.6.4 AEROBIC RICE PRODUCTION

In aerobic rice production system, irrigation water is added only when required, aiming to keep soils moist but avoiding water standing in the field. Methane production can be drastically minimized or nearly eliminated in this system. However, in general, aerobic paddy production has lower yields compared to traditional methods. Some farmers are still getting good yields by cultivating on raised beds or in ditches that restrict standing water to furrows.

3.7 CONSERVATION TILLAGE (NO TILLAGE/ZERO TILLAGE)

A number of agricultural activities contribute toward emission of GHGs and global warming. Major farm activities, including intensive plow tillage (PT), use of agricultural chemicals or fertilizers and burning of crop residues, are adding to GHGs in the environment. In India, the burning of non-conventional fuel and resultant emission of GHGs gasses is served; in the Indo-Gangetic basin, rice-wheat cropping system is dominant in this region. Where conventional method of land preparation not only disturbs the soil environment but also leads to atmospheric pollution. The no-till system eliminates pre-planting tillage operations, which is the use of seed bed preparation thereby reducing the fossil fuel burning and sequestering carbon in the soil that ultimately reduces GHG emissions. Under no-till system, water requirement is reduced because pre-sowing irrigation is not required.

Detailed knowledge of the effects of agronomical operations on GHG emissions is imperative for the recommendation of low emission practices. Reduced tillage (RT) and no tillage (NT) are widely recommended in the world for crop production to improve soil structure, reduce soil erosion and enhance soil organic matter as compared with conventional tillage (CT). However, the effect of NT/RT on mitigation of climate change has been intensively debated because of the substantial inconsistency in individual field experiments (Abdalla *et al.*, 2016).

In recent past, we are concentrated totally on no-till system rather than other agricultural practices as a means of mitigating climate change because the study concluded that, long-term no tillage reduced the GHG emissions by improving the soil properties (Nawaz et al., 2016). This was deliberate because current uptake of no-till and its probable rate of adoption in the medium term is far greater than for the other practices, such as agroforestry and biochar applications to soil. No-till agriculture can provide significant benefits for farmers and sustainability in many situations: reduced GHG emissions are a small but important additional benefit (Neufeldt et al., 2015).

A recent theoretical study concluded that C sequestration in agricultural soils through changed management practices, including no-till, could provide "only a humble contribution to solving the climate problem of the up-coming decades". Of course, even small contributions are welcome-put colloquially, every little helps. But it is important that scientists are realistic when making statements about the relative magnitudes of mitigation through different options. This is mandatory for assisting policymakers (Powlson et al., 2014).

3.8 MULCHING

Kasirajan and Ngouajio (2012) define mulch as a coating material spread over the soil surface. It can also be defined as a technique of covering the land surface around the plants with an organic or synthetic material to build a favorable condition for the growth of plant and proficient crop production (Chakraborty et al., 2008). Mulching protects the soil to safeguard organisms and crop roots from varying meteorological conditions. It is also an important water conservation practice in modern agricultural production particularly in arid and semi-arid environments. The material used for mulch also protects soil surface from sunlight and leads to less evaporation by preserving soil water and adjusting soil temperature. Mulching also helps in saving water resources and increasing water use efficiently by use of water within soil root zone. In addition, mulching is not only a water-saving technique but also responsible for the beautification of farmlands. Different factors responsible for selection of mulching materials are availability of local materials, climate, durability and cost-effectiveness. It is also essential that the material should be selected, which is environment friendly and economically viable. Therefore, it is important that mulching material is selected and used, which can save the water and increase crop yield under rainfed cultivation. Mulch application over the soil also decreases soil temperature (0–5°C) and penetration resistance (0–60 cm). Mulching helps in maintaining soil fertility and soil moisture conservation and controlling soil erosion. Prosdocimi et al. (2016) conducted a 3-year study and reported that mulch application significantly improved the soil's physical and chemical properties depending upon the mulch application rate and decreased in soil bulk density and soil erosion and improvement in total soil porosity. Some other studies have concluded that barley straw mulch and jujube branch mulch reduced the runoff and thus prevented the soil losses by decreasing soil erodibility and nutrient erosion.

3.9 SITE-SPECIFIC NUTRIENT MANAGEMENT (SSNM)

Site-Specific Nutrient Management (SSNM) developed in Asia is a plant-based approach to "feeding" rice with nutrients as and when needed. Site-Specific Nutrient Management (SSNM) enables rice-producing farmers to optimally apply essential nutrients required by the rice crop. The optimum supply of nutrients to the rice crop may vary from field to field depending on crop and land management, previous use of nutrients, crop residue management and organic materials and crop cultivar. Therefore, the SSNM approach provides principles and guidelines to the farmers, which enables the farmers to apply the essential nutrients optimally that match the needs of their rice crop in a specific field and season.

The SSNM approach aims at applying nutrients at optimal rates and right times in order to achieve high rice yield and high nutrient use efficiency by the rice crop, leading to high cash value of the crop harvest per unit of fertilizer invested. Researchers developed the concept of SSNM in the mid-1990s; the initial concept was systematically transformed to offer farmers and extension workers with simplified plant-need-based N, P and K management. The SSNM approach is combined with the leaf color chart (LCC) developed by IRRI, which gives two balancing and equally effective options for improved nitrogen management in rice using the LCC. Nitrogen management in the "real-time" option, farmers monitor the rice leaf color regularly and apply nitrogen fertilizer based on leaf color change (i.e. leaves become more yellowish-green than the critical threshold value indicated on the LCC). In the "fixed-time/adjustable dose" option, the time for N fertilization is preset at critical growth stages of the crop and farmers adjust the dose of nitrogen upward or downward based on the leaf color. The selection of a suitable decision for using the LCC may be dependent on farmer preferences and location-specific factors.

Adequate quantities of P and K fertilizations are applied in the plant-based SSNM approach to overcome deficiencies and maintain soil fertility. The need for phosphorus fertilizer for the rice crop in a given field or location is estimated based on targeted yield and a P-limited yield. The need of the rice crop for potassium fertilizer was similarly obtained from an estimate of an attainable yield target and a K-limited yield. The yield target is the rice grain yield attainable by farmers with good crop and nutrient management under average climatic conditions. It provides an estimate of the total amounts of P and K needed by the rice crop because the amounts of P and K fertilizers taken up by a rice crop are directly related to crop yield.

3.10 NANOTECHNOLOGY

Nano-materials in soils comprise clay minerals as well as metal oxides. Natural nano-particles (<100 nm) occur widely in the environment, especially in soils (Calabi-Floody et al., 2012; Parfitt et al., 1999; Wada et al., 2013). Pan and Xing (2012) reported that nano-clays could be effective in improving soil water, carbon (C) and nutrient storage capacities because of their large surface area. The use of nano-materials improves C stabilization in soil because of their unique electronic, kinetic, magnetic and optical properties (Calabi-Floody et al., 2012; Khedr et al.,2006; Monreal et al., 2010).

A nano-sand is water repellent, developed to prevent water drainage in dry land, which care for the release of nutrients and molecules to care for plant progress (Davidson and Gu, 2012). The nano-sand is able to block water drainage below the depth of the plant roots zone and sustains a subsurface water table, providing vegetation with constant water supply. Many other nano-materials that have been recently developed are nano-membranes for water purification and desalination that would be several times more efficient and nano-clay able to fix the sand and increase available water for plants by 25%. Nano-clay works as a binder agent and keeps moisture in the sand. Revegetation of deserts with nano-clay could lead to mitigation of wind erosion, generate aggregate in the soil, increase the amount in the soil, increase the amount of available water and thus improve the growth of plants and trees.

Nano-clay is composed of clay minerals divided into their smallest components that are 0.7–1.5 nm thick, with a diameter of 20–300 nm and mixed with water. The study results of nano-clay, applied in hot, dry, sandy soil in Egypt, increase in yield of up to 416% while using only one third of the normal consumption of irrigation water (Olesen, 2010).

3.11 CONCLUSIONS

The rapid development, industrialization and intensive agriculture practices and unsystematic destruction of natural resources have resulted in global warming due to change in atmospheric gases. Climate-related parameters, such as temperature, rainfall, soil moisture and sea level, can be changed due to global warming and may result in extensive and frequent occurrence of climatic extremes, such as droughts, heat waves and floods, in different parts of the world, particularly in developing countries. Soil moisture in most agricultural soils plays a significant role in the release and consumption of GHGs that can be controlled by adopting the suitable irrigation methods. Advanced and efficient methods of irrigation such as drip and sprinkler along with proper irrigation scheduling are feasible solutions to boost agricultural production and save a significant quantity of irrigation water. Management of land and water resources along with adoption of on-farm water management practices is essential for long-term agricultural sustainability. The holistic development of land, water and other natural resources plays a vital role in sustaining these resources and also helps in reducing the global warming effect.

REFERENCES

Abdalla, K., Chivenge, P., Ciais, P. and Chaplot, V. (2016). No-tillage lessens soil CO_2 emissions the most under arid and sandy soil conditions: Results from a meta-analysis. *Biogeosciences* 13: 3619–3633.

Ahmed, I. and Umar, R. (2009). Groundwater flow modelling of Yamuna Krishni interstream, a part of Central Ganga Plain Uttar Pradesh. *Journal of Earth System Science* 118(5): 507–523.

Anonymous. 2004. Central Ground Water Board: Dynamic ground water resources of India (as on March 2004), Central Ground Water Board, Ministry of Water Resources, Government of India, http://cgwb.gov.in/documents/DGWR2004.pdf, last access: 9 April 2010, Faridabad, India, 126 pp., 2006.

Arce, L., Sanchez-Martín, A., Benito, A., Garcia-Torres, L. and Vallejo, A. (2008). Influence of drip and furrow irrigation systems on nitrogen oxide emissions from a horticultural crop. *Soil Biology and Biochemistry* 40(7): 1698–1706.

Basche, A. D., Miguez, F. E., Kaspar, T. K. and Castellano, M. J.. 2014. Do cover crops increase or decrease nitrous oxide emissions? A meta-analysis. *Journal of Soil and Water Conservation* 69(6): 471–482. DOI 10.2489/jswc.69.6.471.

Bond-Lamberty, B. and Thomson, A. (2010). Temperature-associated increases in the global soil respiration record. *Nature* 464: 579.

Borken, W. and Matzner, E. (2009). Reappraisal of drying and wetting effects on c and n mineralization and fluxes in soils. *Global Change Biology* 15: 808–824.

Bowden, R. D., Nadelhoer, K. J., Boone, R. D., Melillo, J. M. and Garrison, J. B. (1993). Contributions of aboveground litter, belowground litter, and root respiration to total soil respiration in a temperate mixed hardwood forest. *Canadian Journal of Forest Research* 23: 1402–1407.

Butterbach-Bahl, K., Baggs, E. M., Dannenmann, M., Kiese, R. and Zechmeister-Boltenstern, S. (2013). Nitrous oxide emissions from soils: How well do we understand the processes and their controls? *Philosophical Transactions of the Royal Society B: Biological Sciences* 368: 20130122.

Calabi-Floody, M., Velásquez, G., Gianfreda, L., Saggar, S., Bolan, N., Rumpel, C. and Mora, M. L. (2012). Improving bioavailability of phosphorous from cattle dung by using phosphatase immobilized on natural clay and nanoclay. *Chemophere* 89: 644–655.

Chakraborty, D., Nagarajan, S., Aggarwal, P., Gupta, V. K., Tomar, R. K., Garg, R. N., Sahoo, R. N., Sarkar, A., Chopra, U. K., Sarma, K. S. S. and Kalra, N. (2008). Effect of mulching on soil and plant water status, and the growth and yield of wheat (*Triticum aestivum* L.) in a semi-arid environment. *Agricultural Water Management* 95: 1323–1334.

Ciais, P., Sabine, C. and Bala, G.. 2013. Carbon and Other Biogeochemical Cycles. In *Climate change 2013: the physical science basis. contribution of working group I to the fifth assessment report of the intergovernmental panel on climate change*, Stocker, T.F., Qin, D., Plattner, G.K. (Eds.) (pp. 465–570). Cambridge University Press, Cambridge and New York, NY.

Davidson, D. and Gu, F. X. (2012). Material for sustained and controlled release of nutrient and molecules to support plant growth. *Journal of Agriculture and Food Chemistry* 60: 870–876.

De Silva, C. S., Weatherhead, E. K., Knox, J. W. and Rodriguez-Diaz, J. A. (2007). Predicting the impacts of climate change – A case study on paddy irrigation water requirements in Sri Lanka. *Agricultural Water Management* 93(1–2): 19–29.

Doll, P. (2002). Impact of climate change and variability on irrigation requirements: A global perspective. *Climatic Change* 54: 269–293.

Edwards, K. P., Madramootoo, C. A., Whalen, J. K., Adamchuk, V. I., Su, A. S. M. and Benslim, H. (2018). Nitrous oxide and carbon dioxide emissions from surface and subsurface drip irrigated tomato fields. *Canadian Journal of Soil Science* 98: 389–398.

Elgaali, E., Garcia, L. A. and Ojima, D. S. (2007). High resolution modeling of the regional impacts of climate change on irrigation water demand. *Climate Change* 84: 441–461.

FAO. 2014. AQUASTAT. Food and Agriculture Organization of the United Nations (FAO). online:http://www.fao.org/nr/water/aquastat/didyouknow/index3.stm (accessed on 27 May 2019).

Fares, A., Bensley, A., Bayabil, H., Awal, R., Fares, S., Valenzuela, H. and Abbas, F. (2017). Carbon dioxide emission in relation with irrigation and organic amendments from a sweet corn field. *Journal of Environmental Science and Health, Part B* 52: 387–394.

Feyereisen, G. W., Wilson, B. N., Sands, G. R., Strock, J. S. and Porter, P. M. (2006). Potential for a Rye cover crop to reduce nitrate loss in southwestern Minnesota. *Agronomy Journal* 98(6): 1416–1426.

Foster, S. and Loucks, D. P.. 2006. Non-Renewable Groundwater Resources, Series on Groundwater No. 10, UNESCO, Paris,

Frederic, K. D. (1997). Adapting to climate impact in the supply and demand for water. *Climate Change* 37: 141–156.

Garland, G. M., Suddick, E., Burger, M. and Horwath, W. R. (2011). Direct N_2O emissions following transition from conventional till to no-till in a cover cropped mediterranean vineyard (Vitis vinifera). *Agriculture, Ecosystems & Environment* 144(1): 423–428.

Hanson, P., Edwards, N., Garten, C. T. and Andrews, J. (2000). Separating root and soil microbial contributions to soil respiration: A review of methods and observations. *Biogeochemistry* 48: 115–146.

Islam, A., Ahuja, L. R., Garcia, L. A., Ma, L., Saseendran, A. S. and Trout, T. J. (2012). Modeling the impact of climate change on irrigated corn production in the Central Great Plains. *Agricultural Water Management* 110: 94–108.

Islam, S. F. U., van Groenigen, J. W., Jensen, L. S., Sander, B. O. and de Neergaard, A. (2018). The Effective mitigation of greenhouse gas emissions from rice paddies without compromising yield by early-season drainage. *Science of The Total Environment* 612: 1329–1339.

Itoh, M., Sudo, S., Mori, S., Saito, H., Yoshida, T., Shiratori, Y., Suga, S., Yoshikawa, N., Suzue, Y., Mizukami, H., Mochida, T. and Yagi, K. (2011). Mitigation of methane emissions from paddy fields by prolonging midseason drainage. *Agriculture, Ecosystems and Environment* 141: 359–372.

Kallenbach, C. M., Rolston, D. E. and Horwath, W. R. (2010). Cover cropping affects soil N2O and CO_2 emissions differently depending on type of irrigation. *Agriculture, Ecosystems and Environment* 137(3–4): 251–260.

Kasirajan, S. and Ngouajio, M. (2012). Polyethylene and biodegradable mulches for agricultural applications: A review. *Agronomy for Sustainable Development [Internet]* 32: 501–529.

Khedr, M. H., Omar, A. A. and Abdel-Moaty, S. A. (2006). Reduction of carbon dioxide into carbon by freshly reduced $CoFe_2O_4$ nanoparticles. *Materials Science and Engineering: A* 432: 26–33.

Lal, R. (2004). Soil carbon sequestration impacts on global climate change and food security. *Science* 304: 1623.

Li, J., Zhang, J. and Ren, L. (2003). Water and nitrogen distribution as affected by fertigation of ammonium nitrate from a point source. *Irrigation Science* 22(1): 19–30.

Maag, M. and Vinther, F. P. (1996). Nitrous oxide emission by nitrification and denitrification in different soil types and at different soil moisture contents and temperatures. *Applied Soil Ecology* 4: 5–14.

Maier, C. A. and Kress, L. W. (2000). Soil CO_2 evolution and root respiration in 11-year-old loblolly pine (*Pinus taeda*) plantations as affected by moisture and nutrient availability. *Canadian Journal of Forest Research* 30: 347–359.

Mall, R. K., Gupta, A., Singh, R., Singh, R. S. and Rathore, L. S. (2006). Water resources and climate change: An Indian perspective. *Current Science* 90: 1610–1626.

Mascheleyn, P. H., DeLaune, R. D. and Patrick, W. H. (1993). Methane and nitrous oxide emissions from laboratory measurements of rice soil suspension: Effect of soil oxidation-reduction status. *Chemosphere* 26: 251–260.

Monreal, C. M., Sultan, Y. and Schnitzer, M. (2010). Soil organic matter in nano-scale structures of a cultivated Black Chernozem. *Geoderma* 159: 237–242.

Mukerrji, R. 2009. Vulnerability and adaptation experiences from Rajasthan and Andhra Pradesh: water resource management. SDC V&A Programme, India.

Nawaz, A., Lal, R., Shrestha, R. K. and Farooq, M.. 2016. Mulching affects soil properties and greenhouse gas emissions under long-term no-till and plough-till systems in alfisolotcentral Ohio Land Degrad. Develop. Published online in Wiley Online Library (wileyonlinelibrary.com) DOI: 10.1002/ldr.2553.

Nealson, K. H. and Saarini, D. (1994). Iron and manganese in anaerobic respiration: Environmental significance, physiology, and regulation. *Annual Review of Microbiology* 48: 311–343.

Neftel, A., Andreas, B., Martin, S., Bernhard, L. and Sergei, V. T. (2000). An experimental determination of the scale length of N_2O in the soil of a grassland. *Journal of Geophysical Research: Atmospheres* 105(D10): 12095–12103.

Nesheim, M. C., Oria, M. and Yih, P. T. 2015. Food System. In *A framework for assessing Effects of the food system*. Washington, DC: National Academies Press, Available online: https://www.ncbi.nlm.nih.gov/books/NBK305182/ (accessed on 17 June 2015).

Neufeldt, H., Kissinger, G. and Alcamo, J. (2015). No-till agriculture and climate change mitigation. *Nature Climate Change* 5: 488–489.

Oertel, C., Matschullat, J., Zurba, K., Zimmermann, F. and Erasmi, S. (2016). Greenhouse gas emissions from soils—A review. *Geochemistry* 76: 327–352.

Olesen, K. P. 2010. Turning sandy soil to farmland: 66% water saved in sandy soil treated with Nano Clay. Desert Control Institute Inc. Vassoy, P. 10. Available from: http://www.desertcontrol.com

Orchard, V. A. and Cook, F. J. (1983). Relationship between soil respiration and soil moisture. *Soil Biology and Biochemistry* 15: 447–453.

Pan, B. and Xing, B. (2012). Applications and implications of manufactured nanoparticles in soils: A review. *European Journal of Soil Science* 63: 437–456.

Parfitt, R. L., Yuan, G. and Theng, B. K. (1999). A ^{13}CNMR study of the interactions of soil organic matter with aluminium and allophane in podzols. *European Journal of Soil Science* 50: 695–700.

Powlson, D. S., Stirling, C. M., Jat, M., Gerard, B. G., Palm, C. A. and Sanchez, P. A. (2014). Limited potential of no-till agriculture for climate change mitigation. *Nature Climate Change* 4: 678–683.

Prosdocimi, M., Jordán, A., Tarolli, P., Keesstra, S., Novara, A. and Cerdà, A. 2016. The immediate effectiveness of barley straw mulch in reducing soil erodibility and surface runoff generation in Mediterranean vineyards. *Science of the Total Environment* 547: 323–330. DOI:10.1016/j.scitotenv.2015.12.076.

Rastogi, M., Singh, S. and Pathak, H. (2002). Emission of carbon dioxide from soil. *Current Science* 82: 510–517.

Ratering, S. and Schnell, S. (2001). Nitrate-dependent iron (II) oxidation in paddy soil. *Environmental Microbiology* 3: 100–109.

Ray, D. K., Mueller, N. D., West, P. C. and Foley, J. A. (2013). Yield trends are Insufficient to double global crop production by 2050. *PLoS ONE* 8: e66428.

Rehana, S. and Mujumdar, P. P. 2012. Regional impacts of climate change on irrigation water demands hydrological processes Hydrol. Process. Published online in Wiley Online Library (wileyonlinelibrary.com) DOI: 10.1002/hyp.9379.

Sanchez-Martín, L., Meijide, A., Garcia-Torres, L. and Vallejo, A. (2010). Combination of drip irrigation and organic fertilizer for mitigating emissions of nitrogen oxides in semiarid climate. *Agriculture, Ecosystems and Environment* 137(1–2): 99–107.

Schellenberg, D. L., Alsina, M. M., Muhammad, S. and Stockert, C. M. (2012). Yield-scaled global warming potential from N_2O emissions and CH4 oxidation for almond (Prunus dulcis) irrigated with nitrogen fertilizers on arid land. *Agriculture, Ecosystems & Environment* 155: 7–15.

Scialabba, N. E.-H. and Müller-Lindenlauf, M. (2010). Organic agriculture and climate change. *Renewable Agriculture and Food Systems*, Special Issue: "Sustainable Agriculture Systems in a resource Limited Future" (June 2010). 25(2): 158–169.

Sharma, P., Shukla, M. K., Sammis, T. W. and Adhikari, P. 2012. Nitrate-nitrogen leaching from onion bed under furrow and drip irrigation systems. *Applied and Environmental Soil Science* 2012: 1–17. DOI:10.1155/2012/650206

Shipley, P. R., Meisiner, J. J. and Decker, A. M. (1992). Conserving residual corn fertilizer nitrogen with winter cover crops. *Agronomy Journal* 84: 869–876.

Singh, C. B., Aujla, T. S., Sandhu, B. S. and Khera, K. L. (1996). Effects of transplanting data and irrigation regime on growth, yield and water use in rice (*Oryza sativa*) in Northern India. *Indian Journal of Agricultural Sciences* 66: 137–141.

Skopp, J., Jawson, M. D. and Doran, J. W. (1990). Steady-state aerobic microbial activity as a function of soil water content. *Soil Science Society of America Journal* 54: 1619–1625.

Smith, K. A. and Conen, F.. 2004. Impacts of land management on fluxes of trace greenhouse gases. *Soil Use Manage* 20: 255–263. DOI:10.1079/SUM2004238

Smith, P., Martino, D., Cai, Z., Gwary, D., Janzen, H., Kumar, P., McCarl, B., Ogle, S., O'Mara, F. and Rice, C. 2007. Agriculture. In *Climate change 2007: mitigation. contribution of working group III to the fourth assessment report of the intergovernmental panel on climate change*, Netz, B., Davidson, O.R., Bosch, P.R., Dave, R., Meyer, L.A. (Eds.). Cambridge and New York, NY: Cambridge University Press.

Stres, B., Stopar, D., Mahne, I., Hacin, J., Resman, L., Pal, L., Fuka, M. M., Leskovec, S., Danevčič, T. and Mandic-Mulec, I. (2008). Influence of temperature and soil water content on bacterial, archaeal and denitrifying microbial communities in drained fen grassland soil microcosms. *FEMS Microbiology Ecology* 66: 110–122.

Sulkava, P., Huhta, V. and Laakso, J. (1994). Impact of soil fauna structure on decomposition and N-mineralisation in relation to temperature and moisture in forest soil. *Pedobiologia* 40: 505–513.39.

Trost, B., Prochnow, A., Drastig, K., Meyer-Aurich, A., Ellmer, F. and Baumecker, M. (2013). Irrigation, soil organic carbon and N$_2$O emissions. A review. *Agronomy for Sustainable Development* 33: 733–749.

Wada, Y., Wiser, D., Eisner, S, Flore, M., Gerten, D., Haddeland, I., Hanasaki, N., Masaki, Y., Portmann, T., Stacke, T., Tesseler, Z. and Schewe, J. (2013). Multi-model projections and uncertainties of irrigation water demand under climate change. *Geophysical Research Letters* 40: 4626–4632.

Wu, J., Guo, W., Feng, J.,; Li, L., Yang, H., Wang, X., and Bian, X. (2014). Greenhouse gas emissions from cotton field under different irrigation methods and fertilization regimes in arid northwestern China. *The Scientific World Journal*, pp. 1–10, http://dx.doi.org/10.1155/2014/407832.

Yang, H., Rong, S., Zhenxing, Z., Ling, W., Qing, W. and Wenxue, W. (2016). Responses of nitrifying and denitrifying bacteria to flooding-drying cycles in flooded rice soil. *Applied Soil Ecology* 103: 101–109.

4 Managing Groundwater Resources and Its Quality for Sustainable Development of Agriculture

G. T. Patle

4.1 INTRODUCTION

Water resource development and management is one of the most important measures of a country's economic prosperity. Increasing population, expanding irrigated agricultural areas, and economic development are mainly responsible to increase in global water demand. Surface water and groundwater are the world's most important water resources, but they are under significant stress owing to overexploitation and poor management in many parts of the world (Boretti & Rosa, 2019; Nhemachena *et al.*, 2020; Simonovic, 2002). Water crises in many regions of the world are caused by the uneven distribution of water resources over time and location, as well as its change by human use and misuse. Increased demand from various sectors, especially residential, agriculture, and industrial, is increasing the pressure on water supplies, resulting in tensions, conflict among users, and excessive environmental pressure (CWC, 2010; Velmurugan *et al.*, 2020). The total volume of water on the planet is around 1.4 billion km^3. However, the majority of water (approximately three-quarters) is available in the ocean, and fresh water forms a very minor proportion (35 million km^3) of the massive quantity available on the planet, accounting for around 2.5% of the overall volume. About 24 million km^3 (68.9%) of these are in the form of ice and permanent snow cover in mountainous regions, the Antarctic, and the Arctic regions, with the remaining 29.9% present as groundwater (shallow and deep groundwater basins up to 2000 m). The remaining 0.3% is found in lakes and rivers, with the remaining 0.9% found in soil moisture, swamp water, and permafrost atmosphere (Lakshminarasimhan, 2022; Robert, 2001; Sophocleous, 2004). Although surface water availability (water in lakes, rivers, and reservoirs) can meet such demand globally, groundwater is regarded as the most important source of water found in groundwater-rich aquifers. According to estimates, approximately 67% of all groundwater is used for irrigation (food production), 22% for home reasons (drinking water and sanitation), and 11% for industrial usage. The total global groundwater abstraction is 1000 km^3 per year (Bhat, 2014; O'Mara, 1988; Shah *et al.*, 2001; Siebert *et al.*, 2013). Overexploitation or chronic groundwater depletion

DOI: 10.1201/9781003351672-4

can occur if groundwater abstraction exceeds groundwater recharge across large areas and for a long time (Dillon *et al.*, 2019; Shah *et al.*, 2001; Van der Gun, 2012).

4.2   WATER RESOURCES POTENTIAL IN INDIA

The annual precipitation, including snowfall, which is the country's principal supply of water, is estimated to be in the range of 4000 billion cubic meters (BCM). According to the central water commission (CWC), the country's overall water resource potential is approximately 1869 BCM, which includes both surface and groundwater resources (Chavda & Ravenscroft, 2020; Mushtaq *et al.*, 2020). Due to topographic limits and uneven resource distribution over space and time, it is anticipated that only roughly 1123 BCM of the total potential of 1869 BCM can be put to effective use. The surface water potential is 690 BCM, with the remainder being groundwater potential. In 2001, the national per capita annual water availability was 1816 cu m, but it was 1544 cu m in 2011. The country's per capita availability will be 1140 cu m in 2050, down from 1608 cu m in 2010. International agencies regard shortage situations to be any situation in which availability is less than 1000 cu m per capita. Uncontrolled resource development to fulfil rising demand has resulted in diminishing groundwater levels (Chauhan *et al.*, 2022; Chavda & Ravenscroft, 2020; Mushtaq *et al.*, 2020).

4.3   GROUNDWATER DEVELOPMENT AND UTILIZATION PROSPECTS IN INDIA

India, followed by China and the United States, is the largest user of groundwater (Siebert *et al.*, 2010) and contributes 20% of the global agricultural water withdrawal (Zektser & Everett, 2013). Groundwater has played a crucial part in maintaining India's economy, environment, and way of living, as well as in satisfying the water requirements of India's numerous user sectors. It is also the greatest and most productive source of irrigation water, in addition to being the principal source of water for home and many industrial applications. As a source of water supply, groundwater has a number of significant advantages over surface water, including higher quality, greater protection from possible contamination including infection, less sensitivity to seasonal and perennial fluctuations, and a much more uniform distribution across large regions. The Central Ground Water Board (CGWB) is charged with assessing and managing the country's groundwater resources via groundwater management research, exploration, evaluation, and monitoring of the groundwater regime (Chatterjee & Purohit, 2009; Hooda *et al.*, 2000). According to the report, the annual replenishable groundwater resource for the entire country is 433 BCM, the net annual groundwater availability is estimated to be 399 BCM, and the annual groundwater draught for irrigation, domestic, and industrial use was 231 BCM (Jha & Sinha, 2009; Saha *et al.*, 2018; Sinha *et al.*, 2020). The groundwater development stage for the entire country is 58%. Two major sources contribute to the annual replenishable groundwater resource: precipitation and other sources, such as canal seepage return flow from irrigation, seepage from water bodies, and artificial recharge owing to water conservation structures. The contribution of precipitation to the country's annual replenishable groundwater resource is 67%, while the contribution of all other sources is 33%. In India, around four states account for more than 90% of groundwater potential. In terms of replenishable groundwater resources, Uttar Pradesh ranks highest (17.84%), followed by Andhra Pradesh (8.29%), Madhya Pradesh (8.10%), Maharashtra (7.85%), Bihar (6.78%), West Bengal (6.76%), and Assam (6.59%). According to the groundwater development stage, the CGWB has classed the country's groundwater resources as either safe, semi-critical, critical, or overexploited (Majumder & Sivaramakrishnan, 2014; Shankar *et al.*, 2011).

- **Safe:** If the groundwater development stage is between 0 and 70%, the water is considered safe. These are regions with groundwater development potential.
- **Semi-critical:** If the groundwater development stage is between 70 and 90%, it falls into the semi-critical group. These are regions in which cautious groundwater development is advised.
- **Critical:** When the groundwater development stage is between 90 and 100%, it is categorized as critical. For groundwater development, these regions require intensive monitoring and evaluation.
- **Overexploited:** If the groundwater development state is larger than 100, it falls into the overexploited group. These are the regions where future groundwater development will be coupled with water conservation measures.

The following states' groundwater development phases are described here. Andhra Pradesh 37%, Arunachal Pradesh 0%, Assam 14%, Bihar 44%, Chhattisgarh 35%, Delhi 137%, Goa 28%, Gujarat 67%, Haryana 133%, Himachal Pradesh 71%, Jammu & Kashmir 21%, Jharkhand 32%, Karnataka 64%, Kerala 47%, Madhya Pradesh 57%, Maharashtra 53%, Manipur 1%, Meghalaya 0%, Mizoram 3%, Nagaland 0%, Odisha 28% (*Ground Water Year Book – India 2017-18*, n.d.)

4.3.1 USE OF GROUNDWATER IN AGRICULTURE AND FOOD SECURITY IN INDIA

India is the greatest user of groundwater in the world, consuming an estimated 230 km^3 annually, or more than a quarter of the global total. Groundwater is an essential resource for rural areas in India, as more than 60% of irrigated agriculture and 85% of drinking water supplies rely on it. India's agricultural output and food security have benefited greatly from the availability of groundwater (Harikumar, 2013; Mukherjee *et al.*, 2015). About 70% of India's food supply is derived from irrigated agriculture, with a substantial contribution from groundwater (Gandhi and Namboodiri, 2009). About 80% of the nation's water is used in the irrigation sector, making agriculture the greatest consumer of water (Oza, 2007). There has been a spectacular increase in irrigation development in India. Net irrigated area grew from 21 million hectares in 1950–51 to 62.29 million hectares in 2007–08 (CWC, 2010; Gandhi and Namboodiri, 2009). Its contribution to the country's net irrigated area has reached 60.7% (CWC, 2010).

4.3.2 Impact of Climate Change on Groundwater Resources

Variation in hydro-meteorological factors is anticipated to generate future hydrologic cycle imbalances as a result of climate change. This will impact the recharging and availability of groundwater in arid and semi-arid regions. Approximately 29% of the nation's groundwater blocks are classified as semi-critical, critical, or overexploited, and the situation is developing rapidly (Harikumar, 2013; Mukherjee *et al.*, 2015; Thampi and Raneesh, 2012). Recent decades have recognized global warming and climate change as a major threat to fresh water resources (Green *et al.*, 2011). Variations in hydro-meteorological factors and, thus, the underlying groundwater supplies are expected to be impacted by climate change. Changes in spatial and temporal variation in rainfall patterns, an increase in extreme rainfall events in terms of increased rainfall intensity, an increase in surface air temperature, and alterations to the hydrologic cycle are anticipated to have the greatest influence on groundwater supplies. These factors may influence the quantity of water available for groundwater recharge (Scibek & Allen, 2006). Changes in rainfall patterns can have an impact on the natural recharging process, while an increase in temperature can increase crop evapotranspiration and irrigation demand. It is anticipated that groundwater levels will continue to fall as a result of the combined effects of rising agricultural, domestic, and industrial demands and climate change. However, the precise extent of climate change's impact on groundwater resources is still unknown. Compared to surface water resources, it is difficult to quantify the influence of climate change on groundwater resources due to uncertainty in climate projections (Jyrkama & Sykes, 2007). Due to long-term changes in climate variables, such as air temperature, precipitation, humidity, wind speed, duration of sunshine hours, and evaporation, climate change has a direct impact on surface water resources. However, due to the complex relationship between climate parameter and groundwater recharge, it is difficult to quantify the impact of climate change on groundwater resources. Groundwater recharge is governed by numerous complicated characteristics and processes, which are impacted by a number of variables, including rainfall and its intensity, evapotranspiration, infiltration, soil moisture storage in the vadose zone, hydraulic property of aquifer, and water table depth.

4.3.3 Groundwater Depletion and Water Quality Issues in India

Many regions of the country may experience a decline in groundwater levels within the next few decades as a result of growing groundwater consumption for irrigation, household, and industrial purposes. Similarly, poor management methods exacerbated the issue. The groundwater level is decreasing mostly as a result of an increase in cropland devoted to water-intensive crops and an increase in water demand from other sectors. The effects of declining water levels on agricultural production are significant. This will undoubtedly affect the groundwater development and utilization policy (Scott & Shah, 2004; Singh *et al.*, 2015; Sishodia *et al.*, 2016). Similarly, the persistent deterioration of groundwater quality is a serious concern that has imposed additional restrictions on the development and use of groundwater for drinking, irrigation, and industrial purposes. The majority of problems with groundwater quality are due to contamination, overexploitation, or a mix of the two. The majority of groundwater quality issues are difficult to detect and resolve. The quality of groundwater is declining mostly owing to geological and human activity. In many regions of the country, the quality of groundwater is unfit for drinking and irrigation, and such water may be harmful (Laghari *et al.*, 2012; Singh & Singh, 2002). An increase in groundwater salinity is a key concern for agricultural use. In numerous Indian states, excessive quantities of fluoride, nitrate, iron, arsenic, total hardness, and other harmful metal ions have been detected (Laghari *et al.*, 2012; Singh & Singh, 2002).

The purity of groundwater determines its usefulness for various uses. The purpose of water quality tests is to determine the acceptability of water for various usage without adverse impacts. The biological, chemical, and physical aspects of water comprise its quality (Lerner & Harris, 2009; Tripathi & Singal, 2019; Wutich *et al.*, 2020). Therefore, numerous scientific measurements are employed to characterize water quality, and these measurements are often based on the water's intended usage. To protect specific water uses, water quality recommendations give basic scientific knowledge regarding water quality measures and ecologically appropriate toxicological threshold values. The most common criteria used to evaluate water quality are to drinking water, irrigation water, ecosystem health, human contact safety, etc. (Voulvoulis *et al.*, 2017).

4.4 WHO DRINKING WATER QUALITY CRITERIA

In 1984 and 1985, WHO published its first version in three volumes. The primary objective of these rules is to protect public health by providing potable water. In 1988, revised drinking water recommendations were begun and published in three volumes. Several nations utilize these principles to establish national standards to assure the safety of public water sources. The last edition of WHO water quality guidelines was issued in 1993 (WHO, 2009; Yamamura, 2001).

Water quality index (WQI) labeling: The Canadian Council of Ministers of the Environment (CCME) has established the classification index for water quality as bad, marginal, fair, good, or exceptional. The number ranges from 1 to 100, with 1 indicating the lowest water quality and 100 indicating the highest. Table 4.1 provides the following designations.

4.4.1 DRINKING WATER STANDARDS IN INDIA

The drinking water quality standards approved in India are listed below. These standards are referenced in IS: 10500: 1991, the Indian standard. The comparison

TABLE 4.1
Water Quality Index Designations

Excellent	95–100
Good	80–94
Fair	65–79
Marginal	45–64
Poor	0–44

TABLE 4.2
Indian Standards & WHO Guideline for Drinking Water

Sl. No.	Parameter	BIS, Indian Standards (IS 10500:1991)		World Health Organization (WHO Guideline)
		Desirable Limit	Permissible Limit	Maximum Allowable Concentration
1	Colour	5 Hazen Units	25 Hazen Units	15 True Colour Units
2	Turbidity	5.0 NTU	10 NTU	5.0 NTU
3	PH	6.5–8.5	No relaxation	6.5–8.5
4	Total Hardness (as $CaCO_3$)	300 mg/L	600 mg/L	500 mg/L
5	Chlorides (as Cl)	250 mg/L	1000 mg/L	250 mg/L
6	Dissolved Solids	500 mg/L	2000 mg/L	1000 mg/L
7	Calcium (as Ca)	75 mg/L	200 mg/L	–
8	Sulphate (as SO_4^{2-})	200 mg/L	400 mg/L	400 mg/L
9	Nitrate (as NO^{3-})	45 mg/L	100 mg/L	10 mg/L
10	Fluoride (as F^-)	1.0 mg/L	1.5 mg/L	1.5 mg/L
11	Anionic Detergent (as MBAS)	0.2 mg/L	1.0 mg/L	–
12	Mineral Oil	0.01 mg/L	0.03 mg/L	–
13	Alkalinity	200 mg/L	600 mg/L	–
14	Boron	1.0 mg/L	5.0 mg/L	–
15	Zinc (as Zn)	5.0 mg/L	15 mg/L	5.0 mg/L
16	Iron (as Fe)	0.3 mg/L	1.0 mg/L	0.3 mg/L
17	Manganese (as Mn)	0.1 mg/L	0.3 mg/L	0.1 mg/L
18	Copper (as Cu)	0.05 mg/L	1.5 mg/L	1.0 mg/L
19	Arsenic (as As)	0.05 mg/L	No relaxation	0.05 mg/L
20	Cyanide (as CN)	0.05 mg/L	No relaxation	0.1 mg/L
21	Lead (as Pb)	0.05 mg/L	No relaxation	0.05 mg/L
22	Chromium (as Cr^{6+})	0.05 mg/L	No relaxation	0.05 mg/L
23	Aluminum (as Al)	0.03 mg/L	0.2 mg/L	0.2 mg/L
24	Cadmium (as Cd)	0.01 mg/L	No relaxation	0.005 mg/L
25	Selenium (as Se)	0.01 mg/L	No relaxation	0.01 mg/L
26	Mercury (as Hg)	0.001 mg/L	No relaxation	0.001 mg/L
27	Total Pesticides	Absent	0.001 mg/L	–

between Indian drinking water standards and WHO norms is provided in Table 4.2. The Bureau of Indian Standards (BIS) has established and released values as per Table 4.2 with desired and permitted limits.

The comparison of WHO drinking water guidelines for selected parameters against guidelines from the European Union (EU), United States Environmental Protection Agency (USEPA), and India is presented in Table 4.3.

The presence of arsenic and fluoride in excess of the limit suggested for safe drinking water constitutes India's principal groundwater quality problems. In 1980, the first report of arsenic in groundwater was made in West Bengal, India. In West

TABLE 4.3

Comparison of Drinking Water Guidelines for Selected Parameters

Parameters	WHO	EU	USEPA	India
pH	6.5–8	6.5–9.5	6.5–8.5	6.5–8.5
Chloride	250 mg/L	250 mg/L	250 mg/L	250
Iron	0.3 mg/L	0.2 mg/L	0.3 mg/L	0.3
Lead	0.01 mg/L	0.01 mg/L	0.015 mg/L	0.05
Arsenic	0.01 mg/L	0.01 mg/L	0.01 mg/L	0.05
Copper	2.0 mg/L	2.0 mg/L	1.3 mg/L	0.05
Faecal coliform bacteria	0 Counts/100 mL	0 Counts/100 mL	0 Counts/100 mL	

Bengal, 79 blocks in 8 districts exceeded the allowed level of 0.01 mg/L for arsenic. Arsenic in groundwater is primarily found in the middle aquifer between 20 and 100 m deep. The deeper aquifers are not contaminated with arsenic. The poisoning of groundwater with arsenic has also been discovered in the states of Bihar, Chhattisgarh, Uttar Pradesh, and Assam, in addition to West Bengal (Basu *et al.*, 2015; Roychowdhury *et al.*, 2005; Samanta *et al.*, 2004).

The available options for eliminating arsenic from water are:

- Groundwater extraction from arsenic-free aquifers.
- Provision of piped water from surface water sources.
- Diluted surface water with groundwater.
- Treatment of groundwater for arsenic removal by adsorption (Activated alumina/Granulated ferric hydrated oxide) or precipitation and coagulation process.
- Rainwater collection.

4.4.2 FLUORIDE

The bulk of India's rural population drinks and uses groundwater for residential reasons. The presence of fluoride in groundwater poses the greatest health risks associated with fluorosis. The allowable concentration of fluoride in groundwater is 1.5 mg/L. Andhra Pradesh (16), Assam (2), Bihar (5), Chhattisgarh (2), Delhi (7), Haryana (11), Gujarat (18), Jammu and Kashmir (1), Jharkhand (4), Kerala (2), Karnataka (14), Maharashtra (8), Madhya Pradesh (13), Orrisa (18), Punjab (9), Rajasthan (32), Tamil Nadu (8), Uttar Pradesh (7), and West Bengal (7) have also detected fluoride in groundwater, which is above the allowable concentration (Basu *et al.*, 2015; Roychowdhury *et al.*, 2005; Samanta *et al.*, 2004).

4.4.3 IRRIGATION WATER QUALITY

Water is one of the most critical inputs for sustainable agricultural production, which depends on the availability of sufficient quantities of high-quality water

(Vidhya, 2021). The quality of irrigation water is vital for increasing crop output and quantity, maintaining soil productivity, and protecting the environment. Three indicators are typically used to evaluate the quality of irrigation: total salt concentration, sodium adsorption ratio (SAR) of water, and boron content. However, the presence of trace elements in water can restrict its usage for irrigation (Aulakh & Malhi, 2005; Panigrahi *et al.*, 2021; Sharma *et al.*, 2015; Singh *et al.*, 2022).

4.4.4   SALINITY HAZARDS

Highly saline water is toxic to plants and poses a salinity issue. Soils with significant total salinity are referred to as saline soils. High soil salt concentrations can lead to a "physiological" drought situation. In other words, despite the field's seeming plenty of moisture, the plants wilt because their roots cannot absorb it (Avliyakulov *et al.*, 2020; El-Mahdy *et al.*, 2018; Shilev, 2020). Typically, TDS (total dissolved solids) or EC (electric conductivity) is used to quantify the salinity of water (EC). Electrical conductivity measures salt concentration in irrigation water (EC). Conventionally, saline water has been defined as possessing more than 1.5 mmhos/cm of total dissolved salts. Saline waters are those whose primary salt is sodium chloride (El-Shamy *et al.*, 2021; Ibrahim *et al.*, 2019; Mohanavelu *et al.*, 2021; Zaman *et al.*, 2018). Table 4.4 provides a classification of irrigation water based on its total salt level.

4.4.5   SODIUM ADSORPTION RATIO

In determining the quality of irrigation water, the SAR and the Residual Sodium Carbonate (RSC) are also major factors (Bhatti *et al.*, 2019; Kumar & Dabral, 2017; Uzen, 2017). Table 4.5 provides evaluation criteria for irrigation water quality.

More than 3 ppm of boron in irrigation water is detrimental to crops, particularly on light soils. The classification of irrigation water based on its boron content is shown in Table 4.6.

TABLE 4.4

Classification of Irrigation Water Based on Total Salt Content

Class	EC (ds/m)	Quality Characterization	Soils for Which Suitable
C1	<1.5	Normal waters	All soils
C2	1.5–3	Low salinity waters	Light and medium textured soils
C3	3–5	Medium salinity waters	Light and medium textured soils
C4	5–10	Saline waters	for semi-tolerant crops
C5	>10	High salinity waters	Light and medium textured soils for tolerant crops
			Not suitable

Source: Zaman *et al.* (2018).

TABLE 4.5

Guidelines for Evaluation of Quality of Irrigation Water

Water Class	Sodium (Na) (%) SAR	Electrical Conductivity (μmhos/cm) at 25°C	RSC (meq/1)	Alkalinity Hazards
Excellent	<20	<250	<10	<1.25
Good	20–40	250–750	10–18	1.25–2.0
Medium	40–60	750–2250	18–26	2.0–2.5
Bad	60–80	2250–4000	>26	2.5–3.0
Very bad	>80	>4000	>26	>3.0

Source: Kumar & Dabral (2017); Zaman *et al.* (2018).

TABLE 4.6

Classification of Irrigation Water Based on Boron Content

Class	Boron (ppm)	Characterization	Soils Suitable
B1	3	Normal waters	All soils
B2	3–4	Low boron waters	Clay soils and medium textured soils
B3	4–5	Medium boron waters	Heavy textured soils
B4	5–10	Boron waters	Heavy textured soils
B5	>10	High boron waters	Not suitable

Source: Johnson & Zhang (2017); Zaman *et al.* (2018).

4.5 CONCLUSION

Surface and groundwater resources experiencing the threat in terms of the scarcity of water resources due to increasing demands from many sectors, especially irrigation need, drinking, and industrial water demand due to continuously growing populations. Consequence of climate change is a major concern for both surface and groundwater resources due to spatiotemporal variation in precipitation, elevating temperatures, and overall variation in water cycle causing the imbalance in the storage and availability of water. Degradation in the water quality of groundwater sources due to physical, chemical, and bacteriological pollution needs to be examined before its use for drinking, domestic, and irrigation use. Water quality standards and norms are helpful for correct categorization of groundwater use and also suggest the ameliorative measures for the improving the quality of water. Water quality norms for the use of agriculture need to be adopted for the good health of the soil, crop, and sustainability of agriculture.

REFERENCES

Aulakh, M. S. and Malhi, S. S. (2005). Interactions of nitrogen with other nutrients and water: Effect on crop yield and quality, nutrient use efficiency, carbon sequestration, and environmental pollution. *Advances in Agronomy* 86: 341–409.

Avliyakulov, M. A., Kumari, M., Rajabov, N. Q. and Durdiev, N. K. (2020). Characterization of soil salinity and its impact on wheat crop using space-borne hyperspectral data. *Geoinformation Support of Sustainable Development of Territories* 26(Part 3): 271–285.

Basu, A., Sen, P. and Jha, A. (2015). Environmental arsenic toxicity in West Bengal, India: A brief policy review. *Indian Journal of Public Health* 59(4): 295–298. https://www.ijph.in/article.asp?issn=0019-557X

Bhat, T. A. (2014). An analysis of demand and supply of water in India. *Journal of Environment and Earth Science* 4(11): 67–72.

Bhatti, E. U. H., Khan, M. M., Shah, S. A. R., Raza, S. S., Shoaib, M. and Adnan, M. (2019). Dynamics of water quality: Impact assessment process for water resource management. *Processes* 7(2). https://doi.org/10.3390/pr7020102

Boretti, A. and Rosa, L. (2019). Reassessing the projections of the world water development report. *NPJ Clean Water* 2(1): 1–6.

Chatterjee, R. and Purohit, R. R. (2009). Estimation of replenishable groundwater resources of India and their status of utilization. *Current Science* 96(12): 1581–1591.

Chauhan, N., Paliwal, R., Kumar, S. and Kumar, R. (2022). Watershed prioritization in lower Shivaliks region of India using integrated principal component and hierarchical cluster analysis techniques: A case of upper Ghaggar watershed. *Journal of the Indian Society of Remote Sensing* 50(6): 1051–1070.

Chavda, D. and Ravenscroft, P. (2020). Groundwater monitoring for sustainable water management. *NDCWWC Journal* 9(1 and 2): 54–59.

CWC (2010). Water and Related Statistics. Information System Organization. Water Planning and Project Wing, Central Water Commission, 1–264.

Dillon, P., Stuyfzand, P., Grischek, T., Lluria, M., Pyne, R. D. G., Jain, R. C., Bear, J., Schwarz, J., Wang, W. and Fernandez, E. (2019). Sixty years of global progress in managed aquifer recharge. *Hydrogeology Journal* 27(1): 1–30.

El-Mahdy, M. E., Abbas, M. S. and Sobhy, H. M. (2018). Investigating the water quality of the water resources bank of Egypt: Lake Nasser. *Conventional water resources and agriculture in Egypt, The handbook of environmental chemistry*. Springer Nature (pp. 639–655). DOI: 10.1007/698_2018_331

El-Shamy, A. M., Abdo, A., Gad, E. A. M., Gado, A. A. and El-Kashef, E. (2021). The consequence of magnetic field on the parameters of brackish water in batch and continuous flow system. *Bulletin of the National Research Centre* 45(1): 1–13.

Gandhi and Namboodiri (2009). Groundwater Irrigation in India: Gains, Costs and Risks. Indian Institute of Management, Ahmedabad (India), 1–38.

Green, T. R., Taniguchi, M., Kooi, H., Gurdak, J. J., Allen, D. M., Hiscock, K. M., Treidel, H. and Aureli, A. (2011). Beneath the surface of global change: Impacts of climate change on groundwater. *Journal of Hydrology* 405: 532–560.

Ground Water Year Book – India 2017-18. (n.d.). Central Ground Water Board, Ministry of Water Resources, River Development and Ganga Rejuvenation, Government of India, Faridabad page 1–109.

Harikumar, P. S. (2013). Water quality and sanitation. *Science and Technology in Transforming Women's Lives* 1: 70–78.

Hooda, P. S., Edwards, A. C., Anderson, H. A. and Miller, A. (2000). A review of water quality concerns in livestock farming areas. *Science of the Total Environment* 250(1–3): 143–167.

Ibrahim, W., Qiu, C., Zhang, C., Cao, F., Shuijin, Z. and Wu, F. (2019). Comparative physiological analysis in the tolerance to salinity and drought individual and combination in two cotton genotypes with contrasting salt tolerance. *Physiologia Plantarum* 165(2): 155–168.

Jha, B. M. and Sinha, S. K. (2009). Towards better management of ground water resources in India. *Quarterly Journal of the Royal Meteorological Society* 24(4): 1–20.

Johnson, G. and Zhang, H. (2017). Classification of Irrigation Water Quality. Oklahoma State University, Division of Agricultural Sciences and Natural Resources, Cooperative Extension Service, 2–3. http://dc.library.okstate.edu/cdm/singleitem/collection/AgCoop/id/667/rec/361

Jyrkama, M. I. and Sykes, J. F. (2007). The impact of climate change on spatially varying groundwater recharge in the Grand River watershed (Ontario). *Journal of Hydrology* 338: 237–250.

Kumar, M. and Dabral, N. (2017). Water quality of shallow groundwater: Case study. *Engineering Practices for Agricultural Production and Water Conservation: An Interdisciplinary Approach, November,* 167–204. https://doi.org/10.1201/9781315365954

Laghari, A. N., Vanham, D. and Rauch, W. (2012). The Indus basin in the framework of current and future water resources management. *Hydrology and Earth System Sciences* 16(4): 1063–1083.

Lakshminarasimhan, R. (2022). Solar-Driven Water Treatment: The Path Forward for the Energy–Water Nexus. In *Solar-Driven Water Treatment* (pp. 337–362). Elsevier.

Lerner, D. N. and Harris, B. (2009). The relationship between land use and groundwater resources and quality. *Land Use Policy* 26: S265–S273.

Majumder, A. and Sivaramakrishnan, L. (2014). Ground water budgeting in alluvial Damodar fan delta: A study in semi-critical Pandua block of West Bengal, India. *International Journal of Geology, Earth & Environmental Sciences* 4(3): 23–37.

Mohanavelu, A., Naganna, S. R. and Al-Ansari, N. (2021). Irrigation induced salinity and sodicity hazards on soil and groundwater: An overview of its causes, impacts and mitigation strategies. *Agriculture* 11(10): 983.

Mukherjee, A., Saha, D., Harvey, C. F., Taylor, R. G., Ahmed, K. M. and Bhanja, S. N. (2015). Groundwater systems of the Indian sub-continent. *Journal of Hydrology: Regional Studies* 4: 1–14.

Mushtaq, B., Bandh, S. A. and Shafi, S. (2020). Management of water resources. In *Environmental management* (pp. 1–46). Springer.

Nhemachena, C., Nhamo, L., Matchaya, G., Nhemachena, C. R., Muchara, B., Karuaihe, S. T. and Mpandeli, S. (2020). Climate change impacts on water and agriculture sectors in Southern Africa: Threats and opportunities for sustainable development. *Water* 12(10): 2673.

O'Mara, G. (1988). *Efficiency in irrigation: The conjunctive use of surface and groundwater resources*. The World Bank.

Oza, A. (2007). Irrigation: Achievements and Challenges Part I. http://www.iitk.ac.in/3inetwork/html/reports/IIR2007/07-Irrigation.pdf, assessed on 23 July, 2013.

Panigrahi, N., Thompson, A. J., Zubelzu, S. and Knox, J. W. (2021). Identifying opportunities to improve management of water stress in banana production. *Scientia Horticulturae* 276: 109735.

Robert, F. (2001). The origin of water on Earth. *Science* 293(5532): 1056–1058.

Roychowdhury, T., Tokunaga, H., Uchino, T. and Ando, M. (2005). Effect of arsenic-contaminated irrigation water on agricultural land soil and plants in West Bengal, India. *Chemosphere* 58(6): 799–810. https://doi.org/10.1016/j.chemosphere.2004.08.098

Saha, R., Kumar, G. P., Pandiri, M., Das, I. C., Rao, P. N., Reddy, K. and Kumar, S. N. K, V. (2018). Knowledge guided integrated geo-hydrological, geo-mathematical and GIS based groundwater draft estimation modelling in Budhan Pochampalli watershed, Nalgonda district, Telangana state, India. *Earth Science India* 11(4): 216–231.

Samanta, G., Sharma, R., Roychowdhury, T. and Chakraborti, D. (2004). Arsenic and other elements in hair, nails, and skin-scales of arsenic victims in West Bengal, India. *Science of the Total Environment* 326(1): 33–47. https://doi.org/10.1016/j.scitotenv.2003.12.006

Scibek, J. and Allen, D. M. (2006). Modeled impacts of predicted climate change on recharge and groundwater levels. *Water Resources Research* 42: W11405. doi:10.1029/2005WR004742

Scott, C. A. and Shah, T. (2004). Groundwater overdraft reduction through agricultural energy policy: Insights from India and Mexico. *International Journal of Water Resources Development* 20(2): 149–164.

Shah, T., Molden, D., Sakthivadivel, R. and Seckler, D. (2001). Global groundwater situation: Opportunities and challenges. *Economic and Political Weekly* 36(43): 4142–4150.

Shankar, P. S. V., Kulkarni, H. and Krishnan, S. (2011). India's groundwater challenge and the way forward. *Economic and Political Weekly* 46(2): 37–45.

Sharma, B., Molden, D. and Cook, S. (2015). Water Use Efficiency in Agriculture: Measurement, Current Situation and Trends.

Shilev, S. (2020). Plant-growth-promoting bacteria mitigating soil salinity stress in plants. *Applied Sciences* 10(20): 7326.

Siebert, S., Burke, J., Faures, J. M., Frenken, K., Hoogeveen, J., Doll, P. and Portmann, F. T. (2010). Groundwater use for irrigation – A global inventory. *Hydrology Earth System Sciences* 14: 1863–1880.

Siebert, S., Henrich, V., Frenken, K. and Burke, J. (2013). Update of the Digital Global Map of Irrigation Areas to Version 5. Rheinische Friedrich-Wilhelms-Universität, Bonn, Germany and Food and Agriculture Organization of the United Nations, Rome, Italy.

Simonovic, S. P. (2002). World water dynamics: Global modeling of water resources. *Journal of Environmental Management* 66(3): 249–267.

Singh, V. K., Rajanna, G. A., Paramesha, V. and Upadhyay, P. K. (2022). Agricultural water footprint and precision management. *Sustainable Agriculture Systems and Technologies* 251–266. https://doi.org/10.1002/9781119808565

Singh, S., Raju, N. J. and Ramakrishna, C. (2015). Evaluation of groundwater quality and its suitability for domestic and irrigation use in parts of the Chandauli-Varanasi region, Uttar Pradesh, India. *Journal of Water Resource and Protection* 7(7): 572.

Singh, D. K. and Singh, A. K. (2002). Groundwater situation in India: Problems and perspective. *International Journal of Water Resources Development* 18(4): 563–580.

Sinha, A. R., Prakash, S., Singh, K., Kumar, S. and Murari, K. (2020). Study on scarcity and remediation of ground water recharge in Patna region. 10(6): 24–32. https://doi.org/10.9790/9622-1006072432

Sishodia, R. P., Shukla, S., Graham, W. D., Wani, S. P. and Garg, K. K. (2016). Bi-decadal groundwater level trends in a semi-arid south Indian region: Declines, causes and management. *Journal of Hydrology: Regional Studies* 8: 43–58.

Sophocleous, M. (2004). Global and regional water availability and demand: Prospects for the future. *Natural Resources Research* 13(2): 61–75.

Thampi and Raneesh (2012). Impact of anticipated climate change on direct groundwater recharge in a humid tropical basin based on a simple conceptual model. *Hydrological Processes* 26: 1655–1671.

Tripathi, M. and Singal, S. K. (2019). Use of principal component analysis for parameter selection for development of a novel water quality index: A case study of river Ganga India. *Ecological Indicators* 96: 430–436. https://doi.org/10.1016/j.ecolind.2018.09.025

Uzen, N. (2017). Use of wastewater for agricultural irrigation and infectious diseases. Diyarbakir example. *Journal of Environmental Protection and Ecology* 17(2): 488–497.

Van der Gun, J. (2012). Groundwater and Global Change: Trends, Opportunities and Challenges.

Velmurugan, A., Swarnam, P., Subramani, T., Meena, B. and Kaledhonkar, M. J. (2020). Water demand and salinity. *Desalination-Challenges and Opportunities.*

Vidhya, V. (2021). Farmers: Issues, challenges and future strategies. In P. D. T, A. Kumar, S. Mukherjee and S. Mandal (Eds.), *Indian agriculture, farmer and labour: Issues and reforms* (p. 104). Bharti Publications.

Voulvoulis, N., Arpon, K. D. and Giakoumis, T. (2017). The EU water framework directive: From great expectations to problems with implementation. *Science of the Total Environment* 575: 358–366. https://doi.org/10.1016/j.scitotenv.2016.09.228

WHO. (2009). Boron in Drinking-Water. WHO, Geneva, Switzerland, 1–36. http://scholar.google.com/scholar?hl=en&btnG=Search&q=intitle:N-Nitrosodimethylamine+in+drinking-water+Background+document+for+development+of+WHO+Guidelines+for+Drinking-water+Quality#0

Wutich, A., Rosinger, A. Y., Stoler, J., Jepson, W. and Brewis, A. (2020). Measuring human water needs. *American Journal of Human Biology*, *32*(1), 1–17. https://doi.org/10.1002/ajhb.23350

Yamamura, S. (2001). Drinking Water Guidelines and Standards. United Nations Synthesis Report on Arsenic in Drinking Water, 18.

Zaman, M., Shahid, S. A. and Heng, L. (2018). Irrigation water quality. In *Guideline for salinity assessment, mitigation and adaptation using nuclear and related techniques* (pp. 113–131). Springer.

Zektser, I. S. and Everett, L. G. (2013). Groundwater Resources of the World and Their Use, UNESCO IHP-VI Series on Groundwater No 6, UNESCO, Paris, France, http://unesdoc.unesco.org/images/0013/001344/134433e.pdf, last access: 28 June 2013.

5 Climate-Resilient Agriculture System
An Approach to Integrated Farming System (IFS)

Ananya Mishra, Rohitashw Kumar,
and Munjid Maryam

5.1 INTRODUCTION

The agricultural sector in India faces various and dynamic challenges, including maintaining the protection of the natural resources, the negative impact on climate, decreasing factor production, various nutrient and nutrient mining shortages, over-exploitation of groundwater resources, soil erosion due to heavy tillage process, and reduces the organic carbon and a shrinking pattern in landholding size, both of which are expected to worsen with time, which are some of the more common issues across the world, resulting in a stagnation in system productivity. India's population is about 1.30 billion people and more than 70% of the Indian population lives in countryside areas, where agriculture practice is the key source of revenue. With an average farm size of 1.57 ha, Indian agriculture is defined by small farm holdings (Jayanthi, 2018). The Indian economy is primarily based on agriculture. Minor and marginal agriculturalists are the backbone of the Indian economy, accounting for 85% of the overall agricultural population but just 44% of total effective land (GoI, 2014). India's active farm holdings are also shrinking. Due to immense demographic pressure on the limited land assets available for agriculture, the normal extent of land holding has decreased by three times in Bihar and Kerala over the last four decades, whereas it has decreased by more than two times in Madhya Pradesh, Andhra Pradesh, Maharashtra, and Karnataka (NABARD Rural Pulse, 2014). As a result of these threats, both farmers and the scientific community have established a wide variety of agricultural methods that, in theory, could improve agricultural systems' resilience to climate change (Wezel *et al.,* 2014). Climate change has direct and indirect consequences on agronomic production, such as shifting rainfall levels, drought, floods, and insect and disease regional displacement (IPCC [Intergovernmental Panel on Climate Change], 2013). In addition to these immediate challenges, these countries struggle for lack of proper tools to plan and deal with natural hazards. Furthermore, due to its dependency on weather and the fact that farmers appear to be poorer than their urban counterparts, the agricultural segment is the most vulnerable to climate change (Rao *et al.,* 2019). Given the ramifications of climate change, increasing and maintaining agricultural production are crucial for securing adequate food and

DOI: 10.1201/9781003351672-5

nourishment stability for upcoming generations in India (Rao *et al.*, 2016). Climate challenges are better tackled by growing adaptation capability and tolerance, which reduce the negative effects of climate variation (FAO, 2013). Changing environments can cause farmers to abandon traditional farming practices in favour of climate-resilient agriculture (CRA). CRA activities and technology, in general, are low-emission techniques that seek to increase food quality and resilience (Sain *et al.*, 2017). To ensure protections, CRA activities should be based on conventional practices while integrating modern science, and farmers should be involved in the production process. Investment risk, access to capital, extension programmes, and farm supplies, as well as enhanced facility governance by farmer organisations, should all be considered while developing CRA practice (Gentle and Maraseni, 2012; Issaka *et al.*, 2016). The arrangement and operation of a farming method can be used to explain and comprehend it. In a broader context, the system covers land use patterns, supply ties, land tenures, holding scale and delivery, irrigation, marketing (including transportation and storage), lending unions, capital services, science, and education (Walia *et al.*, 2019). The components of farming systems communicate in a chain, making it impossible to work with such complex interconnected systems. The integrated farming system (IFS) method has been broadly encouraged for enhancing production, productivity, income, and soil quality in various ecological surroundings across India (Balamatti and Shamaraj, 2017). Improving annual crop yield will continue to be the goal of any farming system's productivity growth. Effective rainwater harvesting, appropriate sowing and tillage activities, improved variety selection, suitable intercropping and seed rotations, productive soil fertility control, proper tree conservation steps, and crop preparation are the main elements for improving crop productivity in the area (Tanwar *et al.*, 2018a,b). Micro and marginal farmers find it impossible to maximise farm productivity by spending large sums of money on concentrated farm operations. In this case, IFS plays a critical role in enhancing farm productivity while requiring less capital (Reddy *et al.*, 2020). This can be accomplished by making the best use of energy, recycling garbage, and using domestic labour. Waste is used as a supply in IFS, which removes waste in the environment while also increasing farm productivity and decreasing production costs (Gupta *et al.*, 2012). Many studies on IFS have been practiced in numerous countries and in several parts of India, and they have discussed issues relating to increasing the efficiency of agricultural production systems (Rathore *et al.*, 2019), enhancing total production, profitability, gainful jobs, effective resource recycling, and resource use efficiencies by leveraging complementarities, and creating interconnections between various agricultural subsystems and entrepreneurs. It is a concept of environmental reliability that leads to sustainable cultivation, as well as a dependable way of obtaining a reasonably high output with significant manure pricing. Furthermore, small increases in land production are no longer adequate to meet the needs of poor farmers' limited resources. As a result, intelligent resource management, including optimal resource distribution, is critical to reducing the possibility of land survival. The Ministry of Agriculture and Farmers' Welfare, Government of India, has put a strong focus on the IFS as part of its mission to double farmers' income by 2022 (Kumar 2018). Furthermore, a thorough understanding of the components' relationships and interconnections would enhance food security.

5.2 CONCEPTS OF INTEGRATED FARMING SCHEME (IFS)

The scientific combination of various dependent and interrelating farm venture for the productive use of soil, manpower, and other capital of a farm household that offer farmers all-year income is known as an integrated farming method. Within a working farm unit (or a group of units), the IFS comprises an entire compound of construction, resource allocation, and management as well as decisions-making and actions, which are the outcomes as agricultural products processing, and endorsement of the farm products. It's a holistic approach to agricultural management that balances the ecological importance of a classified and balanced ecosystem with agriculture's economic demands to maintain a steady supply of nutritious and inexpensive food. IFS combines poultry, fish, forestry, farming, and agro-based industry in a cooperative system in which wastes from a single process are used as inputs for another processes, perhaps without or with care, to offer production means, like fuel, for maximum productivity, use fertiliser, and feed at the lowest possible cost. The integrated farming method is a complex philosophy that must be adaptable to any farm, in any region, and be open to change and technical advancements at all times. A general feature of these schemes is that they combine crop and livestock businesses, and in some situations, they can involve a mix of both trees and aquaculture (Kumar 2018). A farming system is a group of agricultural businesses in which owners allocate money for the most productive use of the individual enterprises in order to boost farm production and profitability. It is a part of farming systems that considers concepts such as mitigating risk, the overall output and income through recycling and minimising external inputs, and optimising the use of biodegradable waste and agricultural wastes. The distinction between integrated farming and mixed farming may be that in integrated farming systems, businesses communicate ecologically, in time and space, are equally supportive, and depend on one another (Kumar *et al.*, 2018). Above all, the IFS is a holistic approach to agriculture that benefits society as a whole, not just those that practise it. Table 5.1 describes the steps of farm operations, that is the inputs, processes, and outputs of a farm make it a network.

Crops are grouped in a tier structure in the space concept, mixing at least two crops of different field durations intercropping them by appropriately changing the plantation process. For the majority of marginal farmers, income from arable crops is inadequate. The aim of sustainability is to use inputs in a way that does not degrade the condition of the ecosystem in which it communicates. As a result, it is clear that

TABLE 5.1

Steps of Farm Operations

Steps	Details
Inputs	These are the items that are used on the field and are classified into two groups: Physical (e.g. rainfall) and human (e.g. work, money, and labour)
Processes	These are the actions that occur on the farm in order to convert inputs into outputs (e.g. sowing, irrigation, and harvesting)
Outputs	These would be the farms' commodities (i.e. wheat, barley, and cattle)

farming is a method, the aim of which is to achieve long-term production sustainability. The ultimate goal is to develop theoretically feasible and commercially viable farming system models for irrigated, hilly, costal, and rainy area by combining cropping with allied enterprises in order to produce income and jobs from the field.

5.3 THE GUIDING PRINCIPLES: INTEGRATED FARMING SCHEME

Agriculture's development mechanism must be sustainable and resilient, with increased efficiency, higher wages, job creation, risk reduction, resource use and protection, and climate change mitigation and adaptation both becoming priorities.

Increased productivity: As the world's population grows, so does the need for fruit, meat, fodder, and fibre, but the amount of land available for agriculture shrinks (Rathore *et al.*, 2019). Dairy + crop, horticulture + crop + goats, crop + goats + dairy + horticulture, crop + horticulture + pigs, poultry +dairy + vegetables, dairy + vegetables + horticulture, dairy + vegetables, and dairy + crop + other farm animals are the main components of IFS (Tripathi and Rathi, 2011). CAZRI conducted a survey in a hot arid area that revealed productive crops (pearl millet, maize, moth bean, and sorghum) are combined with *Hardwickia binata*, *Prosopis cineraria*, and Z. In comparison to 1.77 t/ha under single cropping, mauritiana produced 2.50, 2.95, and 7.55 t/ha (pearl millet equivalent yield). As a result, the IFS had a 1.4 to 1 ratio. Yields are 4.3 times greater than single cropping (Tanwar *et al.*, 2018a,b). Under such systems, pigs are restricted in crop fields long before cultivation and the field is ploughed by excavating the roots. After harvesting for fertilising the soil, chickens are used to remove rotting fruit and weeds from orchards and vineyards. In farms with both cropland and pasture, cattle or other animals are permitted to graze crops in crops centre. Water-dependent agriculture systems allow for the productive and effective farm nutrient recycling, resulting in the produce of diesel, nutrients, and mineralised irrigation water made from compost tea. Animal houses were built over a reservoir, allowing animal waste to fall straight into the pond, where fish drink (Kumar *et al.*, 2018). Waterway irrigation (from the IGNP, Bhakra, and Gang canals), electricity to rural areas (which boosted groundwater irrigation), field automation, new variety of crops and processing methods, road building, and increased marketing opportunities have all resulted in significant shifts in land use and farming methods in arid regions (Kar, 2014). Jholmal, a type of organic fertiliser and organic pesticide derived from excrement, faeces, water, and additives (which include locally available organic matter), was familiarised in Nepal as element of CRA operations (Katovich, 2014). The choice of these specific practices was inspired by established signals that small land holder adaption of CRA exercise was difficult due to the large investment and expensive expanding requirements, business network, and regional structures of authority (Issaka *et al.*, 2016). Rural regions now suffer from labour scarcities as a consequence of migration. Due to the high labour requirements of vegetable farming, the formation of farmers' groups can result in labour being shared between farm households (Rai *et al.*, 2018).

Generation of employment opportunities: Crop-based agriculture is extremely seasonal, with labour demand peaks at any point in the year, and farmers do not have a sufficient job in the remainder of the year. The IFS has the capability of generating new jobs and a fairer allocation of jobs across the year, while ensuring that local workers are constantly sinking. The IFS is a work-intensive scheme, creating on-farm jobs, and the farmers and their family members contribute much of the work needed for the development process (Dasgupta *et al.*, 2015). Almost 80% of the world population living in rural areas is dependent on agriculture alone for revenue and nourishment demands. Steps must be taken in the agricultural field to tackle the issues of food shortage, hunger, and ecology in sustainability. Returns have to be increased as environmental protection and profitability, economic equity and higher production must be preserved (Iram *et al.*, 2020). Not only are healthy soils crucial for agricultural production, but they can also supply important environmental resources through carbon sequestration, contaminants neutralisation, and erosion deterrence. If all these requirements are met, agriculture is said to be a sustainable agriculture practice (Nakajima, 2020; Williams *et al.*, 2020). Without additional labour, a family with children can manage 0.75 acre. One hectare of mulberry garden provides year-round employment to 13 persons in irrigated conditions. Regular practice takes 607-man days and 827-women days to complete the same task (Rana and Chopra, 2013). The focus should be on improving the status of various elements of the current farming system, developing their compatibility with the prevailing agricultural system, reducing the reliance on outside production, carrying high risks, and being able to produce more incomes and jobs every day.

Reduction of risk: A sole commodity-based agriculture is more vulnerable in comparison with IFS to climate change, biotic and economic changes (relative input and export prices). Strong pluvial variability in more marginal areas will also raise the risk of development, and the variability of rainfall is expected to increase in the future (Thornton *et al.*, 2014). In the agricultural system, no systematic assessment has yet been made of the impacts and various reactions by livestock and crop farmers to climate change. The variety of risk-management techniques that farmers currently use is well known, but there are growing evidence that these might not be adequate in the long term (Thornton and Herrero, 2015). Increase in price led farmers and other participants in the distribution network to rearrange risk assessment of their prices (Assefa *et al.*, 2017). Swarnam and Velmurugan (2020) describe the IFS role in providing entrepreneurs with the chance to diversify and improve their resources (Fig. 5.1).

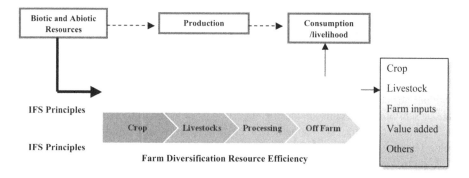

FIGURE 5.1   IFS role in providing entrepreneurs with the chance to diversify and improve their resources. (Swarnam and Velmurugan, 2020.)

To improve adaptability that means ability to modify input, output, in reaction to shocks and demands, marketing and risk management are used, without modifying the farming system's processes and feedback mechanisms (Meuwissen *et al.*, 2019). The application also demonstrates that evaluation of agricultural system resilience needs to take into account all the problems and not dwell on one particular issue, as is usually the case for risk assessment studies.

5.4   IFS OBJECTIVES

- A combination of forestry, fisheries, livestock, dairy, champagne and beekeeping, and all these operations can be carried out on marginalised and swamps, too.
- By minimising cost of production by processing waste and by-products of one company, IFS enhances land efficiency and profitability as inputs to other companies.
- In IFS, waste materials and by-products, such as cultivable residuals and animal wastes, are effectively recycled. There is also less dependency on external inputs, leading to a more robust manufacturing environment. Soil productivity restructuring is possible by means of organic manure, processing of compost, use of vegetables in a cultivation scheme, etc.
- The IFS creates a dependable and sustainable productive system through varied crops and companies in order to decrease risks and reduce resistance to climate change.
- Protecting the atmosphere and improving stability over a longer term to preserve the basis of natural resources.
- Provide sustainable and economically viable technology which involves the reasonable use of the regional resources.
- The IFS solution effectively recycles waste materials and nutrients, which allows productive use of capital by connecting relevant companies and components and less reliant on external inputs, while eliminating environmental emissions due to high use of external inputs.
- Throughout the year, the integrated farming system provides farmers with a supply of money through the selling of diverse farms, including milk, egg, mushroom, tomatoes, fruit, and food grains.
- To ensure steady and secure income and to improve production/rejuvenate the system, the aims of integrated agriculture systems (IFS) are as follows:
  - To maintain agro-ecological balance through reducing pests and pathogens, by managing the natural crop system and by reducing the usage of (in-organic fertilisers and pesticides) chemical products.
  - Various elements are interconnected to produce varieties of items and provide the agriculture family a healthy diet.

5.5   CLIMATE CHANGE NEXUS WITH AGRICULTURE

Climate change is the most severe ecological concern confronting human, having ramifications for natural ecosystems, agriculture, and water resources (Senapati, 2009). Climate change includes rising temperatures, changes in rainfall pattern, and higher

CO_2 concentration levels in the atmosphere. According to general circulation models (GCMs), growing levels of greenhouse gases (GHGs) are anticipated to raise global mean surface temperature by 1.5–4.5°C within the next 100 years (Senapati *et al.*, 2013). This will cause sea levels to rise, climate zones to migrate towards pole, and soil moisture levels to diminish. Climate change is expected to have an impact on the agricultural sector as it may alter precipitation patterns, posing a serious threat to soil health (Sharma *et al.*, 2015). These changes could impact the increased CO_2 concentration in the atmosphere that can affect the rate at which plant species and wild weeds develop. Concentration levels have impact on yield (both amount and quality), rate of growth, chlorophyll content and evapotranspiration rates, moisture content, and farm requirements such as pesticides, fungicides, and fertilisers, among other things. Carbon dioxide in the environment impacts agriculture, which can affect plant and animal productivity by changing precipitation, temperature, and radiation levels. Inundation of agriculture and high salinity of groundwater on coastlines may result from rising sea levels (Mahato, 2014). Productivity may be affected by environmental factors such as frequency of soil drainage and severity (resulting in soil leaching), soil erosion, soil conditions, and crop diversity loss. In India, precipitation is expected to increase by 10–15% in several locations, with a simultaneous precipitation reduction of 5–25% in drought-prone central India and a severe decline in winter rainfall in northern India (Senapati *et al.*, 2013). Soil erosion becomes a severe issue, particularly India is vulnerable to desertification due to its arid and semi-arid climate. Soil erosion is exacerbated by the removal of natural flora and cultivation on sandy land and marginal lands.

5.6   CLIMATE CHANGE IMPACT ON WATER RESOURCES

Rise in temperature often leads the changes in hydrological cycle which intensify evaporation and rainfall rates. That is, rising temperatures may alter patterns of precipitation, runoff distribution (both spatially and temporally) groundwater reserves, soil moisture, as well as occurrence of droughts and floods events (Schulze, 2011). Consequently, changes in temperature have a direct impact on the availability of water resources in India. Climate variability and change studies are particularly important for the Indian subcontinent because rainfall is essential for economic and social advancement. The functioning of water resources in the country is heavily influenced by climatic circumstances. Around 80% of rainfall occurs during four monsoonal months and may not be evenly distributed across the country, causing scarcity in some areas. As a result, substantial water storage facilities are required to supply demand during the dry season (Mehrotra and Mehrotra, 1995). The impact of climate change on the water balances of the basin has been identified. Land use and water usage practices, on the other hand, play a role in determining if and how communities are impacted in those altering areas (Hagemann *et al.*, 2013).

5.7   CLIMATE RESILIENCE THROUGH IFS

India is sensitive to climate change due to its large population's reliance on agriculture sector, excessive demand for natural resources, and inadequate coping strategies. One of the primary issues facing modern agriculture today is producing enough food for a growing population while limiting negative effects on ecosystems. Human intervention has a significant impact on natural ecosystems, altering biogeochemical cycles and changing biotic communities, resulting in a loss of biodiversity and ecosystem health. This shift in focus corresponds to the requirement for site-specific analyses in order to identify appropriate agricultural technologies and methods for climate-smart agriculture. In modern agriculture practices, traditional rice intensification systems are subjected to heavy doses of various fertilisers, adversely affecting ecosystem quality in general, thus promoting the adoption of integrated systems such as rice-duck, rice-fish, and rice-duck-fish method with significantly reduced chemical fertiliser and pesticides, resulting in minimal environmental health degradation. IFS' main goal is to recycle garbage in order to save money on manures and fertilisers while also providing an environmentally pleasant working environment (Panwar *et al.*, 2021). Plantation crops, dairy, sheep, orchards, goats, fish, pigs, poultry, mushroom, and biogas are integrated with the goal of meeting household demand for the 4Fs (fodder, food, fuel, and feed), as well as upcycling to meet the nutrient requirements of various environmental components. Future mechanised agricultural production trends may also adopt the integrated farming system, where environmental pollution and climate change effects are significantly reduced, as well as agricultural operational cost reduction/ minimisation, which may be advantageous factors for farmers to adopt the integrated farming system in the future. This sort of agriculture involves economic, social, and ecological components that must all be considered at the same time. Individual farm productivity and social systems are immensely variable, implying that small-scale farmers ability to adjust to climate change and variability will be similarly vary, resulting in diverse farm-scale climate consequences (Antwi-Agyei et al., 2012). Rice-duck-fish integration is mutually beneficial and sustainable since the integrated system uses less input energy and agricultural waste while simultaneously serving as a storage for optimal nutrient recycling (Nayak *et al.*, 2016). Integrated system may be one way to boost the sustainability of tropical agroecosystems by bringing more variability of vegetation inside the farm, which is an essential driver of soil bacterial response to ecological changes. Integrated systems have the ability to contribute to land conservation by minimising the need to open new areas while avoiding the negative consequences of agricultural production. Land conservation through agriculture strengthening is frequently linked to higher use of agrochemicals and single culture systems in order to maximise production, potentially resulting in biodiversity decline both within the production area and in natural habitat as a side effect, not necessarily protecting agriculture expansion. As a result of these and faster migration, climate change has the potential to reduce population biological variation, which could have an impact on ecosystem functions and resilience (Mishra *et al.*, 2021). To operate farm equipment with resilience, solar energy can be used as a power source for green operations, including food-related integrated activities (Richa *et al.*, 2021). Considering IFS has a larger recycling potential, small and marginal farmers adapting IFS systems are less reliant on purchased inputs. As a result of the greater diversity of agricultural by-products, small land holder mixed farms may be less susceptible to climate change and crop failure (Paramesh et al., 2022).

5.8   SUSTAINABILITY THROUGH IFS

In agriculture, sustainable growth must provide a comprehensive and economy-friendly farming method (IFS) with productive soil, aquatic planting and pest control practices. The system's nutrient recycling promotes the system's self-sustainability which not only reduces its dependency on external inputs, seeds/fertilisers etc. but also provides the farm family with equivalent and rich food supplies with a lower agricultural cost and a larger profit margin for the same land that is a key factor in maintaining sustainability (Kumar *et al.*, 2018). Agricultural system study and implementation are recognised as a promising method to achieve the multiple objectives of hunger, food stability, prosperity, and sustainability. By integrating additional businesses in an integrated agriculture system, the efficiency and viability of the dominant agricultural systems are improved to ensure the safety of marginal and small farmers and, at the same time, the sustainable agricultural ecosystem and environment. In the IFS model, the average total NPK need was 285.5, 116.3, and 109.9 kg/ha. Moreover, fish pond silt was excavated once every three years and mixed into the soil. An excavation containing a 15-cm deep terrestrial surface area of 800 m^2 pond areas saved about 950 rupees also contributed a total of 18.56 kg nitrogen, 6.21 kg phosphorus, and 74.24 kg potassium. With an average value of 0.95, the OC (Organic Cotton) percentage of pond soil was 1.20. The budgeting of nutrients explicitly promotes the independence of the system, which not only reduces the dependence on foreign supplies but also spares money wasted on expensive chemical fertilisers (Singh *et al.*, 2012). The IFS solution is therefore highly sensitive to the complexities of organic farming and the entire method can be turned into organic farming by IFS operation in subsequent years. While the meaning of rice-fish-duck co-cultivation is well established, the ecological processes underlying sustainability of the system have not been studied in depth and studies on many aspects of integrated rice-based agro-systems in particular are limited. The establishment of joint indices for better evaluations of habitat efficiency and biodiversity was highlighted in modern studies (Che-Salmal et al., 2017). The fundamentals of the integrated rice-duck-fish method consist of the productive use by others of waste and the activities of one part, turned into useful goods, thus improving the sustainability, efficiency, and profitability of farms. Moving to integrated systems will increase the quality of the environment, reliability and machine viability. An evaluation of soil and water quality physico-chemical and biological variables will help describe the ecological process and evaluation of agro-ecosystems qualities (Nayak *et al.*, 2018a,b). Biodiversity in soil and water is important in order to contribute to a broad variety of ecological resources that can work in natural and maintained environments (Wagg *et al.*, 2014). Rice is combined with other resources such as fish and duck is important to ensure soil quality, the ecology of organisms, and sustainability of production in order to solve this problem. In this protocol, the fertilisation and plant safety standards are the most common feature and additional constraint relating to the labour practices, crop rotation, etc. is provided by a set of principles and practices which are standardised and codified in the related protocols (Vlahos *et al.*, 2017).

5.9 CONCLUSIONS

The findings of this study indicate that CRA practices will improve farmers' resilience, increase farm incomes, and create jobs at local level. CRA has also received assistance from the farmers involved. It does, however, take more labour and more resources than traditional farming. Farmers who want to supplement their income by cultivating seasonal vegetables can do so under this system. The effect on agriculture that will vary with climate change severity is likely to be substantial. Climate change was highly sensitive to agriculture, particularly crops. Food safety at global level will probably not be endangered, but certain areas will certainly suffer from food shortages and starvation. The adoption of an independent agricultural company by itself can't support the agricultural family, but the IFS solution promises to tackle the issues of Indian farming communities' sustainable economic development. The integrated agriculture system is considered an important instrument for managing natural and human resources as well as for solving small- and medium-scale farmer problems in developing nations like India very effectively. This multidisciplinary whole-farm strategy aims to boost the income and jobs of smallholders through the integration into the farm of numerous agricultural companies and the processing of crop residues and by-products. The most adequate and effective model for farming systems gives the highest productivity of irrigated agro-ecosystems, comprising plant components, milk production, poultry, and fishery. In participatory studies of multilevel approaches in the fields of farmers, the IFS models for various environment-friendly landscapes and subsystems can be tuned. Real location-based technological expertise, multidimensional position to satisfy domestic necessity, avenues for jobs, the use of capital in a reasonable way, sustained competitiveness, investment ability, and framework are required for economic capacity. IFS contributes to conservation of capital through agricultural and animal waste created in a system are effectively recycled by means of compost, thereby mitigating environmental hazards in inorganic fertiliser and agrochemicals. Furthermore, external input spending would be reduced. This means that the scheme is commercially and environmentally sustainable.

ACKNOWLEDGEMENT

We wish to express our heartfelt gratitude to the division of Soil and Water Engineering, and Division of Irrigation and Drainage, College of Agricultural Engineering and Technology, Sher-e-Kashmir University of Agricultural Sciences and Technology.

REFERENCES

Antwi-Agyei, P., Fraser, E.D., Dougill, A.J., Stringer, L.C. and Simelton, E. 2012. Mapping the vulnerability of crop production to drought in Ghana using rainfall, yield and socio-economic data. *Applied Geography* 32(2): 324–334.

Assefa, T.T., Meuwissen, M.P. and Lansink, A.G.O. 2017. Price risk perceptions and management strategies in selected European food supply chains: An exploratory approach. *NJAS-Wageningen Journal of Life Sciences* 80: 15–26.

Balamatti, A. and Shamaraj, H. 2017. Participatory evaluation of choice and combination of enterprises for integrated farming system under dry-land and irrigated agro-ecosystems. *Indian Journal of Agronomy* 62(1): 8–15.

Che-Salmal, M.R., Sureger, A.Z., Hasan, A.A. and Nasution, Z. 2017. Dynamics of aquatic organisms in a rice field ecosystem: Effects of seasons and cultivation phases abundance and predator-pray interactions. *Tropical Ecology* 58: 177–191.

Dasgupta, P., Goswami, R., Ali, M., Chakraborty, S. and Saha, S. 2015. Multifunctional role of integrated farming system in developing countries. *International Journal of Bioresource and Stress Management* 6(3): 424–432.

FAO. 2013. Climate change guidelines for forest managers. Rome

Gentle, P. and Maraseni, T.N. 2012. Climate change, poverty and livelihoods: Adaptation practices by rural mountain communities in Nepal 2. *Environmental Science & Policy* 1: 24–34.

GoI. 2014. Agricultural statistics at a glance, Directorate of Economics and Statistics, Govt. of India, New Delhi.

Gupta, V., Rai, P.K. and Risam, K.S. 2012. Integrated crop-livestock farming systems: A strategy for resource conservation and environmental sustainability. *Indian Research Journal of Extension Education* 2: 49–54.

Hagemann, S., Chen, C., Clark, D.B., Folwell, S., Gosling, S.N., Haddeland, I., Hanasaki, N., Heinke, J., Ludwig, F., Voss, F. and Wiltshire, A.J. 2013. Climate change impact on available water resources obtained using multiple global climate and hydrology models. *Earth System Dynamics* 4(1): 129–144.

IPCC (Intergovernmental Panel on Climate Change). 2013. *Summary for policymakers. climate change. The physical science basis.* Cambridge: Cambridge University Press.

Iram, S., Iqbal, A., Ahmad, K.S. and Jaffri, S.B. 2020. Congruously designed eco-curative integrated farming model designing and employment for sustainable encompassments. *Environmental Science and Pollution Research* 27(16): 19543–19560.

Issaka, Y.B., Antwi, M. and Tawia, G. 2016. A comparative analysis of productivity among organic and non-organic farms in the West Mamprusi District of Ghana. *Agriculture* 6(2): 13.

Jayanthi, C. 2018. Integrated rice based farming system for enhancing productivity & climate resilience. *ORYZA—An International Journal on Rice* 55(spl): 64–70.

Kar, A. 2014. Agricultural land use in arid Western Rajasthan: Resource exploitation and emerging issues. *Agropedology* 24(02):179–196.

Katovich, E. 2014. Costs and returns of grain and vegetable crop production in Nepal's Mid-Western Development Region.

Kumar, S. 2018. Livelihood improvement through integrated farming system interventions to resource poor farmers: Integrated farming system interventions for poor farmers. *Journal of AgriSearch* 5(1): 19–24.

Mahato, A. 2014. Climate change and its impact on agriculture. *International Journal of Scientific and Research Publications* 4(4): 1–6.

Mehrotra, D. and Mehrotra, R. 1995. Climate change and hydrology with emphasis on the Indian subcontinent. *Hydrological Sciences Journal* 40(2): 231–242.

Meuwissen, M.P., Feindt, P.H., Spiegel, A., Termeer, C.J., Mathijs, E., de Mey, Y., Finger, R., Balmann, A., Wauters, E., Urquhart, J. and Vigani, M. 2019. A framework to assess the resilience of farming systems. *Agricultural Systems* 176: 102656.

Mishra, A., Kumar, R. and Richa, R. 2021. Biodiversity conservation to mitigate the impact of climate change on agro-ecosystems. In *Biological diversity: Current status and conservation policie*s, Volume 1, Kumar., V., Kumar, S., Kamboj, N., Payum, T., Kumar, P. and Kumari, S (Eds.) (pp. 89–107). Haridwar: Agro Environ Media, Publication Cell of Agriculture and Environmental Science Academy.

NABARD Rural Pulse. 2014. Agricultural land holdings in India Issue-I: 1–4.

Nakajima, T. 2020. Soil health and carbon sequestration in urban farmland. In *Recycle based organic agriculture in a city* (pp. 147–158). Singapore: Springer.

Nayak, P.K., Lal, B., Panda, B.B., Poonam, A. and Nayak, A.K. 2016. Rice-fish based integrated farming system: A potential for climate change resiliency, adaptation and mitigation strategies. *Extended summaries, 4th International Agronomy Congress* 237: 238.

Nayak, P.K., Nayak, A.K., Panda, B.B., Lal, B., Gautam, P., Poonam, A., Shahid, M., Tripathi, R., Kumar, U., Mohapatra, S.D. and Jambhulkar, N.N. 2018a. Ecological mechanism and diversity in rice based integrated farming system. *Ecological Indicators* 91: 359–375.

Nayak, P.K., Tripathi, R., Panda, B.B., Poonam, A., ShahidMd, M.S. and Nayak, A.K. 2018b. Crop-Livestock-agroforestry based Integrated farming system: An eco-efficient sustainable practice for food and nutritional security.

Panwar, A.S., Prusty, A.K., Shamim, M., Ravisankar, N., Ansari, M.A., Singh, R. and Modipuram, M. 2021. Nutrient recycling in integrated farming systems for climate resilience and sustainable income. *Indian Journal of Fertilisers* 17(11): 1126–1137.

Paramesh, V., Ravisankar, N., Behera, U., Arunachalam, V., Kumar, P., Solomon Rajkumar, R., Dhar Misra, S., Mohan Kumar, R., Prusty, A.K., Jacob, D. and Panwar, A.S. 2022. Integrated farming system approaches to achieve food and nutritional security for enhancing profitability, employment, and climate resilience in India. *Food and Energy Security* 321, 1–16. https://doi.org/10.1002/fes3.321

Rai, R.K., Bhatta, L.D., Acharya, U. and Bhatta, A.P. 2018. Assessing climate-resilient agriculture for smallholders. *Environmental Development* 27: 26–33.

Rana, S.S. and Chopra, P. 2013. Integrated farming system. *Department of Agronomy, College of Agriculture, CSK Himachal Pradesh Krishi Vishvavidyalaya: Palampur, India.*

Rao, C.S., Gopinath, K.A., Prasad, J.V.N.S. and Singh, A.K. 2016. Climate resilient villages for sustainable food security in tropical India: Concept, process, technologies, Institutions, and impacts. *Advances in Agronomy* 140: 101–214.

Rao, C.S., Kareemulla, K., Krishnan, P., Murthy, G.R.K., Ramesh, P., Ananthan, P.S. and Joshi, P.K. 2019. Agro-ecosystem based sustainability indicators for climate resilient agriculture in India: A conceptual framework. *Ecological Indicators* 105: 621–633.

Rathore, V.S., Tanwar, S.P.S., Kumar, P. and Yadav, O.P. 2019, February. Integrated farming system: Key to sustainability in arid and semi-arid regions. ICAR.

Reddy, G.K., Govardhan, M., Kumari, C.P., Pasha, M.L., Baba, M.A. and Rani, B. 2020. Integrated farming system a promising farmer and eco-Friendly approach for doubling the farm income in India—A review. *International Journal of Current Microbiology* 9(1): 2243–2252.

Richa, R., Shahi, N.C., Lohani, U.C., Kothakota, A., Pandiselvam, R., Sagarika, N., Singh, A., Omre, P.K. and Kumar, A. 2021. Design and development of resistance heating apparatus-cum-solar drying system for enhancing fish drying rate. *Journal of Food Process Engineering* 45(2): 13839.

Sain, G., Loboguerrero, A.M., Corner-Dolloff, C., Lizarazo, M., Nowak, A., Martínez-Barón, D. and Andrieu, N. 2017. Costs and benefits of climate-smart agriculture: The case of the dry corridor in Guatemala. *Agricultural Systems* 151: 163–173.

Schulze, R.E. 2011. Approaches towards practical adaptive management options for selected water-related sectors in South Africa in a context of climate change. *Water SA* 37(5): 621–646.

Senapati, M.R. 2009. Vulnerabilities to climate change. *Kurukshetra, A Journal published by Ministry of Rural Development, Govt. of India.*

Senapati, M.R., Behera, B. and Mishra, S.R. 2013. Impact of climate change on Indian agriculture & its mitigating priorities. *American Journal of Environmental Protection* 1(4): 109–111.

Sharma, A.P.M., Sharma, S.K. and Patil, R.J. 2015. Estimation of rainfall erosivity factor (R) of universal soil loss equation for soil erosion modelling using GIS techniques in Shakkar River watershed. *Research Paper Crop Residue Management With Conservation Agriculture for Sustaining Natural Resources* 49(2): 234–238.

Singh, J.P., Gangwar, B., Kochewad, S.A. and Pandey, D.K. 2012. Integrated farming system for improving livelihood of small farmers of western plain zone of Uttar Pradesh, India. *SAARC Journal of Agriculture* 10(1): 45–53.

Swarnam, T.P. and Velmurugan, A. 2020. Prospects for entrepreneurship development in integrated farming systems: Island perspective. *Journal of the Andaman Science Association* 25(1): 9–14.

Tanwar, S.P.S., Bhati, T.K., Singh, A., Patidar, M., Mathur, B.K., Kumar, P. and Yadav, O.P. 2018a. Rainfed Integrated Farming System in Arid Zone of India: Resilience unmatched.

Tanwar, S.P.S., Singh, A., Kumar, P., Mathur, B.K. and Patidar, M. 2018b. Rainfed integrated farming systems for arid zone agriculture diversification for resilience. *Indian Farming* 68(9): 403–414.

Thornton, P.K., Ericksen, P.J., Herrero, M. and Challinor, A.J. 2014. Climate variability and vulnerability to climate change: A review. *Global Change Biology* 20(11): 3313–3328.

Thornton, P.K. and Herrero, M. 2015. Adapting to climate change in the mixed crop and livestock farming systems in sub-Saharan Africa. *Nature Climate Change* 5(9): 830–836.

Tripathi, S.C. and Rathi, R.C. 2011, May. Livestock farming system module for hills. In *Souvenir. National symposium on technological interventions for sustainable agriculture, 3rd–5th May, GBPUAT, hill campus, Ranichuri*: 103–104.

Vlahos, G., Karanikolas, P. and Koutsouris, A. 2017. Integrated farming in Greece: A transition-to-sustainability perspective. *International Journal of Agricultural Resources, Governance and Ecology* 13(1): 43–59.

Wagg, C., Bender, S.F., Widmer, F. and Van Der Heijden, M.G. 2014. Soil biodiversity and soil community composition determine ecosystem multifunctionality. *Proceedings of the National Academy of Sciences* 111(14): 5266–5270.

Walia, S.S., Dhawan, V., Dhawan, A.K. and Ravisankar, N., 2019. Integrated farming system: Enhancing income source for marginal and small farmers. In *Natural Resource Management: Ecological Perspectives* (pp. 63–94). Cham: Springer.

Wezel, A., Casagrande, M., Celette, F., Vian, J.F., Ferrer, A. and Peigné, J. 2014. Agroecological practices for sustainable agriculture. A review. *Agronomy for Sustainable Development* 34(1): 1–20.

Williams, H., Colombi, T. and Keller, T. 2020. The influence of soil management on soil health: An on-farm study in southern Sweden. *Geoderma* 360: 114010.

6 Climate-Resilient Agriculture for Sustainable Livelihoods and Food Security

Y. B. Vala and S. H. Malve

6.1 INTRODUCTION

The United Nations Food and Agriculture Organization (FAO) estimates that feeding the world population will require a 60% increase in total agricultural production. With many of the resources needed for sustainable food security already stretched, the food security challenges are huge. At the same time, climate change is already negatively impacting agricultural production globally and locally. Climate risks to cropping, livestock and fisheries are expected to increase in coming decades, particularly in low-income countries where adaptive capacity is weaker. Impacts on agriculture threaten both food security and rural livelihoods resulting in affecting the broad-based development.

Climate-resilient agriculture (CRA) is an integrative approach to address these interlinked challenges of food security and climate change that explicitly aims at three objectives:

1. Sustainably increasing agricultural productivity, to support equitable increases in farm incomes, food security and development.
2. Adapting and building resilience of agricultural and food security systems to climate change at multiple levels.
3. Reducing greenhouse gas emissions from agriculture (including crops, livestock and fisheries).

CRA invites to consider these three objectives together (i) at different scales – from farm to landscape; (ii) at different levels – from local to global; and (iii) over short- and long-time horizons, taking into account national and local specificities and priorities.

What is new about CRA is an explicit consideration of climatic risks that are happening more rapidly and with greater intensity than in the past. New climate risks require changes in agricultural technologies and approaches to improve the lives of those still locked in food insecurity and poverty and to prevent the loss of gains already achieved. CRA approaches entail greater investment in managing climate risks, understanding and planning for adaptive transitions that may be needed, for example into new farming systems or livelihoods, exploiting opportunities for reducing or removing greenhouse gas emissions where feasible.

DOI: 10.1201/9781003351672-6

6.2 WHAT IS CLIMATE-RESILIENT AGRICULTURE?

"Agriculture that sustainably increases productivity, enhances resilience, reduces/ removes GHGs where possible, and enhances achievement of national food security and development goals" (FAO, 2013). The principal goal of CRA is identified as food security and development, while productivity, adaptation and mitigation are identified as the three interlinked pillars necessary for achieving this goal.

6.3 IMPACT OF CLIMATE CHANGE ON AGRICULTURE

- Reduction in crop yield.
- Shortage of water.
- Irregularities in onset of monsoon, drought, flood and cyclone.
- Decline in soil fertility.
- Incidence of pest and diseases.
- Loss of biodiversity.
- Adverse impact on coastal agriculture due to rise in sea levels (17.5–57.5 cm) and sea-water intrusion, by 2100 another 10–20 cm rise if polar ice melting continues (IPCC, 2014).

Fig. 6.1 presents the global atmospheric concentrations of carbon dioxide, methane, nitrous oxide, and certain manufactured greenhouse gases have all risen significantly over the last few hundred years. Historical measurements show that the current global atmospheric concentrations of carbon dioxide, methane and nitrous oxide are unprecedented compared with the past years. Carbon dioxide concentrations are rising mostly

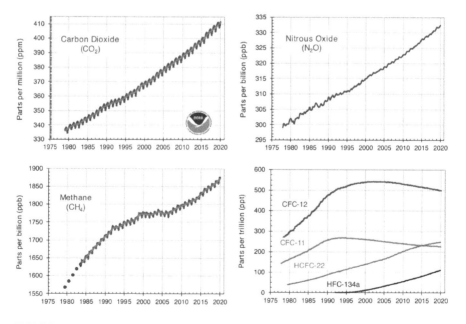

FIGURE 6.1 Rapidly increasing concentration of greenhouse gases in atmosphere. (*Source*: National Oceanic and Atmospheric Administration, 2019.)

because of the fossil fuels that people are burning for energy. Fossil fuels like coal and oil contain carbon that plants pulled out of the atmosphere through photosynthesis over many millions of years; we are returning that carbon to the atmosphere in just a few hundred. Since the middle of the 20th century, annual emissions from burning fossil fuels have increased every decade, from an average of 3 billion tons of carbon (11 billion tons of carbon dioxide) a year in the 1960s to 9.5 billion tons of carbon (35 tons of carbon dioxide) per year in the 2010s, according to *the Global Carbon Update 2021*.

6.3.1 CLIMATE CHANGE AND FOOD SECURITY

Climate change has added to the enormity of India's food security challenges. While the relationship between climate change and food security is complex, most studies focus on one dimension of food security. This chapter provides an overview of the impact of climate change on India's food security, keeping in mind three dimensions: availability, access and absorption. It finds that ensuring food security in the face of climate change will be a formidable challenge and recommends, among others, the adoption of sustainable agricultural practices, greater emphasis on urban food security and public health, provision of livelihood security and long-term relief measures in the event of natural disasters. At the heart of the Sustainable Development Goals are targets to end hunger, achieve food security and improve nutrition. For India, food security continues to be high on its list of development priorities because the country's relatively high rates of economic growth have not led to a reduction in hunger and under nutrition.

Fig. 6.2 indicates the vulnerability of Indian agriculture to climate change in the year of 2021–2050, which clearly indicates the vulnerability of agriculture due to climate change in Indian different regions in the term of vulnerability index.

Fig. 6.3 indicates climate change is perceptible through a rise in all India's mean temperature and increased frequency of extreme rainfall events in the last three decades. This causes fluctuation in production of major crops in different years. Impact of climate change on Indian agriculture was studied under the National Innovations in Climate Resilient Agriculture (NICRA). Rainfed rice yields in India are projected to reduce marginally (<2.5%) in 2050 and 2080 and irrigated rice yields by 7% in 2050 and 10% in 2080 scenarios. Further, wheat yields are projected to reduce by 6–25% in 2100 and maize yields by 18–23%. Future climates are likely to benefit chickpeas with increase in productivity (23–54%).

6.3.2 THREE PILLARS OF CRA

1. **Productivity:** CRA aims to sustainably increase agricultural productivity and incomes from crops, livestock and fish, without having a negative impact on the environment.
2. **Adaptation:** CRA aims to reduce the exposure of farmers to short-term risks, while also strengthening their resilience by building their capacity to adapt and prosper in the face of shocks and longer term stresses.
3. **Mitigation:** Adapting agriculture to climate change and maintaining food production could help to solve the current problems. Important aspect of mitigation is the uptake of carbon in plants and soils, which can help to reduce the concentration of carbon dioxide in our atmosphere.

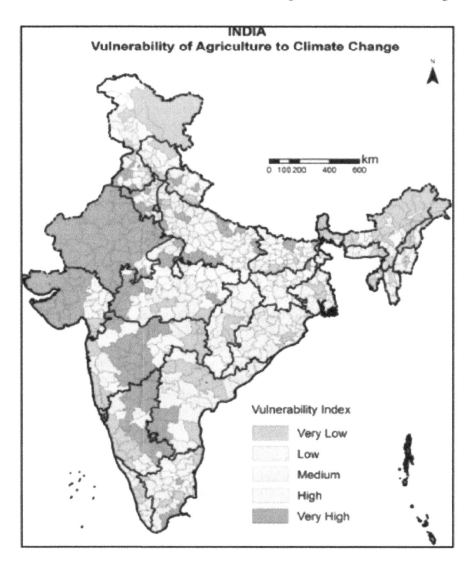

FIGURE 6.2 Vulnerability of Indian agriculture to climate change (2021–2050), Chakrabarty (2022).

6.3.3 WHY CLIMATE-RESILIENT AGRICULTURE?

- Changing climate
- Increasing population
- Desertification
- Looming water crisis
- Bio-diversity crisis
- Food crisis

FIGURE 6.3   Rainwater harvesting and recycling.

6.4   STRATEGIES TO ACHIEVE THE OBJECTIVE OF CRA

A set of strategies are there to achieve the objectives of CRA. Use of integrated renewable energy sources, resource conservation technologies, developing new cultivars and weather-tuned farming practices are some of the potential tactics to achieve these objectives. Foremost important strategy is to adopt "weather tuned agricultural practices". Frequency of extreme weather events is increasing and crops are highly susceptible to these weather aberrations. Timely forecasts of weather and following weather-based agro advisories may solve this problem to an extent. Popularizing the use of renewable energy sources is important in forming an energy-resilient farming practice. A few such technologies are wind mills, solar panels, biogas extraction units, bio-oil mining and purification and bio energy-operated water pumps. Optimum utilization of all forms of agricultural inputs and natural resources can be attained by various resource conservation technologies, such as zero tillage, crop rotation and residue recycling. These techniques also reduce greenhouse gas emissions by means of carbon sequestration. Climate change has a detrimental effect on crop production by imposing drought and temperature stress due to global warming. The crop yield in south and central Asia is expected to decrease by 5–30%. The introduction of genetically modified or improved crop varieties that have the tolerance to these abiotic stresses will serve as a solution for this.

6.5   KEY CHARACTERISTICS OF CRA

CRA integrates climate change into the planning and development of sustainable agricultural systems. CRA produces triple-win outcomes, increased productivity, enhanced resilience and reduced emissions. CRA adopts a landscape approach that builds upon the principles of sustainable agriculture but goes beyond the narrow sectoral approaches that result in uncoordinated and competing land uses, to integrated planning and management.

CRA has multiple entry points, ranging from the development of technologies and practices to the elaboration of climate change models and scenarios, information technologies, insurance schemes, value chains and the strengthening of institutional and political enabling environments. Interventions must take into account how different elements interact at the landscape level, within or among ecosystems and as a part of different institutional arrangements and political realities. To achieve food security goals and enhance resilience, CRA approaches must involve the poorest and most vulnerable groups.

6.5.1 How CRA is Different from Conservation Agriculture?

Conservation agriculture is not universally applicable and innovative approaches for promotion among small-scale farmers are required. Possible constraints include:

- Insufficient quantities of residues and need for crop residues as livestock feed.
- Fertilizers are sometimes necessary as a compliment to legume residues in order to increase crop yields.
- Weeds are major challenge in smallholder cropping systems. Many adaptations of conservation agriculture use herbicides to control weeds.
- Conservation agriculture practices together with best management practices in rice and wheat-based cropping systems.

6.5.2 Climate Smart Villages: Way Forward to CRA

Stakeholders and various institutions are responsible for disseminating the knowledge of various technologies among grass root level such as farmer's field schools and farm radio. Climate Resilient Villages (CSVs) are the concept that takes a bridge between the scientific knowledge and local actions so that the novel strategies of CRA can be effectively implemented among the farmers. With the cooperation of farmers, the scientists test and implement the CRA practices. According to the CCAFS (2017), the idea of climate Resilient villages is to incorporate CRA into village development plans. CSVs have been implemented in Indian states like Haryana and Bihar. Let us take an example of Haryana. Haryana is a semiarid region where irrigation intensity is more than 175%. Based on the survey conducted among eight randomly selected villages in Haryana, various climate-resilient agricultural practices have been selected. Village committees have been constituted comprising farmers, researchers and local planners. The CRA strategies which are relevant for the communities are implemented. Practices like laser land leveling, stress-tolerant crop varieties, zero tillage, crop diversification and legume integration, information and communication technologies to access weather forecast, site-specific nutrient management, direct seeded rice, nutrient expert decision support tools for maize and wheat, alternate wetting and drying (AWD) of rice, residue mulching and plant health monitoring using green seeker are some of the CRA strategies adopted in the village. Weather alert and weather-based agro advisories are disseminated by the meteorology department to the farmers, which helps them to take timely decisions. The farmers could save up to 30% of water by adopting direct sowing and AWD of crops (CCAFS 2017). Due to the success of these CSVs, the government of Haryana has decided to launch additional 500 CSVs in the state.

6.6   CLIMATE RESILIENT AGRICULTURAL PRACTICES

1. Rainwater harvesting and recycling (Fig. 6.3)
2. Greenhouse gas reduction
3. Watershed development
4. Adapted cultivars and cropping systems
5. Alternative land use system
6. Conservation agriculture
7. Small-farm mechanization through custom hiring centers
8. Timely and accurate weather forecasting

6.6.1   RAINWATER HARVESTING AND RECYCLING

There are two main techniques of rainwater harvesting. Storage of rainwater on surface for future use is a traditional technique and structures used were underground tanks, ponds, check dams etc. Recharge to ground water is a new concept of rainwater harvesting. The different methods are used for groundwater recharge. The *in-site* water harvesting is one of the important methods for conserving rainwater where it falls. Farmers normally preserve rainwater to sustain crops by adopting various methods of soil moisture conservation and subsequently recharge ground water. People's participation plays an important role in water conservation. There is a need to impart knowledge and build necessary skills of water conservation among the stakeholders (Fig. 6.4).

FIGURE 6.4   Training of stakeholders.

FIGURE 6.5   Contour trench for water conservation.

6.6.2   WATER HARVESTING: CONTOUR TRENCHING FOR RUNOFF COLLECTION

Contour trench construction is an extension of the practice of plowing fields at a right angle to the slope. Contour trenches are ditches dug along a hillside in such a way that they follow a contour and run perpendicular to the flow of water. The soil excavated from the ditch is used to form a berm (a narrow shelf) on the downhill edge of the ditch. Contour trenching is used for runoff collection and water harvesting (Figs. 6.5 and 6.6).

FIGURE 6.6   Contour trench in hilly area.

TABLE 6.1

Methane Emission from Different Rice Ecosystems in India

Ecosystem	Water Regime	Rice Area (M ha)	Methane (Mt)
Irrigated	Continuous flooding	6.78	1.10
	Single aeration	8.98	0.59
	Multiple aerations	9.39	0.19
Rainfed	Flood-prone	3.05	0.58
	Drought-prone	8.22	0.54
Deep water		1.29	0.25
Upland		5.16	0.00
Total		42.86	3.25

6.6.3 GREENHOUSE GAS REDUCTION

The irrigated system especially continuous flooding is releasing more methane gas as compared to rain- and deep-water cultivation. The upland paddy system does not release methane gas and results in less methane emission from paddy fields. Table 6.1 covers different ecosystems of paddy cultivation w.r.t. methane emission gas.

The different methods of paddy cultivation (i.e. dry seeding, AWD, aerobic rice production and mid-season drawdown) are shown in Fig. 6.7.

FIGURE 6.7 Methods of paddy cultivation.

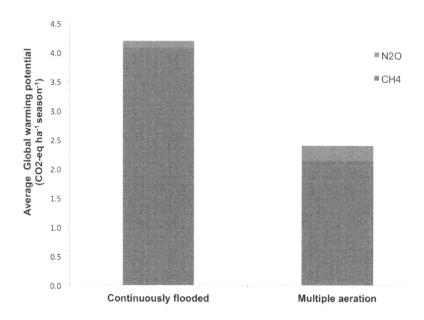

FIGURE 6.8   Global warming potential of different rice growing practices.

Fig. 6.8 indicates the results of global warming potential under different rice growing practices, i.e. continuously flooded and multiple aeration systems. The analysis of graph clearly indicates that the continuously flooded system in rice field increased nitrous oxide and methane gas emissions from the rice growing field.

6.6.4   WATERSHED DEVELOPMENT

Watershed development refers to the conservation regeneration and the judicious use of all resources natural (like land, water plants and animals) and human – within the watershed area. Watershed management tries to bring about the best possible balance in the environment between natural resources on the one side and man and animals on the other.

Watershed Development Programs
- IWDP: Integrated Watershed Development Program
- DPAP: Drought Prone Area Development Program
- DDP: Desert Development Programme

Objectives
- To mitigate the adverse effects of drought on crops and livestock
- To control desertification
- To encourage restoration of ecological balance and
- To promote economic development of village community

TABLE 6.2
Impact of Watershed Development on Soil Erosion Reduction in Gujarat State across the Different Schemes

	Reduction in Soil Erosion Percent (%)	
Schemes	>50%	Up to 50%
IWDP	70	30
DPAP	30	65

Table 6.2 indicates the impact of watershed development on soil erosion reduction in Gujarat state across the different schemes. Reduction in soil erosion percent where the IWDP and DPAP schemes are adopted in Gujarat state.

Fig. 6.9 indicates the meta-analysis of 311 watersheds and results of graph show that runoff is totally decreased and cropping intensity of particulars is increased. Fig. 6.10 shows the change in groundwater level in the state of Madhya Pradesh (MP), Gujarat, Andhra Pradesh (AP), Tamil Nādu (TN) and Utter Pradesh (UP) after adoption of Watershed Development Programme.

6.6.5 ADAPTED CULTIVARS AND CROPPING SYSTEMS

The use of adapted crops and varieties (including both herbaceous and tree crops) as suggested by the United Nation's FAO among the climate smart practices for risk reduction, soil and water conservation and efficient water management.

Crops and varieties (either annual or perennial) help reduce the negative impacts of climate change on agricultural systems and at the same time to ensure stable

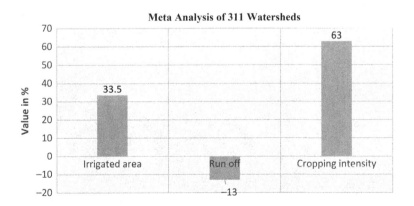

FIGURE 6.9 Benefits from watershed.

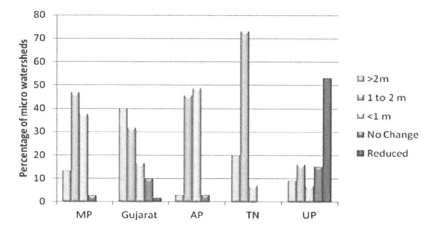

FIGURE 6.10 Change in groundwater level after WDPs in different states.

agricultural production. Chickpea crop cultivation at farmer's field under NICRA project, KVK, Gujarat (Fig. 6.11).

The term cropping system refers to the crops, crop sequences and management techniques used on a particular agricultural field over a period of years. It includes all spatial and temporal aspects of managing an agricultural system. Fig. 6.12 shows the front line demonstration of cropping system at farmer's field.

FIGURE 6.11 Crop cultivation at farmer's field under NICRA project, KVK, Gujarat.

FIGURE 6.12 Front line demonstration at farmer's field.

Historically, cropping systems have been designed to maximize yield, but modern agriculture is increasingly concerned with promoting environmental sustainability in cropping systems.

Important cropping system

i. Mono-cropping
ii. Intercropping
iii. Strip cropping
iv. Relay cropping

6.6.6 Alternative Land Use System

6.6.6.1 Classification of Alternative Land Use System

The land use system can be divided as alley cropping, agri-horticulture stem, agri-silviculture ley farming, tree farming etc. Fig. 6.13 gives the classification of alternative land use system.

1. **Agri-silvicultural Systems**
   - Field crops: Sorghum/Safflower/Chickpea/Sunflower
   - Tree: Anjan/Acacia albida/Khejri
2. **Silvipasture Systems**
   - Tree: Neem/Sissoo/Raintree/Anjan
   - Forage crops: Anjan grass and Stylo
3. **Agri-Silvipasture Systems**
   - Field crops: Sorghum, groundnut, Finger millet Maize, Pulses
   - Tree: Neem/Sissoo/Raintree/Anjan
   - Forage crops: Anjan grass and Stylo

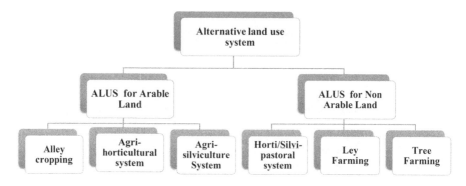

FIGURE 6.13   Flow chart of classification of alternative land use system.

4. **Agri-Horti Systems**
   • Field crops: Pulses, Minor millets, wheat
   • Horticulture crops: Mango, Sapota, Citrus
5. **Silvi-Horti Systems**
   • Tree crops: Subabul, Casuarina
   • Horticulture crops: Mango, Sapota, Citrus
6. **Conservation Agriculture**

6.6.6.2   Practices of Conservation Agriculture
   i. *In-situ* moisture conservation and Mulches
   ii. Conservation tillage
   iii. Crop residues management
   iv. Land configuration
   v. System of rice intensification (SRI)

Mulching: covering of the soil with crop residues, such as straw, maize stalks, palm fronds or standing stubble to protect it from raindrop impact and to reduce the velocity of runoff and wind called mulching (Fig. 6.14).

Reasons for applying mulch include conservation of soil moisture, improving fertility and health of the soil, reducing weed growth and enhancing the visual appeal of the area. Mulches of different kinds, such as leaves, straws, plastic, stubbles etc., minimize evaporation and increase the absorption of moisture and protect the surface of the land against the beating action of raindrops. Later on, the decay forms humus, which improves the physical condition of soil.

6.6.7   SMALL-FARM MECHANIZATION THROUGH CUSTOM HIRING CENTRES

Custom Hiring Centres (CHCs) help in hiring the need-based implements/machinery by poor farmers at affordable cost and carrying out field operations like land preparation, sowing, interculture etc. It covers large areas in short time and other agricultural operations with high energy efficiency.

FIGURE 6.14   Mulching under different crops.

Most Popular Implements in CHCs

- Laser land leveller
- Zero till drill
- Drum seeder
- Rotavator
- Happy seeder
- Ridge and furrow planter
- Multi-crop thresher
- Power tiller

6.6.7.1   Laser Land Leveling

Laser land leveling is leveling the field within certain degree of desired slope using a guided laser beam throughout the field. Fig. 6.15 shows the laser land leveler.

FIGURE 6.15   Laser land leveler.

Happy seeder

FIGURE 6.16   Happy seeder.

Advantages

- It reduces the water logging at any particular place within the field.
- It helps in even germination of seed.
- Helps in maintaining plant population.
- Approx. 40–55% saving in irrigation water.
- Reduce salinity problem.
- Improvement in nutrient use efficiencies.

6.6.7.2 Happy Seeder

Happy Seeder (Fig. 6.16) is a tractor-mounted machine that cuts and lifts rice straw, sows wheat into the soil and deposits the straw over the sown area as a mulch. Happy seeder can reduce air pollution and greenhouse gas emissions.

6.6.8 Timely and Accurate Weather Forecasting

Timely sowing with the collaboration of weather gives better germination, acceptable yield, higher net income, ability to reduce attack of plant pests and avoid negative impacts due to climate change. Weather forecasting and early warning systems will be very useful in minimizing risks of adverse climatic conditions.

Short- and medium-range forecast for agriculture is useful for the following activities:

- Preparatory activities, including land preparation and preparation of plant material.
- Planting or sowing, management of crops, fruit trees and vines.
- Application of fertilizer, irrigation; thinning, pest and disease control.
- Harvesting, on-farm post-harvest processing.

6.7   IMPORTANT STEPS IN CLIMATE-RESILIENT AGRICULTURE

- Community paddy nursery is taken as a contingency measure for delayed planting to combat the problems experienced by the farmers during deficit rainfall in lowlands.

- Large number of community tanks/ponds with substantial water storage capacity is essential as a means of augmentation and management of village level water resources for storing and using properly the surplus rainfall or runoff during kharif season.
- Crop diversification encourages livelihood security and resilience to climate variability to overcome the constraints of low yields or even crop failure due to erratic monsoon rainfall and skewed distribution associated with the practice of sole cropping.
- Intercropping is a feasible and more profitable option to minimize risk in crop production, ensure reasonable returns, improve soil fertility with a legume intercrop and it is a key drought coping strategy.
- Direct seeded rice is vital for promoting water use efficiency in unpuddled field to cope with water shortages.
- Drought-tolerant paddy cultivars, such as Sahbhagidhan, Naveen, Anjali, BirsaVikasDhan 109, and Abhishek, should be promoted to tackle deficit rainfall situations.
- Drum seeding of rice needs to be encouraged for water saving and timeliness in planting to overcome the challenges of water shortages because of deficit rainfall, declining groundwater table due to insufficient recharge, late and limited release of irrigation water from canals or poor inflows into tanks associated with the transplanted rice in irrigated and rainfed areas.
- Flood tolerant high yielding rice varieties, viz. Swarna Sub 1, MTU-1010, MTU-1001, and MTU-1140, impart resilience to farmers in flood-prone areas.
- Rejuvenation of farming in cyclone- and flood-prone coastal agro-ecosystems through land shaping should come into practice for rainwater harvesting, utilization and integration of farm enterprises.
- Check dams are established for ex-situ storage of excess runoff water in seasonal streams at suitable sites in different rainfall zones for direct use or for improving the groundwater availability.
- Short duration crop varieties of pulse and oilseed crops are predominantly suitable for late sowing under rainfed conditions.
- Enhancing resilience through improvement in conveyance efficiency as one of the reasons for gap between potential created and utilized under canal irrigation systems is the lack of maintenance of conveyance channels which became silted up.
- Rainwater harvesting and recycling was demonstrated by the construction of temporary check dams to help in groundwater recharge and rising water table in the area, for life saving irrigation in rabi and summer crops, improvement of crop stand (70–75%), time-saving (25–30%) in irrigation (1.5 hr/ha/irrigation), water saving (25–30%), higher yield (10–15%), lesser seed rate (20–25%) etc.
- Recharging of tube wells to improve shallow aquifers was taken up as a major intervention because farmers were not aware of in-situ soil and moisture conservation techniques; hence, 35–40% of total rainfall was being

lost as runoff and the rabi crops often suffered due to moisture stress affecting productivity.

- Adequate supply of fodder, either green or dry, is crucial to the livelihoods of farmers involved in animal husbandry to tackle fodder scarcity, for example Sorghum (Pusa Chari Hybrid-106, CSH-14, CSH-23, CSV-17), Bajra (CO-8, TNSC-1, APFB-2, AvikaBajra Chari), Maize (African tall, APFM-8), Berseem (Wardan, UPB-110), Lucerne (CO-1, LLC-3, RL-88), and perennial fodders like APBN-1, CO-3 and CO-4.

- Improving the resilience of poor farmers by reclaiming cultivable wastelands or land improvement through leveling and bunding in undulated upland areas to arrest the unabated land degradation.

- Small-scale water harvesting structures like farm ponds at individual farm level enable the farmers to reuse the harvested water during critical growth stages; provide pre-sowing irrigation to rabi crops for improving livelihoods of small farmers; increase irrigated area in the villages, crop productivity and groundwater level; take up at least two assured crops in a year and also shift to vegetable cultivation.

- Direct rainwater harvesting and its judicious use for crop production through water catch ponds/pits, i.e. Jalkund (a low cost rainwater harvesting structure) can be highly beneficial to the farmers for providing protective irrigation to the crops for successful cultivation during moisture scarcity conditions during dry seasons; for animal husbandry activities like piggery, poultry and duckery; taking up fish rearing in the harvested water; and cultivating high value vegetable crops, such as brinjal, chilli, tomato, radish, amaranthus, coriander, and cowpea.

- Small farm mechanization through Custom Hiring Centres (CHCs) for farm implements, such as rotavator, zero till drill, drum seeder, multi-crop planter, power weeder, and chaff cutter, can successfully empower farmers to tide over the shortage of labor and improve efficiency of agricultural operations. Mechanization provides access to small and marginal farmers to costly farm machinery, facilitates timeliness and precision in agricultural operations and efficient use of resources and applied inputs, promotes adoption of climate resilient practices and technologies cultivation, provides work opportunities to skilled labor and small artisans, conserves available soil moisture under stress conditions and provides adequate drainage of excess rain and floodwaters.

- Identification of suitable varieties of main and alternate crops is needed so that participatory seed production of short duration, drought and flood tolerant varieties of rice, soybean, groundnut, green gram, finger millet, foxtail millet, pigeon pea etc. can be taken up well in advance and kept in the village level seed banks for use in contingency situations at the local level to combat seed shortages.

- Zero till drill wheat not only escapes terminal heat stress but also saves irrigation water up to 10–15% during first irrigation; facilitates two days early and uniform germination, better plant stand than traditional method and no crust formation after rains; improves crop yield, soil structure and fertility and causes no lodging of crops at the time of maturity in case of heavy rains.

- There is a need for improved planting methods like broad bed and furrow (BBF), furrow irrigated raised bed (FIRB) and ridge and furrow method for in-situ soil and water conservation and proper drainage technology in deep black soils; for enhancing water use efficiency and crop productivity (5–10%); facilitates less moisture stress during non-rainy days, 20–25% lower seed rate, better weed management; saves time (25–30%) and water (25–30%) in irrigation as well as reduces crop lodging and compaction of soil.
- In-situ incorporation of biomass and crop residues is mandatory for improving soil health.
- Integrated farming system modules with a combination of small enterprises, such as crop, livestock, poultry, piggery, fish and duck rearing are demonstrated to minimize risk from a single enterprise in the face of natural calamities and diversified enterprises bring in the much-needed year-round income to farmers in mono-cropped paddy growing areas and improve their livelihoods and resilience to extreme weather events.

6.8 RESEARCH REVIEWS AND CASE STUDIES

6.8.1 RAINWATER HARVESTING AND RECYCLING

Anonymous (2019) concluded that in villages of Banaskantha district of Gujarat under deep loamy sand to sandy loam soil type intervention of ridge and furrow moisture conservation practices gave higher yield (kg/ha), RWUE, Net returns and B:C over flat bed. Effect of *in-situ* moisture conservation (ridge and furrow) on yield and economics of castor in Banaskantha District of Gujarat are given in Table 6.3 (Anonymous, 2019).

TABLE 6.3

Effect of *In-Situ* Moisture Conservation (ridge and Furrow) on Yield and Economics of castor in Banaskantha District of Gujarat

Village	Farming Situation/ Soil Type	Intervention	Yield (kg/ha) Seed	Yield (kg/ha) Stalk	RWUE (kg/ ha-mm)	Net Returns (Rs./ha)	B:C Ratio
Kalimati/ Dholiya	Deep loamy sand to sandy loam	Supplemental irrigation	1434	2841	1.62	52,282	3.03
		Farmer's practice	993	2268	1.12	33,948	2.37
		% Yield increase	44.41	25.26	–	–	–
Ghanghu	Deep loamy sand to sandy loam	Supplemental irrigation	1404	2947	1.59	50,910	2.95
		Farmer's practice	1020	2304	1.15	35,248	2.46
		% Yield increase	37.65	27.91	–	–	–

TABLE 6.4

Performance of Ridge and Furrow System on Growth and Yield of Rainfed Soybean

Parameters	Ridge and Furrow System	Normal Flatbed Sowing	% Change Over Control
Plant population (No./m²)	48.2	40.6	18.72
Root length (cm) at harvest	70.2	60.8	15.46
Plant height at 60 DAS	25.6	20.2	26.73
Number of root nodules/plant at 60 DAS	31.2	26.6	17.29
No. of pods/plant	50.2	37.3	34.35
Seed yield (kg/ha)	1510	1395	8.24
Straw yield (kg/ha)	1705	1596	6.83

Basediya et al. (2018) reported that intervention of ridge and furrow moisture conservation practices gave higher growth and yield parameters over flatbed sowing (Table 6.4).

6.8.2 Greenhouse Gases Reduction

Chidthaisong *et al.* (2018) studied the effects of AWD on methane and nitrous oxide emissions from a paddy field in Thailand they revealed that AWD is effective as it reduced the methane emission from field. Table 6.5 describes the effects of AWD on methane and nitrous oxide emissions from a paddy field (Chidthaisong *et al.*, 2018).

6.8.3 Adapted Cultivars and Cropping Systems

Chaudhari *et al.* (2017) reported that cropping system of ground nut + pigeon pea gave significantly highest LER, Groundnut pod equivalent yield (kg/ha), net returns and B:C ratio under rainfed condition (Table 6.6).

Vaghela *et al.* (2019) revealed that castor equivalent yield gave significantly highest in groundnut + castor (2:1) cropping system as compared to other cropping

TABLE 6.5

Evaluating the Effects of Alternate Wetting and Drying (AWD) on Methane and Nitrous Oxide Emissions from a Paddy Field

Treatments	CH_4 (kg/ha)	N_2O (kg/ha)	Grain Yield (t/ha)
T_1 – CF	17.3	0.785	4.50
T_2 – AWD	8.8	0.979	4.19
T_3 – AWDS	21.0	0.851	4.44

TABLE 6.6

LER, Groundnut Pod Equivalent Yield and Economics of Groundnut-based Cropping System under Rainfed Condition

Treatments	Land Equivalent Ratio	Groundnut Pod Equivalent Yield (kg/ha)	Net Return (Rs/ha)	B:C Ratio
T_1 – Sole Groundnut	1.00	1489	46,467	3.23
T_2 – Groundnut + Pearl millet	1.02	991	28,365	2.60
T_3 – Groundnut + Black gram	1.04	1241	37,742	3.18
T_4 – Groundnut + Soybean	1.01	1186	34,906	2.97
T_5 – Groundnut + Cotton	1.27	1766	53,215	3.38
T_6 – Groundnut + Castor	1.42	1794	56,599	3.76
T_7 – Groundnut + Pigeon pea	1.45	1802	57,392	3.82
S.Em±	0.04	73	3131	0.17
CD (P= 0.05)	0.13	226	9584	0.51

systems as well as groundnut + castor (2:1) cropping system shows that highest WUE in kg/ha-mm. Table 6.7 presents the castor equivalent yield and Water use efficiency (WUE) as influenced by different castor-based cropping systems.

6.8.4 ALTERNATIVE LAND USE SYSTEM

Anonymous (2014) revealed that Subabul + Pearlmillet + Horsegram (3:1) silvipasture system under rainfed agro-ecosystem gives higher yield (q/ha) as compared to

TABLE 6.7

Castor Equivalent Yield and Water Use Efficiency (WUE) as Influenced by Different Castor-Based Cropping Systems

Treatments	Castor Equivalent Yield (kg/ha)	WUE (kg/ha-mm)
T_1 – Greengram + Castor (2:1)	3769	7.5
T_2 – Cowpea + Castor (2:1)	3310	6.6
T_3 – Groundnut + Castor (2:1)	5145	10.3
T_4 – Sesamum + Castor (2:1)	2987	6.0
T_5 – Castor + Summer Pearlmillet	4569	6.3
T_6 – Castor + Summer Greengram	4681	7.5
T_7 – Castor + Summer Sesamum	4661	6.2
T_8 – Castor + Summer mothbean	4305	6.9
T_9 – Castor sole	3777	10.0
S.Em±	281.9	0.6
CD (P = 0.05)	14.1	1.8

TABLE 6.8

Intensive Forage Production through Silvipasture System under Rainfed Agro-ecosystem

Treatments	Subabul (q/ha)	Intercrop Yield (q/ha)
T₁ – Subabul + *Cenchrus ciliaris* + Stylosenthes (3:1)	36.9	169.3
T₂ – Subabul + Pearlmillet + Horsegram (3:1)	37.0	488.3
T₃ – Subabul (Sole)	34.0	–

other alternate land use systems. Intensive forage production through silvipasture system under rainfed agro ecosystem is given in Table 6.8.

6.8.5 CONSERVATION AGRICULTURE

Rajashekarappa *et al.* (2013) reported that among the different organic mulches, *in-situ* green manuring was found to be most effective treatment as it recorded highest grain yield (kg/ha) and infiltration rate (cm/hr) and lower soil temperature. Table 6.9 describes the effect of different organic mulches on soil moisture content, soil temperature, infiltration rate and yield of *Kharif* maize (rainfed).

TABLE 6.9

Effect of Different Organic Mulches on Soil Moisture Content, Soil Temperature, Infiltration Rate and Yield of *Kharif* Maize (Rainfed)

Treatments	Soil moisture Content (%) 15 cm depth	30 cm depth	Soil Temperature (°C) at 10 cm Depth	Infiltration Rate (cm/hr)	Grain Yield (kg/ha)
T₁ – No mulch	12.90	14.16	27.20	9.06	4189
T₂ – Additional FYM @ 10 t/ha	11.88	13.50	28.40	9.20	3946
T₃ – Coconut coir pit mulching@ 5 t/ha	12.03	13.33	27.83	9.36	4178
T₄ – Straw mulching @ 5 t/ha	12.53	13.90	27.40	9.00	4292
T₅ – Tank silt application @ 10 t/ha	12.37	13.70	27.60	8.58	4081
T₆ – Mulching with coconut frond	13.83	14.16	**26.83**	9.55	4651
T₇ – *In-situ* green manuring with sunhemp	13.13	**15.83**	**26.66**	**9.68**	**5269**
T₈ – Green leaf mulching with Gliricidia	13.00	14.36	26.96	9.53	4607
T₉ – Intercropping of horsegram	12.07	13.13	28.33	8.73	3416
CD (P=0.05)	NS	2.54	4.55	1.43	723.0

Bold values show the highest value among that row for particular treatment.

TABLE 6.10
Effect of Transplanting Methods on Yield of Rice

Treatments	Grain Yield (t/ha)	B:C Ratio
Transplanting methods (M)		
T$_1$ – System of rice Intensification	6.1	3.12
T$_2$ – Normal Transplanting	4.0	2.14
T$_3$ – Mechanized transplanting	4.6	2.64
CD (P = 0.05)	0.16	0.16

Shantappa (2014) revealed that system of rice intensification gave higher grain yield (kg/ha) and B:C ratio over to other methods. Effect of transplanting methods on yield of rice are discussed in Table 6.10.

6.9   CONCLUSIONS

Food security is the all-time concern of people across the globe. Agriculture is in the midst of threat due to climate change and decline in area under crop cultivation. Climate change and climatic variability are likely to affect sustainability of agricultural production thereby affecting national food security. Adoption of climate Resilient technologies can help in coping up with the challenge of climate change. Some climate-resilient technologies like rainwater harvesting and recycling, efficient cropping systems, greenhouse gases reduction, alternate land use systems, conservation agriculture and weather forecasting, growing heat/drought tolerant crop varieties, changes in crop management practices and development of improved farm machineries help in agricultural adaptation to the changing climate. Exchanging information and providing technical advice on improving efficiency, productivity and resilience of agriculture at regional and national scales should be considered. Besides, capacity building and awareness on multiple advantages of climate Resilient, sustainable agricultural technologies should be promoted. Farmers should be ensured with better support prices of agricultural produce to enable them to cope with higher adaptation costs of cultivation under changing climatic scenarios.

REFERENCES

Anonymous. 2014. Annual report ICAR-CRIDA 2014–15, Hyderabad, Telangana.
Anonymous. 2019. Annual report NICRA (2019–20), AICRPDA, SDAU, S. K. Nagar.
Basediya, A. L., Mishra, S., Gupta, R., Kumar, P. and Basediya, S.S. 2018. Performance of Ridge and furrow system on the growth and yield attribution of soybean in Barwani district of M.P. India. *International Journal of Current Microbiology and Applied Sciences* 7(8): 499–505.

CCAFS. 2017. Progressing towards climate resilient agriculture; top ten success stories from CCAFS in South Asia. CGIAR Research Program on Climate Change, Agriculture and Food Security (CCAFS) of South Asia, New Delhi, India.

Chakrabarty, M. 2022. Climate change and food security in India. 10.4324/9781003272656-11.

Chaudhari, D. T., Vekariya, P. D., Vora, V. D., Talpada, M. M. and Sutaria, G. S. 2017. Enhancing productivity of groundnut based intercropping systems under rainfed conditions of Gujarat. *Legume Research* 40(3): 520–525.

Chidthaisong, A., Nittaya, C., Rossopa, B., Buddaboon, C., Kunuthai, C., Sriphirom, P., Towprayoon, S., Tokida, T., Agnes, T. P. and Minamikawa, K. 2018. Evaluating the effects of alternate wetting and drying (AWD) on methane and nitrous oxide emissions from a paddy field in Thailand. *Soil Science and Plant Nutrition* 64(1): 31–38.

FAO, 2013. FAO Success stories on Climate Smart Agriculture. *FAO of the United Nations*, pp. 28.

IPCC. 2014. Climate Change 2014: Synthesis Report. Contribution of Working Groups I, II and III to the Fifth Assessment Report of the Intergovernmental Panel on Climate Change [Core Writing Team, R.K. Pachauri and L.A. Meyer (eds.)]. IPCC, Geneva, Switzerland, 151 pp.

Rajashekarappa, K. S., Basavarajappa, E. and Puttaiah, E.T. 2013. Effect from different organic mulches and *in situ* green manuring on soil properties and yield and economics of maize in south-eastern dry zone of Karnataka. *Global Journal of Biology, Agriculture & Health Sciences* 2(3): 236–240.

Shantappa, (2014). Studies on establishment techniques, irrigation water levels and weed management practices on productivity and emission of greenhouse gasses (GHGS) in rice (Oryza sativa L.). Ph.D. Thesis. University of Agricultural Sciences, Raichur, Karnataka.

Vaghela, S.J., Patel, J.C., Patel, D.G. and Dabhi, J. S. 2019. *Journal of Pharmacognosy and Phytochemistry* 7(3): 748–751.

7 Climate-Resilient Agriculture to Double the Farmer's Income
A Case Study of Chamba (HP)

Kehar Singh Thakur and Rajeev Raina

7.1 INTRODUCTION

7.1.1 DISTRICT PROFILE OF CHAMBA

The Chamba district in its present form came into existence on 1 November 1966, surrounded by Jammu and Kashmir on the north-west, the Ladakh area of Jammu and Kashmir state on the north-east and east, and Lahaul and Bara-Bhangal area of Himachal Pradesh on the south-east, and on south surrounded by the District Kangra of Himachal Pradesh and Gurdaspur District of Punjab. The district is situated between north latitude 32°11′30″ and 33° 13′ 06″, and east longitude 75°49′00″ and 77°03′30″. The area of the district is 6,522 sq. km with Chamba as its headquarters. There are 1,591 villages in the district, out of which 1,110 are inhabited and 481 are uninhabited (2011 census). The district has been divided into six subdivisions (Chamba, Churah, Pangi, Bharmaur, Dalhousie, and Chowari). The Ravi is the main river of Chamba district and is the heart and soul of the Chambyals. With its tributaries, it drains the whole of Chamba valley proper between the Dhauladhar and Pangi ranges and thus commands the largest and most important part of the district (Thakur, 2018). The river originates from the Bara Bhangahal area of Dhauladhar. Another river that flows in the district is Chandrabhaga mainly covers the Pangi region of the district. The irrigation is mainly done either by tapping natural springs or by lifting surface water from perennial khads and rivers. Kuhl irrigation is prevalent in many valleys, which covers an area of 3,545 ha by tapping natural springs and by lift irrigation schemes.

7.1.2 CLIMATE AND RAINFALL

The climate of the district varies from semi-tropical to semi-arctic. Winter varies from December to February and summer extends from March to June while July to September is rainy months. The maximum rainfall in the district occurs between July and September. The average annual temperature is 20.7°C and the average rainfall of the district is 2,213 mm. Snowfall is received in the higher reaches in the winter months.

DOI: 10.1201/9781003351672-7

7.1.3 Geomorphology and Soils

Chamba district presents an intricate mosaic of mountain ranges, hills, and valleys. It is primarily a hilly district with altitudes ranging from 600 m AMSL to 6,400 m AMSL. Physiographically, the area forms part of the middle Himalayas with high peaks ranging in height from 3,000 to 6,000 m AMSL. It is a complex folding region that has undergone many orogeneses (structural deformations). The topography of the area is rugged with high mountains and deeply dissected by river Ravi and its tributaries. Physiographically, the district can be divided into two units: (i) high hills, which cover almost the entire district, and (ii) a few valley fills. Three types of soils observed in the district are sandy loam, loam, and sandy clay loam.

7.1.4 Agriculture

Agriculture is the main occupation of the people of Himachal Pradesh and has an important place in the economy of the state. Himachal Pradesh is the only state in the country, where 89.96% of the population (2011 census) lives in rural areas. Therefore, dependency on agriculture/horticulture is imminent as it provides direct employment to about 70% of the total workers of the state. Agriculture is the premier source of state income (GSDP). About 14% of the total GSDP comes from agriculture and its allied sectors. Out of the total geographical area of 55.67 lakh ha, the area of operational holdings is about 9.55 lakh ha and is operated by 9.61 lakh farmers. The average land-holding is about 1.00 ha. Distribution of landholdings according to Agricultural Census 2010–2011 shows that 87.95% of the total holdings are of small and marginal farmers. About 11.71% of holdings are owned by semi-medium and medium farmers and only 0.34% by large farmers. About 80% of the total cultivated area in the state is rainfed. Rice, wheat, and maize are important cereal crops in the state. Groundnut, soybean, sunflower in kharif, and rapeseed/mustard and toria are important oilseed crops in the Rabi season. Urad, bean, moong, and rajmah in the kharif season and gram lentil in rabi are the important pulse crops of the state. The main occupation of the community of the area is agriculture. The economy of the district is mainly agrarian as the most population of the district is directly dependent on agriculture and its related activities. Due to diversity in the soil character and agro-climate conditions which are quite best for the growth of various types of vegetables, fruits, cash crops, etc., different types of agriculture, as well as horticulture products, are produced here. The main crops are maize, wheat, paddy, barley, millets, pulses (black gram and rajmash), and oilseed (brown sarson and raya). Besides, many types of vegetables like potato, tomato, cabbage, and cauliflower are also produced. A large population of the area also depends upon agricultural activities for livelihood and a considerable proportion of the population of the area is being supported by agricultural activities also.

7.1.5 Demography

According to 2011 census, the district has a population of 5, 18,844, population density of 338 person/sq.km, sex ratio of 977, and literacy rate of 87.23% (Table 7.1). The decadal population growth rate was almost 13%. In the district, people belonging to various communities live in harmony. The population consists of the majority of

TABLE 7.1
Population (As per 2011 Census)

Total population	5,18,844
Male population	2,60,857
Female population	2,57,987
Sex ratio	950/1,000 female/male
The density of population (per square km)	80 (per square km)

TABLE 7.2
Literacy Rate Aggregate (As per 2011 Census)

Male literacy	89.53%
Female literacy	73.51%
Aggregate	82.80%

Hindus. The marginal population of Sikhs and Muslims also exists. The literacy rate aggregate of the district is presented in Table 7.2.

7.1.6 GEOGRAPHY

Chamba district presents an intricate mosaic of mountain ranges, hills, and valleys. It is primarily a hilly district with altitudes ranging from 600 to 6,400 m AMSL. Physiographically, the area forms part of the middle Himalayas with high peaks ranging in height from 3,000 to 6,000 m AMSL. It is a region of complex folding, which has undergone many orogeneses (structural deformations). The topography of the area is rugged with high mountains and deeply dissected by the river Ravi and its tributaries. The district headquarter is situated 430 km away from the state capital. It is located at latitude 32.5°, and longitude 76.1°. The district is sharing a border with Kangra district to the South, Doda district to the North, and Kargil district to the West. The district occupies an area of approximately 6,528 square kilometers. The district's total population is 5,18,844 according to the census of 2011. Males comprise 2,60,857, while Females comprise 2,57,987; literate people are 3,43,267 among total. Its total area is 6,528 sq. km. It is the seventh-largest district in the state in terms of population but the second largest district in the state in terms of area. The total area of Chamba district is 6,528 sq. km (as per the report of the Survey General of India.). Table 7.3 gives the geographic coordinates of the district.

Geographical Area: 6,528 sq. km.

TABLE 7.3
Geographic Coordinates of the District

Forest area	2,437
Altitude	996 m
Longitude	75.3.30
Latitude	32.11.6
Major river	Ravi River

TABLE 7.4
Major Farming Systems/Enterprises (Based on the Analysis Made by the KVK)

Sl. No		Farming System/Enterprise
1	Irrigated (borewell)	–
2	Irrigated (canal)	Paddy/vegetable – wheat/vegetables
3	Tank Irrigated	Vegetable/pulses – wheat/vegetables
4	Rainfed	Maize – wheat, Hort-Agri – pastoral system, Silvi – pastoral system
5	Enterprises	Horticulture, off-season vegetables, protected cultivation, mushroom production, bee keeping, sheep and goat rearing

The main source of income is agriculture and 80% population of the district is involved in agricultural activities.

Major farming systems/enterprises (based on the analysis made by the KVK) are covered in Table 7.4: the predominant farming situations in the district, the area covered under each of these farming situations in the district, their general productivity levels, and trends in the productivity.

Table 7.5 presents the area, production, and productivity of major crops cultivated in the district.

TABLE 7.5
Area, Production, and Productivity of Major Crops Cultivated in the District

S. No	Crop	Area (ha)	Production 2019–2020 (MT)	Productivity (MT/ha)
Fruit crops				
1	Apple standard	12,893	28,282.5	2.2
2	Plum	348.13	269.7	0.8
3	Peach	160.64	172.6	1.1
4	Apricot	363.62	321.9	0.9
5	Pear	325.85	320.6	1.0
6	Cherry	7.23	4.4	0.6
7	Kiwi fruit	2.59	5.7	2.2
8	Pomegranate	65.22	6.1	0.1
9	Olive	1.97	0.7	0.4
10	Persimmon	3.03	0.9	0.3
11	Almond	219.3	30.8	0.1
12	Walnut	1,255.2	352.9	0.3
13	Mango	525.72	78.8	0.1
14	Litchi	215.97	17.5	0.1
15	Guava	29.95	24.3	0.8
16	Aonla	44.1	10.4	0.2
17	Kinnow	13.26	14.0	1.1
18	Kagzi lime	417.38	502.5	1.2
19	Galgal	336.38	510.7	1.5

(Continued)

TABLE 7.5 (*Continued*)
Area, Production, and Productivity of Major Crops Cultivated in the District

S. No	Crop	Area (ha)	Production 2019–2020 (MT)	Productivity (MT/ha)
20	Pecanut	62.94	11.2	0.2
21	Papaya	0.9	0.2	0.2
22	Grapes	1.15	1.2	1.0
23	Loquat	0.3	0.4	1.3
24	Musambi	0.08	0.2	2.5
Vegetable crops				
1	Peas	1,650.0	32,801.0	19,879.0
2	Tomato	134.0	4,690.0	35,000.0
3	Bean	451.0	6,259.0	13,878.0
4	Onion	39.0	738.0	18,923.0
5	Garlic	29.0	226.0	7,793.1
6	Cabbage	73.0	1,592.0	21,808.2
7	Cauliflower	21.0	424.0	20,190.4
8	Radish, turnip, and carrot	79.0	1,962.0	24,835.4
9	Bhindi	57.0	850.0	14,912.2
10	Cucurbits	30.0	778.0	25,933.3
11	Capsicum and chillies	26.0	390.0	15,000.0
12	Brinjal	36.0	828.0	23,000.0
13	Potato	900	1,17,000	1,30,000.0
Cereals and pulses				
1	Paddy	2,250	39,800	1,769
2	Maize	23,050	6,03,700	2,619
3	Wheat	21,400	3,72,000	1,738
4	Barley	1,560	2,370	152
5	Black gram	1,850	10,400	562
6	Rajmash	600	5,000	833
7	Oilseed	2,950	11,200	380

7.2 CLIMATE-RESILIENT AGRICULTURE

Climate-resilient agriculture (CRA) is an approach that includes sustainably using existing natural resources through crop and livestock production systems to achieve long-term higher productivity and farm incomes under climate variabilities. This practice reduces hunger and poverty in the face of climate change for forthcoming generations. CRA practices can alter the current situation and sustain agricultural production from the local to the global level, especially in a sustainable manner (Prabhakar, 2020).

Improved access and use of technology, transparent trade regimes, increased use of resource conservation technologies, and an increased adaptation of crops and livestock to climatic stress are the outcomes of climate-resilient practices.

Most countries have been facing crises due to disasters and conflicts; food security, however, is adversely affected by inadequate food stocks, basic food price fluctuations, high demand for agro-fuels, and abrupt weather changes.

7.3 STRATEGIES AND TECHNOLOGIES FOR CLIMATE CHANGE ADAPTATION

7.3.1 TOLERANT CROPS

Patterns of drought may need various sets of adaptive forms. To reach deficient downpour conditions, early maturing and drought-tolerant cultivars of green gram (BM 2002-1), chickpea, and pigeon pea (BDN-708) were brought to selected farmer's fields in the Aurangabad district of Maharashtra (rainfall of 645 millimeters) (Prabhakar, 2020).

This provided a 20–25% higher yield than the indigenous cultivars. In the same way, drought-tolerant, early maturing cultivars of pigeon pea (AKT-8811) and sorghum (CSH-14) were introduced in the villages of Amravati district, Maharashtra (rainfall of 877 mm).

7.3.2 TOLERANT BREEDS IN LIVESTOCK AND POULTRY

Local or indigenous breeds have the notion to forage for themselves. In nomadic systems, the animals show their owners when to move in search of new grasslands.

Indigenous breeds have unique characteristics that are adapted to very specific ecosystems across the world. These unique characteristics include resistance to droughts, thermoregulation, ability to walk long distances, fertility and mothering instincts, ability to ingest and digest low-quality feed, and resistance to diseases (Prabhakar, 2020).

These livestock breeds may not be highly productive in terms of meat or milk production but are highly adaptive to the unpredictable nature and have low resource footprints.

7.3.3 FEED MANAGEMENT

Betterment of feeding systems as an adaptation measure can indirectly improve the efficiency of livestock production. Some feeding methods include altering feeding time or frequency and modification of diet composition, including agroforestry species in the animal diet, and training producers in the production and conservation of feed for various agro-ecological zones.

These measures can decrease the risk from variations of climate by encouraging higher intake or compensating for low-feed consumption, decreasing excessive heat load, reducing animal malnutrition and mortality, and reducing feed insecurity during dry seasons, respectively (Prabhakar, 2020).

7.3.4 WATER MANAGEMENT

Smart water technologies like a furrow-irrigated raised bed, micro-irrigation, rainwater harvesting structure, cover-crop method, greenhouse, laser land leveling, reuse wastewater, deficit irrigation, and drainage management can support farmers to decrease the effect of variations in climate.

Various technologies based on a precision estimation of crop water needs, groundwater recharge techniques, adoption of scientific water conservation methods, altering the fertilizer and irrigation schedules, cultivating less water requiring varieties, adjusting the planting dates, irrigation scheduling, and adopting zero-tillage may help farmers to reach satisfactory crop yields, even in deficit rainfall and warmer years(Prabhakar, 2020).

Hence, many international organizations, national governments' research institutions, farmers' organizations, non-profit organizations, and private agencies across the world have been focusing their efforts on the design and development of cost-effective and environmentally friendly water-conserving devices to enhance water use efficiency.

7.3.5 AGRO-ADVISORY

Response farming is an integrative approach; it could be called farming with advisories taken from the technocrats depending on local weather information. The success of response farming not only decreased danger but also enhanced productivity which has already been visible in Tamil Nadu and many other states.

Response farming can be a viable choice for climate change adoption strategies, as the variations in climate are not sudden. The main causes for the success of response farming are both location and time-specific technologies. It is time to take forward the success of response farming to the entire farming community.

7.3.6 SOIL ORGANIC CARBON

Different farm management practices can increase soil carbon stocks and stimulate soil functional stability. Conservation agriculture technologies (reduced tillage, crop rotations, and cover crops), soil conservation practices (contour farming), and nutrient recharge strategies can refill soil organic matter by giving a protective soil cover.

Integrated nutrient management deals with the application of organic and inorganic fertilizers, in addition to farmyard manure, vermicompost, legumes in rotation, and crop residue for sustaining soil health for the long term.

Feeding the soil instead of adding fertilizers to the crop without organic inputs is the key point for the long-term sustainability of Indian agriculture.

7.4 CLIMATE CHANGE AND AGRICULTURE

- One of the critical challenges for a country's food security is climate change and its impact in the form of extreme weather events.
- The predicted 1–2.5°C temperature rise by 2030 is likely to show serious effects on crop yields.
- High temperatures may reduce crop duration, permit changes in photosynthesis, escalate crop respiration rates, and influence the pest population.
- Climate change accelerates nutrient mineralization, hampers fertilizer uses efficiency (FUE), and hastens the evapotranspiration in soil.

7.4.1 AGRI SUB-SECTORS AND CLIMATE CHANGE

1. Food Grains

- Cultivation practices are completely based on climatic situations.
- For example, in India, an increase in temperature by 1.5°C and a reduction in the precipitation of 2 mm can reduce the rice yield by 3–15%.

2. Horticulture

- High temperature causes moisture stress situation, directing to sunburn and cracking symptoms in fruit trees like apricot, apples, and cherries.
- The temperature increase at the ripening stage causes fruit burning and cracking in the litchi plantation.

3. Animal husbandry

- Dairy breeds are more prone to heat stress than meat breeds.
- An increase in metabolic heat production breeds leads to higher susceptibility to heat stress, while the low milk-giving animals are resistant.
- Poultries, no doubt, are severely sensitive to temperature-associated problems, particularly heat stress.

4. Fisheries

- Increasing environmental temperature may cause seasonal betterment in the growth and development of the fishes.
- But it also enhances the dangers to the populations living away from the thermal tolerance zone.

Details about the initial National Initiative on Climate Resilient Agriculture (NICRA) villages and the villages added subsequently are presented in Table 7.6.

TABLE 7.6

Details about the Initial NICRA Villages and the Villages Added Subsequently

Village Information	NICRA/Village since 2010	Additional Village 1 (2018)	Additional Village 2 (2018)	Additional Village 3 (2018)
Name of the village and district	Lagga (Hathla, Osal, Padhruin, Sakla, Prechha, and Ghasiti)	Banjal, Khalla	Khalnera, Chacho	Kiri
No. of households	102	50+30=80	25+20=45	150
Total cultivated area (ha)	89	13.89	10.29	5.28
Area under rainfed cultivation/ flood-affected area (ha)	89	13.89	10.29	5.28
Involvement of households in the program (%) approx by the end of 2019	80	50	50	40

7.5   A CASE STUDY OF DOUBLING FARMERS' INCOME

7.5.1   BRIEF STATEMENT OF THE PROBLEM

Lagga village in district Chamba lies in elevations ranging between 1,528 and 2,074 m (AMSL), 32°33′N latitude, and 76°15′E longitude. The average annual rainfall is about 1,100–1,800 mm. During winter, this area receives moderate to high snowfall. This zone has a single cropping system. Maize, cabbage, cauliflower, apple, beans, and potato are major crops grown in this area. Farmers in this area have small land-holdings that were too without irrigation facilities. Before starting NICRA in this village, the farmers were earlier producing a small amount of maize to meet their domestic requirements and were very poor (Raina *et al.,* 2019).

7.5.2   NATURAL RESOURCES AVAILABLE/UNAVAILABLE

The area is highly rugged hilly terrain, so most of the rainwater is lost by surface runoff resulting in very limited groundwater storage. At present, there are 11 water harvesting structures, with capacities ranging from 60,000 to 80,000-L capacity, which help the farmers to irrigate various crops in the whole village along with a lift irrigation scheme (Thakur *et al.,* 2021a,b).

7.5.3   THE INTERVENTIONS COVERED UNDER THE COMPONENT ARE BROADLY CLASSIFIED IN FOUR MODULES

7.5.3.1   Module I: Natural Resources

This module consists of interventions related to *in-situ* moisture conservation, water harvesting and recycling for supplemental irrigation, improved drainage in flood-prone areas, conservation tillage where appropriate, artificial groundwater recharge, and water-saving irrigation methods. Fig. 7.1 shows the different interventions related to *in-situ* moisture conservation and water harvesting and recycling for supplemental irrigation.

FIGURE 7.1   Water harvesting and recycling for supplemental irrigation and *in-situ* moisture conservation.

FIGURE 7.2   Spur-type apple plants.

7.5.3.2   Module II: Crop Production

This module consists of introducing drought/temperature-tolerant varieties, advancement of planting dates of rabi crops in areas with terminal winter stress, custom hiring center for timely planting, and location-specific intercropping systems with a high sustainable yield index. Spur-type apple plants (Fig. 7.2).

Crop diversification under rainfed conditions play a very important role in providing nutritional food to human, creating eco-friendly environment, generating employment and increasing the farmers' income, and reducing poverty resulting in overall socio-economic development. Fig. 7.3 presents the crop diversification.

7.5.3.3   Module III: Livestock

Cultivation of fodder crops is essential as it is necessary for feeding the animals that can be either grazed directly or fed after harvest to the livestock. Community lands are used for fodder production (Fig. 7.4) during droughts, preventive vaccination, and improved shelters for reducing cold stress in livestock.

FIGURE 7.3   Crop diversification.

FIGURE 7.4   Fodder production.

FIGURE 7.5   Training program to sensitize the farmers to round-the-year production of the fodder and for the formation of Self-Help Groups.

7.5.3.4   Module IV: Institutional Interventions

This module consists of institutional interventions either by strengthening the existing ones or initiating new ones relating to commodity groups, custom hiring centers, collective marketing, the introduction of weather index-based insurance, and climate literacy through a village-level weather station. Training programs should be conducted to sensitize the farmers for round-the-year production of the fodder and for the formation of Self-Help Groups (Fig. 7.5). Fig. 7.6 demonstrates rabi crop campaign. The celebration of World Soil Health Day at Lagga is portrayed in Fig. 7.7. Farmers celebrate the cabbage/cauliflower day along with Van Mahotsava at village Lagga (Figs. 7.6 and 7.7). Fig. 7.8 presents the training program to sensitize the farmers and students about different interventions.

The shift in cropping area at village Lagga after KVK intervention is presented in Table 7.7.

FIGURE 7.6   Awareness program on rabi campaign.

FIGURE 7.7 Celebration of World Soil Health Day.

FIGURE 7.8 Training program to sensitize the farmers and students about different interventions.

TABLE 7.7
Shift in Cropping Area (ha) at Lagga Village after KVK Intervention

Crops	Before Interventions	After Interventions	% Increase/Decrease
Maize	57.68	35.00	39.30(−)
Potato	7.40	12.60	70.30(+)
Apple	5.12	18.00	251(+)
Cabbage	0.30	8.00	2,566(+)
Cauliflower	0.40	6.00	1,400(+)

7.6 ACHIEVEMENTS

- Spur-type apple cultivation has been introduced on about 25 ha in the village. To scale up this intervention, demonstrations have also been laid out in other areas of the district.
- Intercropping (apple + cabbage/cauliflower) is also one of the major and successful interventions in the NICRA village which is also being

adopted by farmers of other villages in the district with the same climatic conditions.

- Initially, two 100-m^2 playhouses were constructed in the village but now more than 65 polyhouses have been constructed. Ten more polyhouses have been proposed for the financial year 2020–2021. The area under protected cultivation is increasing day-by-day and farmers of other villages are also adopting this technology (Thakur *et al.*, 2021).

Integrated farming system module: IFS module was started in the village in the year 2013–2014 to enhance the farm income of marginal and small farmers against extreme weather conditions, viz. cold and frost. Dependency on single farm enterprise sometimes may not work adequately or fails. IFS involves more than one crop in a limited area in which the residue of one is used as input for another enterprise to minimize the cost of cultivation and to get a sustainable yield.

Components of integrated modules include fruit crops (apple), vegetable crops (cabbage, cauliflower, potato), polyhouses (protected cultivation of capsicum), vermicomposting (earthworms are used for compost preparation), and mushroom and fodder cultivation.

Three programs on doubling farmers' income under IFS were conducted in the village with 20 farmers to grow fruit crops such as an apple with the intercropping of cabbage and cauliflower. This has improved the socio-economic status of rural people (Thakur *et al.*, 2021).

Protected cultivation: Protected cultivation has significantly helped farmers in reducing dependency on rainfall and efficient use of land and water resources. Under protected cultivation, the construction of a polyhouse in the village started in the year 2011 onward for the betterment of the farmers. Only 0.02 ha area was covered by protected cultivation in the year 2013–2014 with two polyhouses. In 2015, 0.03 ha area was under this and increased by 0.13 ha in 2016. At present, there are 65 polyhouses. Capsicum cultivation was taken up in the playhouses with the support of a micro-irrigation system (sprinkler system). The average yield of capsicum under protected cultivation was realized up to 545Q/ha with 3.28 BCR. Two progressive farmers of NICRA village Sh. Dharo Ram and Sh. Kavinder Singh cultivate colored capsicum also and earn handsome money out of it (Raina *et al.*, 2019).

7.7 CROP DIVERSIFICATION

1. Cauliflower: The demonstration on crop diversification with cauliflower was conducted over the past few years in NICRA village. In 2012, only 1.8 ha area was covered by this intervention which increased to about 13.5 ha in 2020.

2. Cabbage: Cabbage was demonstrated over a 2.5 ha area in 2012 indicating the adoption of successful intervention of intercropping in the orchard. In 2020, a 12.5 ha area was under this crop.

Before interventions, only 0.7 ha area was taken up by cabbage and cauliflower, and after interventions, 26.0 ha area was taken up by these two crops. By adopting these interventions of intercropping and off-season vegetable cultivation, farmers of Lagga village were able to increase their socio-economic status and get a high net return per year. Approximately 25 kg of seeds of hybrid varieties of cabbage and cauliflower were sown in the year 2019 (Thakur *et al.*, 2021).

3. Apple: Demonstration of crop diversification with the introduction of spur-type varieties of apple was successfully conducted over an area of 25 ha with 55 numbers of farmers. Before interventions, only a 5.12 ha area was under apple cultivation, but now the area has increased tremendously and reached up to 25 ha. High-density plantation with the latest varieties of apple is also demonstrated by two progressive farmers named Sh. Dharo Ram of Padhruin and Sh. Sanjeev Thakur of Chacho (Raina *et al.*, 2019).

7.8   FARMERS WHO ARE ACHIEVERS

1. **Integrated Farming System (IFS): Sh. Hari Singh**

Farmer's Name	Sh. Hari Singh S/o Sh.Frangu Ram	
Age	57 year	
Address	Village Lagga, PO Kiri, Teh. and District Chamba	
Education	Matriculation	
Landholding (ha/ acre)	2.5 ha	
Detail about livestock	5 (cows – 4, bulls – 2, and calf – 1)	
Farming experience (Years)	42	
Detail of Practice	**Existing Practice (Description)**	**Innovation (Improvement Over Existing)**
	Maize, wheat, barley, potato, etc.	Z-type polyhouses (3), apple orchard, and vermin-composting unit
Description of innovation	The traditional way of cultivation	Growing capsicum, tomato, and cucumber under protected structures (polyhouses), apple, cabbage, cauliflower, French bean, potato, etc. in open fields
Specification of the practice	The traditional way of cultivation. Getting only one crop per annum.	Integrated farming system. Getting multi-crops per annum. Cultivating capsicum, tomato, and cucumber under protected structures (3 z-type polyhouses of 40 sq m), 150 plants of apple, cabbage, cauliflower, French bean, potato, etc. in open fields.
Whether innovation is original innovation or modification of recommended technology	Innovation is a modification of recommended technology	

The practical utility of Innovation	This intervention increases the productivity per unit time per unit area and helps in earning more than Rs 2,50,000.00 per annum from the same piece of land where earlier he can get only Rs 50,000.00	
Utility of the innovation from the climate change perspective	Protected cultivation is the main innovation that reduces the impact of climatic vulnerabilities on some of the crops and farmers can get more profit.	
Adoption of the innovation by other farmers in the village	This innovation is widely adopted by ten farmers of the village and four farmers of the surrounding village.	
Crop yield	Maize (20 qt), wheat (28 qt), barley (20 qt), and potato (2 qt).	Maize (25 qt), wheat (7 qt), potato (15 qt), apple (500 boxes), cabbage (10 qt), cauliflower (40 qt), tomato (2 qt), and cucumber (2 qt).
Expenses incurred (Rs/ha/year)	20,000.00	25,000.00
Net returns (Rs/ha/ year)	50,000.00	2,50,000.00
B:C ratio	1:2.5	1:15
Other benefits	Fodder and grass	Pea and fodder

2. Integrated Farming System (IFS): Sh.Dharo Ram

Farmer's Name	Sh.Dharo Ram S/o Sh. Gian Chand	
Age	35 years	
Address	Village Padhruin, PO Kiri, Teh. and District Chamba	
Education	Graduation	
Landholding (ha/ acre)	1.5 ha	
Detail about livestock	6 (cows – 3, bulls – 2, and horses – 1)	
Farming experience (Years)	19	
Detail of Practice	**Existing Practice (Description)**	**Innovation (Improvement Over Existing)**
	Cabbage, cauliflower, capsicum, maize, wheat, barley, apple, etc.	One 252 m² and three 100 m² polyhouses, a high-density apple orchard (490 plants), and two vermin-composting units of size 30×8 ft and 8×6 ft
Description of innovation	The traditional way of cultivation	Growing colored capsicum under protected structures (polyhouses), apple, cabbage, cauliflower, etc. in open fields

Specification of the practice	The traditional way of cultivation. Getting only one crop per annum.	Integrated farming system. Getting multi-crops per unit of land. Cultivating colored capsicum under protected structures, 200 plants of apple (Royal), high-density apple orchard (490 plants), cabbage, cauliflower, etc. in open fields, and mushroom in lean period
Whether innovation is original innovation or modification of recommended technology	Innovation is a modification of recommended technology	
The practical utility of innovation	This intervention increases the productivity per unit time per unit area and helps in earning more than Rs 4,50,000.00 per annum from the same piece of land where earlier he can get only Rs 40,000.00.	
Utility of the innovation from the climate change perspective	Protected cultivation is the main innovation that reduces the impact of climatic vulnerabilities on some crops and farmers can get more profit	
Adoption of the innovation by other farmers in the village	This innovation is widely adopted by 48 farmers of the village	
Crop yield	Cabbage (10 qt), cauliflower (4 qt), capsicum (1 qt), maize (5 qt), wheat (2 qt), apple (45 boxes), etc.	Cabbage (40 qt), cauliflower (15 qt), capsicum (40 qt), maize (5 qt), wheat (2 qt), apple (400–600 boxes), etc.
Expenses incurred (Rs/ha/year)	20,000.00	40,000.00
Net returns (Rs/ha/year)	40,000.00	4,50,000.00
B:C ratio	1:2	1:11.25
Other benefits	Fodder and grass	Pea, tomato, and fodder

3. Integrated Farming System (IFS):Sh.Kavinder Kumar

Farmer's Name	Sh.Kavinder Kumar S/o Sh. Karam Chand
Age	31 years
Address	Village Banjal, PO Kiri, Teh. and District Chamba
Education	10+2
Landholding (ha/acre)	1.5 ha
Detail about livestock	9 (cows – 3, bulls – 4, and calves – 2)
Farming experience (Years)	10

Detail of Practice	Existing Practice (Description)	Innovation (Improvement Over Existing)
	Cabbage, cauliflower, capsicum, maize, wheat, barley, apple, etc.	One 252 m^2 and four 100 m^2 polyhouses, a high-density apple orchard (420 plants), and two vermin-composting units of size 30X8 ft
Description of innovation	The traditional way of cultivation	Growing colored capsicum and sweet cherry tomato under protected structures (polyhouses), apple, cabbage, red cabbage, cauliflower, etc. in open fields
Specification of the practice	The traditional way of cultivation. Getting only one crop per annum.	Integrated farming system. Getting multi-crops per unit of land. Cultivating colored capsicum under protected structures, 150 plants of apple (royal), high-density apple orchard (420 plants), cabbage, cauliflower, etc. in open fields.
Whether innovation is original innovation or modification of recommended technology	Innovation is a modification of recommended technology	
The practical utility of innovation	This intervention increases the productivity per unit time per unit area and helps in earning more than Rs 5,20,000.00 per annum from the same piece of land where earlier he can get only Rs 40,000.00	
Utility of the innovation from the climate change perspective	Protected cultivation is the main innovation that reduces the impact of climatic vulnerabilities on some crops and farmers can get more profit	
Adoption of the innovation by other farmers in the village	This innovation is widely adopted by two farmers of the village and five farmers of the surrounding village	
Crop yield	Maize (5 qt), wheat (2 qt) and apple (45 boxes), etc.	Cabbage (14 qt), cauliflower (12 qt), capsicum (24 qt), maize (5 qt), wheat (2 qt) and apple (420–580 boxes) etc.
Expenses incurred (Rs/ha/year)	20,000.00	40,000.00
Net Returns (Rs/ha/year)	40,000.00	5,20,000.00
B:C ratio	1:2	1:13
Other benefits	Fodder and grass	Pea, tomato, fodder, and grass

The details of farmers and change in the income due to the above initiatives are presented in Table 7.8.

The change in the socio-economic status of the adopted villages with the interventions of the NICRA project is presented in Table 7.9.

Information about the awards and recognitions received by KVK staff/farmers in the NICRA Village are given in Table 7.10.

TABLE 7.8
Increase in Farmer's Income in Two Years during 2018–2019 and 2019–2020

Sl. No.	Name of Farmer	Contact Info of Farmer with Mobile No.	Landholding (Bigha)	Income of Farmer (Rs) as per Their Statement 2013	2019
1	Sh. Dharo Ram	Village Padhruin, PO Kiri, Tehsil and District Chamba – 8219882759	12.5	40,000.00	6,00,000.00
2	Sh. Kavinder	Village Banjal, PO Kiri, Tehsil and District Chamba – 9805295432	22	25,000.00	4,20,000.00
3	Sh. Hari Singh	Village Lagga, PO Kiri, Tehsil and District Chamba – 9816684157	12	1,00,000.00	4,30,000.00
4	Sh. Ram Singh	Village Lathuin, PO Kiri, Tehsil and District Chamba – 9816077810	36	2,00,000.00	5,20,000.00
5	Smt. Pritmo Devi	W/o Sh. Hem Raj, Village Lagg, PO Kiri, Tehsil, and District Chamba – 9816743585	12	20,000.00	1,20,000.00

TABLE 7.9
Impact of NICRA Project on the Socio-Economic Status of Adopted Village

Sl. No	Particulars	Before the NICRA Project (2011)	After the NICRA Project (2020)
1	Carriage vehicle	Nil	12
2	Cars	Nil	48
3	Bikes	Nil	54
4	Pucca houses	10	92
5	Migration of inhabitants	70 persons	04
6	Power tillers	NIL	10

TABLE 7.10
Awards and Recognitions (Received by KVK staff/Farmers in the NICRA Village)

Name of Award	Award is given by	Year
Best KVK Scientist Award presented to Dr. Kehar Singh Thakur, Scientist (Silviculture)	The Society of Krishi Vigyan	2020
Adarsh Vidya Saraswati Rashtriya Puraskar (National Award of Excellence-2020) was presented to Dr. Kehar Singh Thakur, Scientist (Silviculture)	Glacier Journal Research Foundation, Global Management Council, Ahmedabad, Gujarat	2020
The Best Scientist of the Year Award-2020 is presented to Dr. Kehar Singh Thakur, Scientist (Silviculture)	Glacier Journal Research Foundation, Global Management Council, Ahmedabad, Gujarat	2020

(Continued)

TABLE 7.10 (*Continued*)

Awards and Recognitions (Received by KVK staff/Farmers in the NICRA Village)

Name of Award	Award is given by	Year
Nominated to Editorial Board of Glacier Journal Research Foundation, Global Management Council, Ahmedabad, Gujarat	Glacier Journal Research Foundation, Global Management Council, Ahmedabad, Gujarat	2020
Progressive Farmer Award-2020 to Sh. Kavinder Kumar, Village Banjal, PO Kiri, Chamba	Dr. Y S Parmar UHF, Nauni, Solan on 36th Foundation Day	2020
First Position in implementing Krishi Kalyan Abhiyan-1	Ministry Of Agriculture and Farmers Welfare	2019
Third Position in implementing Krishi Kalyan Abhiyan-2	Ministry Of Agriculture and Farmers Welfare	2019
Best NICRA KVK Award-2019	CRIDA, Hyderabad	2019
Best Poster Presentation award-2019 to Dr.Kehar Singh Thakur, Scientist (Silviculture), KVK Chamba	CRIDA, Hyderabad	2019
Progressive Farmer Award-2019 to Sh.Dharo Ram, Village Padhruin, PO Kiri, Chamba	Dr Y S Parmar UHF, Nauni, Solan	2019
Progressive Farmer Award-2017 for the popularization of climate-resilient agriculture in NICRA village Lagga to a progressive farmer Sh. Hari Singh	Dr. Y. S. Parmar, University of Horticulture and Forestry, Nauni, Solan (HP) on 32th Foundation Day	2017

7.9 CONCLUSIONS

Lagga village has a single cropping system. Maize, cabbage, cauliflower, apple, beans, and potato are major crops grown in this area. Farmers in this area have small landholdings that were too without irrigation facilities. Before the inception of the NICRA project in this village, the farmers were earlier producing a small amount of maize to meet their domestic requirements and were very poor. Spur-type apple cultivation has been introduced on about 25 ha in the village. To scale up this intervention, demonstrations have also been laid out in other areas of the district. Intercropping (apple+cabbage/cauliflower) is also one of the major and successful interventions in the NICRA village which is also being adopted by farmers of other villages in the district with the same climatic conditions. More than 65 polyhouses have been constructed to increase the area under protected cultivation. IFS module was started in the village in the year 2013–2014 to enhance the farm income of marginal and small farmers against extreme weather conditions, viz. cold and frost. Components of integrated modules are fruit crops (apple), vegetable crops (cabbage, cauliflower, potato), polyhouses (protected cultivation of capsicum), vermicomposting (earthworms are used for compost preparation), and mushroom and fodder cultivation. The area under crop diversification with cauliflower has increased. In 2012, only a 1.8 ha area was covered by this intervention which increased to about 13.5 ha in 2020. Cabbage was demonstrated over a 2.5 ha area in 2012 indicating the adoption of successful intervention of intercropping in the orchard. In 2020, a 12.5 ha area was under this crop. Crop diversification with the introduction of spur-type varieties of apple was successfully conducted over an area of 25 ha. Before interventions, only a 5.12 ha area was under apple cultivation, but now the area has increased tremendously and reached up to 25 ha. High-density plantation with the latest varieties of apple is also demonstrated by two progressive farmers named Sh. Dharo Ram of Padhruin and Mr. Sanjeev Thakur of Chacho. The NICRA project is able to improve the socio-economic status of Lagga farmers by increasing the number of carriage vehicles, personal cars, bikes, pucca houses, and power tillers and the reduction of the migration of local inhabitants. Scientists of Krishi Vigyan Kendra, Chamba, and farmers of Lagga village get recognition not only at the state level but also at the national level as a result of the NICRA project. At present, NICRA village Lagga is a CRA model in the district of Chamba.

The increase in agriculture-sector expenditure in recent years has been on account of schemes like PM-KISAN, PMFBY, interest subvention, and price support and loan waivers, with a focus on providing direct monetary benefits. Apart from efforts aimed at helping the agrarian economy recover, the government should enhance expenditure on agricultural infrastructure. A number of reports have highlighted that farm operations suffered due to infrastructure bottlenecks such as supply chain distortions, non-availability of credit, lack of quality inputs, and marketing infrastructure. Instead of cash-based schemes, India needs expenditure-enhancing infrastructure for a climate-resilient future. Different interventions under different modules of the NICRA project are able to double farmers' income.

REFERENCES

http://www.fao.org/3/nc938en/nc938en.pdf

https://theconversation.com/india-protests-farmers-could-switch-to-more-climate-resilient-crops-but-they-have-been-given-no-incentive-154700

https://www.downtoearth.org.in/blog/agriculture/climate-resilient-agriculture-systems-the-way-ahead-75385

https://www.downtoearth.org.in/blog/agriculture/why-india-needs-climate-resilient-agriculture-systems-75381

Prabhakar, M. 2020. Climate resilient agriculture in India feasibility and need for up. Central Research Institute for Dryland Agriculture Indian Council of Agricultural Research (ICAR) Hyderabad, India. *4th Indian Agricultural Outlook Forum 2020.*

Raina, R., Thakur, K. S., Kumari, S., Kapoor, R., Thakur, M. and Sharma, A. 2019. Impact of technology demonstrations on farmer's socio-economic status – an initiative of KVK, Chamba at Saru, Himachal Pradesh under NICRA. *Directorate of Extension Education, UHF, Nauni*, pp. 36.

Thakur, K. S. 2018. Upliftment of socio-economic status of rural people of Lagga village of Chamba district through NICRA. *3rd Himachal Pradesh Science Congress.*

Thakur, K. S., Raina, R., Dhiman, S. and Negi, D. 2021a. Presented Research Paper on Impact of climate resilient Agriculture on socio-economic status of rural people through National Innovation on Climate Resilient Agriculture (NICRA) project at Lagga village of district Chamba. 5th International Agronomy Congress on Agri Innovations to combat food and nutrition challenges. Virtual mode. *The Indian Society of Agronomy.*

Thakur, K. S., Raina, R., Kapoor, R. and Sharma, A. 2021b. Ekikrit Krishi Padati aur Vividhikaranduarasafita ki kahanilikhataChambaJila ka NICRA GaamLagga. Published Pamphlet.

8 Integrated Farming System

A Dynamic Approach toward Increasing the Income of Small and Marginal Farmers

Y. B. Vala and M. H. Chavda

8.1 INTRODUCTION

Despite the increasing rate of economic growth in India, the agriculture sector still runs behind. It has been recorded by the Economic Survey of India 2008 that the growth rate of food grain production decelerated to 1.2% during 1990–2007, lower than the population growth of 1.9%. It has been marked that with increasing trends in population, the population will reach 1,370 million by 2030 and 1,600 million by 2050. It is expected that we require 289 and 349 million tons of food grains to satisfy the hunger demand. But to fulfill these needs, more area should be converted into agricultural fields; on the other hand, it is somehow expected to decrease up to 20% by 2030. It is clear that with increase in population, more area gets converted into nonagricultural land so as to build shelter. Thus, there is no scope for horizontal expansion of land for agriculture. Only vertical expansion can be done to change this scenario, integrating farming components requiring lesser space and time and ensuring reasonable returns to farm families. Integrated farming system (IFS) is defined as integration of various agricultural enterprises, namely, cropping, animal husbandry, fishery, and forestry, which have great potentialities in the agricultural economy. The IFS is found beneficial as it has various advantages, including increased farm production and reduction in environmental degradation and maintaining sustainability. IFS is a unique system that helps to design enterprises in such a way that they are beneficial to both inputs and outputs; therefore, full use of the resource is the main motive behind it. These could be practiced in aquaculture, agroforestry, agri-horticulture, apiculture, and sericulture. Being such a wide range of enterprises effective recycling of wastes and crop residues helps an additional source of income for farmers. IFS also works on other principles that include sustainable management as the impact of fertilizers led to a decrease in soil health. IFS leads to an increase in the fertility of the soil and enriches it with nutrients along with recycling of resources.

DOI: 10.1201/9781003351672-8

8.2 WHY WE NEED TO INCREASE FARMERS' INCOME?

On February 28, 2016, honorable PM Narendra Modi addressed farmers' rally in Bareilly, UP. It is my dream to see farmers double their income by 2022 when the country completes 75 years of its independence. Doubling the real income of farmers by 2022–23 over the base year of 2015–16 requires annual growth of 10.41% in farmers' income (whereas, it was at 3.80% in 2016–17).

This implies that the ongoing and previously achieved rate of growth in farm income has to be sharply accelerated. More than one-fifth of rural households with self-employment in agriculture were having income less than the poverty line. The rate of farmers' suicides increases sharply due to losses from farming, shocks in farm income, and low farm income. A sharp fall in the growth rate of agricultural output also increases. Low and high fluctuating farm income cause a detrimental effect on the interest in farming and farm investments and also forcing younger age groups to leave farming.

8.3 WHAT IS IFS?

Integrated farming (IF) is a whole farm management system that aims to deliver more sustainable agriculture. IF systems include farming of livestock, aquaculture, horticulture, and allied activities. It is sometimes called integrated bio-systems or integrated agriculture. It is a system that comprises an interrelated set of enterprises with crop activity as a base, which will provide ways to recycle produces, and waste from one component becomes an input for another part of the system, which reduces cost and improves soil health and increases production and/or income.

8.3.1 CONCEPT OF IFS

- An arrangement of recycling products/by-products of one component as input to another linked component
- Reduction in cost of production
- Increase in productivity per unit area
- Increase in the total income of farm
- Effective use of family labors around the year

8.3.2 WHY IFS?

- High input costs
- For meeting the rising need for food, feed, fuel, and fertilizer
- Nutritional requirement of family
- Increased demand for soil nutrients
- For increasing the income
- Employment
- Standard of living
- Sustainability

8.4 CHALLENGES FACED BY SMALL AND MARGINAL FARMERS IN INDIA

In an agricultural country like India, there are plenty of challenges and hardships for small and marginal farmers. The National Sample Survey Office (NSSO) 2014 Farmer's survey brought light to many issues which small and marginal farmers face. Some of them are as follows:

- Water problem
- Credit and indebtedness
- Land issue
- Climatic changes
- Globalization changes
- Social groups
- Low-level education
- Diversification
- Women role
- Associated risks

As per agriculture census data 2010–11, more than 70% of the farmers in India are small and marginal farmers owing nearly 44% of the total land (Table 8.1). So, for the betterment of the Indian economy, it is very important to focus on the development of these farmers, then only our economy will stand strong.

8.5 RESTRAINING FORCES IN DOUBLING FARMER'S INCOME

- Fragmentation of land holdings
- Water and labor crises
- Increasing cost of production
- Pest and disease incidence
- Wildlife menace
- Lack of mechanization and small tools and machinery
- Climate change effect-biotic and -abiotic pressure

TABLE 8.1
Land Holding Pattern of Farmers in India (Agricultural Census 2010–11)

Sr. No	Size-Group	Percentage of Number of Operational Holdings to Total	Percentage of Area Operated to Total
1	Marginal (below 1.00 ha.)	67.10	22.50
2	Small (1.00–2.00 ha.)	17.91	22.08
3	Semi-medium (2.00–4.00 ha.)	10.04	23.63
4	Medium (4.00–10.00 ha.)	4.25	21.20
5	Large (10.00 ha. and above)	0.70	10.59

- Lack of quality planting material
- Inadequate farmer field schools
- Non-availability of credit and inadequate crop insurance
- Lack of policy and credit orientation for developing rural non-farm employment

8.6   PROBLEMS OF PRESENT AGRICULTURE

Following are the problems in agriculture

- Decline in factor productivity
- Static or decline in food production
- Increasing malnutrition
- Shrinkage in net cultivation area
- Increasing environmental pollution
- Depleting ground water table
- Increasing cost of production
- Low farm income
- Problems of farm labors due to large-scale migration

8.7   FARMING SYSTEM

A farming system is the scientific integration of different interdependent and inter-acting farm enterprises for the efficient use of land, labor, capital, and other resources of a farm's families to get a year-round income.

A farm is a system that includes inputs, processes, and outputs:

- **Inputs**: These are the things that go into the farm and may be split into physical inputs (e.g., amount of rain and soil) and human inputs (e.g., labor and money).
- **Processes**: These are the things that take place on the farm in order to convert the inputs to outputs (e.g., sowing, weeding, and harvesting).
- **Outputs:** These are the products from the farm (i.e., wheat, barley, and cattle) depending on the type of farming, for example, arable/pastoral, commercial/subsistence, the type and amount of inputs, processes, and outputs will vary.

8.8   GOALS

- To maximize the yield of all components to earn a supplement income.
- Rejuvenation system's productivity to achieve agro-ecological equilibrium.
- To give healthy and chemical-free products and reduce the use of chemicals.
- Natural cropping system management should be done to avoid insectpests and diseases.

8.9 IFS COMPONENTS

8.9.1 CROPS

Crops may have subsystems like monocrop, mixed/intercrop, multi-tier crops of cereals, legumes (pulses), oilseeds, and forage.

Suitable grain crops

	Black soil
Cereals	Sorghum, rice
Pulses	Black gram, chickpea, red gram
Oilseeds	Sunflower, safflower, sesame
Other crops	Coriander, chili, cotton, sugarcane
	Sandy soil
Cereals	Ragi, bajra, sorghum, other millets
Pulses	Lab-lab, green gram, soybean, cowpea
Oilseeds	Groundnut, castor, sesame, sunflower

Suitable forage crops
***Black soils*:** Fodder sorghum, fodder cowpea, Dhaman
***Loamy/sandy soil*:** Fodder bajra, *Cenchrus sp.*, fodder cowpea, fodder sorghum, oats

8.9.2 SUITABLE TREE SPECIES

Black soil: Babool, Neem, *Albizia lebbeck*, *Prosopis juliflora*, White babool, etc.
Sandy loam soils: Indian date, *Albizia lebbeck*, *Acacia tortilis*, Neem, Ber, etc.

8.9.3 SUITABLE ANIMALS

Goats, sheep, cattle, fish, pigs, etc.

8.9.4 SUITABLE BIRDS

Poultry birds, ducks, etc.

8.9.5 MISCELLANEOUS COMPONENTS

Mushroom production, biogas, apiary, etc.

8.10 BENEFITS OF INTEGRATED FARMING SYSTEM

- Improves soil fertility and health
- Increases economic yield per unit area
- Reduces in production costs
- Decreases farm input requirements

- Multiplies income sources
- Builds family income support
- Reduces animal feed requirements
- Minimizes the use of chemical fertilizers
- Solves the energy problems with biogas
- Avoids degradation of forests
- Enhances employment generation
- Creates a pollution-free environment
- Recycles of resources
- Improves the status and livelihood of the farmers

8.11   AGRONOMIC APPROACHES FOR IFS

- Adoption of an improved cropping system according to rainfall and soil moisture availability.
- Selection of grain species and multipurpose tree species that supplies grass/pods/leaves for a longer period or throughout the year.
- The surplus fodder leaves, crop residues, etc. during the rainy season should be preserved as silage/hay for the lean season and also for the preparation of compost.

8.12   CRITERIA FOR SELECTION OF ENTERPRISES

Fig. 8.1 describes various enterprises which may be included in IFS.

1. Soil and climatic features of an area/locality.
2. Social status of the family and social customs prevailing in the locality.
3. Economic condition of the farmer.
4. Resource availability at the farm and present level of use of resources.

FIGURE 8.1   Various enterprises to be included in IFS.

5. Economics of proposed IFS and credit facilities
6. Farmer's managerial skills
7. Household demand
8. Market facilities

8.13 ENTERPRISES TO BE LINKED THROUGH IFS IN DIFFERENT CONDITIONS

Wetland: crop, livestock farming, fish farming, fish farming, Azolla farming, poultry, pigeon growing, and mushroom growing.

Gardenland: crop, dairy, goat rearing, poultry, home gardening, vermicompost, sericulture, piggery, and biogas unit.

Dryland: crop, sheep rearing, goat rearing, agroforestry, silvi-pasture, farm pond, and horticulture fruit tree.

Enterprise Integration

- Livestock is the best complementary enterprise with cropping.
- Installation of a biogas plant in the crop-livestock system will make use of the wastes besides providing manure and gas for cooking and lighting.
- In wetlands, there is a better scope for fishery, duck, and buffalo rearing. Using rice straw, mushroom production can be started.
- Under irrigated condition, sericulture, poultry, and piggery with arable crop develop better.
- In rainfed farming, sheep and goat, sericulture, agroforestry, etc. develop well.
- In the integrated system, enterprises with complementary relationship should be selected to avoid competition.

8.14 BENEFICIAL IFS COMBINATION UNDER VARIED ECOSYSTEM

An IFS is a combination of many systems, which attempts to increase farmers' income using natural resources on a sustainable basis that can be obtained by integrating crop husbandry with allied enterprises. Enterprise in the farm business is defined as the production of a single crop or a kind of stock. Generally, farmers take more than one enterprise on their farms. The main objectives of farming are to get maximum profit with minimum expenditure by combining enterprises. The combination of enterprises on a farm is influenced by the relationship that exists between the enterprises.

In wetland ecosystem

- Crop + Fish + Poultry
- Crop + Fish + Duck
- Cropping + Fish + Pigeon
- Cropping + Fish + Goat
- Crop + Fish + Pigeon
- Rice + Fish + Azolla

- Rice + Fish + Poultry/Pigeon +Mushroom
- Crop + Fish + Mushroom

In dryland ecosystem

- Crop + Goat/Sheep
- Crop + Goat + Agroforestry
- Crop + Goat + Agroforestry + Horticulture
- Crop + Goat + Silvi-pasture+ Pigeon + Farmpond
- Crop + Silvi-pasture + Buffalo + Farm pond
- Crop + Goat + Agroforestry + Horticulture + Farm pond

Island ecosystem

- Coconut + Fodder + Milch cattle
- Coconut + Fodder + Fish/Prawn culture
- Coconut + Fish culture in salt-affected lands
- Fruits + Fodder + Milch cattle

Gardenland ecosystem

- Crop + Dairy + Biogas
- Crop + Goat + Biogas + Fishery
- Crop + Dairy + Biogas + Sericulture
- Crop + Dairy + Biogas + Homestead garden + piggery
- Crop + Dairy + Biogas + Swamp production + Mushroom
- Crop + Sericulture + Biogas unit
- Crop + Dairy + Biogas + Horticulture Garden
- Crop + Dairy + Biogas +Vermicompost

Hill region

- Crop (fodder trees, grasses) + Dairy cattle/piggery/poultry
- Agri + Horti-silvi-pastoral system

Most prominent IFS models

- Crop + Livestock
- Crop (Rice) + Fish culture
- Floriculture + Apiary (Beekeeping)
- Agroforestry + Silvi-pasture
- Fishery +Duckery + Poultry
- Crop (Rice) + Fish + Mushroom
- Crop + Livestock + Poultry

8.15   PRESENT STATUS OF FARMING SYSTEM RESEARCH

From previous investigations, we have come to the conclusion that the integration of agricultural enterprises, that is, crop, livestock, fishery, and forestry, has a high potential for improving the agricultural economy. These enterprises also ensure that the rational use of the resources further helps in reducing the problem of unemployment. The farming system is governed by various aspects, that is, physical environment, socio-economic conditions, political forces under various institutional and operational constraints, and above all government favorable policies, which may keep the livelihood fully protected. In the traditional Chinese system, the animal houses were built over a pond so that animal waste fell directly into the pond and was directly consumed by fish as food. In fields, not only fish are harvested but the pond water is also used for irrigation. For marginal farmers, it starts with small livestock, like ducks and chickens, then a few goats for milk, then a milch cow, then a bullock for plow in co-operation with another buffalo family, then two bullocks used to plow the fields of others. Farmers would add a milch buffalo to reach the highest point of achievement. The main concept is to start with small livestock and then the household will help one get out of poverty gradually. The poorest households kept only poultry and these households were most dependent on common property resources for their living for example use and sale of firewood from the forest. A survey on farming systems in the country revealed that milch animals, cows, and buffaloes irrespective of breed and productivity are the first choice of the farmers as a major part of their farming system. By considering economic value, vegetables and fruits followed by beekeeping, sericulture, and mushroom and fish cultivation were the most enterprising components of any of the farming systems in the country. Diversification of the farming system by integration of enterprises in various farming situations enabled to enhance of total production in terms of rice equivalent yield ranging from 9.2% in the eastern Himalayan region to as high as 36.6% in Western-plain and Ghat region when compared to prevailing farming systems of the region. A number of success stories on IFS models, including Sukhomajri watershed of Chandigarh, Fakot watershed in hilly areas of Uttarakhand, Jayanthi models for almost all the situations of TN, in Punjab Darshan Singh model for irrigated conditions, Project Directorate for Cropping Systems Research (PDCSR) model, for western Uttar Pradesh and many more in different parts of the country, suggest that farmer's income can be increased by the integration of enterprises in a farming system mode for sustainability and economic viability of small and marginal category of farmers.

8.16   RESEARCH REVIEWS OR CASE STUDIES

8.16.1   INTEGRATED FARMING SYSTEM RESEARCH UNDER WETLAND ECOSYSTEM

Jayanthi et al. (2003) reported that integration of crop + fish + goat gave higher net return of Rs.1,26,564/ha than the conventional system (Rs.36,190/ha). Integration also gave higher REGEY (39,610 kg/ha) and returns per day (Rs.493/day) (Table 8.2). The higher net income is the result of the integration of crops with fish and goats, which increases the production for the sale to the farmers and by better use of on-farm resources.

Table 8.3 reveals that the cultivation of rice gave an RGEY of only 4,311 kg/ha, whereas the integration of rice-brinjal + mushroom + poultry gave an RGEY of

TABLE 8.2

Productivity and Economic Analysis of Different Integrated Farming System (1998–2000)

Farming System	RGEY (kg/ha)	Cost of Production (Rs./ha)	Gross Return (Rs./ha)	Net Return (Rs./ha)	Per Day Return (Rs./ha)
Cropping alone	12,222	24,922	61,112	36,190	164
Crop + fish + poultry	31,858	44,627	1,59,292	1,14,665	436
Crop + fish + pigeon	32,354	43,310	1,61,772	1,18,462	443
Crop + fish + goat	39,610	51,483	1,78,047	1,26,564	493

21,487 kg/ha (Manjunath and Itnal, 2003a,b). It is because of the better use of enterprises like mushrooms and poultry and their interrelated benefits.

Manjunath and Itnal (2003a,b) reported that the integration of cashew, coconut + forage + dairy, rice-brinjal + rice-cowpea + mushroom + poultry is beneficial as it gives a net return of Rs.1,44,025/ha as compared to the cultivation of crop alone (Table 8.4). Because in this integration system, there is a better use of land and farm resources and also because of the suitability of fodder with coconut and higher net returns are also fetched from the inclusion of vegetables in the cropping system, which gives higher returns and productivity.

Anonymous (2009) generated a model for a wetland at TNAU and observed that integration of crop + poultry + fish + mushroom gives an additional income of Rs.18,360/ha. Returns from crop and IFS were Rs.33,446/ha and Rs.70,619/ha, respectively (Table 8.5). The integration has included all the enterprises that are beneficial to each other as the by-product of one enterprise can be used as input in the other enterprise. Hence, the cost of cultivation is reduced because of less dependence on purchased inputs.

Channabasavanna et al. (2010) conducted a study and based on the study they stated that the integration of rice + hybrid maize-sunflower + vegetable + fodder +

TABLE 8.3

Productivity of Rice-Based Integrated Farming System

Farming System	Component Productivity (kg/ha)			Rice-Grain-Equivalent Yield (kg/ha)
	Crop	Poultry	Mushroom	
Rice cropping alone	4,311	----	----	4,311
Rice-groundnut + mushroom + poultry	6,557 (39)	6,060 (36)	4,305 (25)	16,922
Rice-cowpea + mushroom + poultry	7,662 (43)	6,060 (34)	4,305 (23)	18,027
Rice-brinjal + mushroom + poultry	11,122(52)	6,060 (28)	4,305 (20)	21,487
Rice-sunnhemp + mushroom + poultry	4,993 (33)	6,060 (39)	4,305 (28)	15,358

TABLE 8.4

Economic Analysis of Different Farming Systems for Small and Marginal Holdings in Different Topographies of Goa (Mean Data 1999–2001)

Production System	Productivity (kg/ha)	Gross Returns (Rs./ha)	Cost of Cultivation (Rs./ha)	Net Returns (Rs./ha)
Monocropping				
Cashew	4,489	31,420	6,400	25,020
Coconut	2,624	18,370	12,150	6,225
Rice	4,943	34,600	15,230	19,370
Total	**12,056**	84,390	33,780	**50,610**
Integrated Farming System				
Cashew	7,521	52,645	16,315	36,330
Coconut + forage + dairy	13,553	94,870	62,535	32,335
Rice-brinjal (0.5 ha) + rice-cowpea (0.5 ha) + mushroom + poultry	21,223	1,48,560	73,200	75,360
Total	**42,297**	2,96,075	1,52,050	**1,44,025**

goat + fish + poultry gave a higher BC ratio (1.97) than the conventional system (1.84) (Table 8.6). It gives better results because of the inclusion of different enterprises on the same piece of land which gave a higher benefit-cost ratio and also higher employment generation as this integration involves more manpower; other benefits of this system are that it also reduces the need for water by the crop.

Table 8.7 describes the system productivity, benefit-cost ratio, and employment generation in IFS as compared to the conventional system of the study by Kalpana et al. (2016). They have reported that the combination of rice + fish + duck gave a system productivity of 16,675 kg/ha/yr, whereas system productivity in the conventional

TABLE 8.5

Wetland Model Generated at TNAU for Small and Marginal Farmers

Sl. No.	System	Gross Income (Rs./ha)	% Share in Income
1.	Crop alone	33,446	100
2.	Crop	41,877	59.3
	Poultry	6,143	8.7
	Fish	5,225	7.4
	Mushroom	17,372	24.6
	Total	70,619	100
3.	Additional income generated	18,360	

TABLE 8.6

Productivity, Profitability, Employment Generation and Water Requirement in Rice-Based Integrated Farming System (Pooled Data of three Years)

Treatment	Area (ha)	Productivity (kg/ha/yr)	Cost of Cultivation (Rs./ha)	Net Returns (Rs./ha)	B:C Ratio	Employment Generation (man days/ ha/yr)	Water Requirement (mm)
Integrated Farming System							
Rice-rice system	0.33	2,175	8,683	7,387	1.84	172	848
Hybrid maize-sunflower	0.20	908	3,697	3,540	1.96	45	82
Vegetable	0.20	2,136	4,712	3,673	2.00	31	95
Fodder + goat	0.21	1,339	6,289	7,060	2.75	9	82
Fish	0.06	203	515	926	2.23	5	105
Poultry	0.005	327	2,145	300	1.13	13	35
Total	1.00	7,088	18,225	22,887	1.97	275	1,247
Conventional Cropping System							
Rice-rice	1.00	5,611	25,503	17,293	1.64	459	2370

system was only 5,000 kg/ha/yr. The water filled in the rice field will be used for rearing the fish in return the fish litter will play the role of manure for the rice crop and the duck will also play an important role in the increased system productivity. Due to this closed resource use system in this integration, the benefit-cost ratio will also increase. BC ratio is highest in the fish-fish system because it is the least input-required system.

Table 8.8 presents the economics of different IFS components based on the study conducted by Anonymous (2017a,b,c). The analysis of Table 8.8 revealed that the highest gross and net returns (Rs.44,100 and Rs.43,900) were achieved from boundary plantation (Horticultural crops included).

TABLE 8.7

The System Productivity, Benefit Cost Ratio and Employment Generation in IFS as Compared to Conventional System

Farming System	System Productivity (kg/ha/yr)	System Productivity (kg/ha/day)	Benefit – Cost Ratio	Employment Generation (man days/ha)
Rice-fish-duck	16,675	45.68	1.55	575
Rice-rice	5,000	13.69	1.47	235
Fish-fish	5,962	16.34	1.68	94.21
Duck-duck	4,953	13.57	1.54	56.61

TABLE 8.8

Economics of Different IFS Components (2017–18)

S.N	Name of IFS Components	Area (ha.)	Cost of Cultivation (Rs)	Gross Return (Rs)	Net Return (Rs)	Employment Generation Man Days
1	Cropping systems	0.180	21,935	32,965	11,031 (14.56%)	63.00
2	Multistoried horticulture fruits and vegetables	0.400 + 0.186	21,474	37,427	15,953 (21.05%)	35.63
3	Boundary plantation (included in horticultural crops)	0.028	1,091	44,100	43,009 (56.76)	2.25
4	Nursery/kitchen gardening	0.015	942	970	28 (0.04%)	1,025
5	Compost	0.005	2,500	3,250	750 (0.99%)	0.5
6	Water recharging(other than IFS model area)	0.060	10,000	15,000	5,000 (6.60%)	-
Total			57,942	1,33,712	75,771	

8.16.2 INTEGRATED FARMING SYSTEM RESEARCH IN GARDEN LAND ECOSYSTEM

Rangaswamy et al. (1995) conducted a study and presented the results of receipt, expenditure, and net income for five years in IFS and CCS (Table 8.9). The analysis of Table 8.9 revealed that crop + dairy + biogas + spawn production + silviculture combined through IFS generated a net income of Rs.34,579/ha, whereas the conventional system gave net returns of Rs.13,877/ha. The study showed that the IFS gave higher net returns continuously for years.

Goverdhan et al. (2018) reported that IFS system comprising crop + dairy + sheep + rabbit + hens + quails + manure unit gave net returns of Rs.6,09,160/ha, whereas rice

TABLE 8.9

Receipt, Expenditure, and Net Income for Five Years in IFS and CCS

Year	Receipt (Rs./ha)		Expenditure (Rs./ha)		Net Income (Rs./ha)	
	IFS	CCS	IFS	CCS	IFS	CSS
1988–89	79,862	34,547	37,161	10,660	42,701	23,887
1989–90	70,252	26,400	36,275	12,490	33,977	13,910
1990–91	77,832	26,740	35,855	12,420	41,977	14,320
1991–92	45,707	37,845	17,117	27,507	28,590	10,338
1992–93	1,01,105	35,615	75,440	28,640	25,665	6,975
Mean	74,952	32,230	40,370	18,343	**34,579**	**13,887**

TABLE 8.10

Cost of Production and Income from IFS Unit and Conventional System

Sl. No.	Component	Cost of Production (Rs./ha)	Returns (Rs./ha)	
			Gross	Net
Integrated Farming System				
1.	Crop	89,600.00	2,50,250.00	1,60,650.00
2.	Dairy	1,73,740.00	3,65,000.00	1,91,260.00
3.	Sheep	73,000.00	1,40,000.00	67,000.00
4.	Rabbits	55,000.00	72,000.00	1,700.00
5.	Hens	1,13,000.00	2,00,000.00	87,000.00
6.	Quails	1,44,000.00	2,16,000.00	72,000.00
7.	Manure			14,250.00
Total		6,48,340.00	12,43,250.00	6,09,160.00
Conventional System (Rice-Maize)				
8.	Rice	55,350	1,08,450	53,100
9.	Maize	25,157	1,10,430	85,273
Total		80,507	2,18,880	1,38,373

+ maize conventional system gave only Rs.1,38,373/ha (Table 8.10). It shows that the net returns from the IFS are more than the conventional system of growing rice and maize. It is because in the IFS, various enterprises are linked with each other which decreases the total dependence of the system on the purchased inputs in fact the by-product or the residue of one enterprise is used in other subsidiary enterprises as an input which reduces the expenditure and increases the net returns.

8.16.3 INTEGRATED FARMING SYSTEM RESEARCH IN DRYLAND ECOSYSTEM

Table 8.11 presents the agri-horticultural system (Ber + Mung bean) (Gupta, 1997) The results of the Table 8.11 describe that net profit per ha was Rs. 4,800/- in sole leguminous crop farming; however, in the case of ber intercropping, the profit was Rs. 8,000/- per ha.

Shekinah and Sankaran (2007) conducted a study on system productivity (sorghum grain equivalent yield), employment generation, and economics in sorghum-based IFSs (Table 8.12). They stated that crop + pigeon + buffalo + agroforestry + farm pond

TABLE 8.11

Agri-Horticultural System (Ber + Mung Bean)

Treatment	Annual Rainfall (mm)	Fruit Yield (kg/ha)	Grain Yield (kg/ha)	Net Profit (Rs/ha)
Leguminous crop	200	-	520	4,800
Intercrop with ber	200	800	200	8,000

TABLE 8.12

System Productivity (Sorghum Grain Equivalent Yield), Employment Generation, and Economics in Sorghum-Based Integrated Farming Systems

Farming System	Productivity (t/ha)		Employment(man days/ha)		System Productivity (t/ha)	Cost of Cultivation Rs./ha	Net Returns (Rs./ha)	B:C Ratio
	2000–01	2001–02	2000–01	2001–02				
Cropping alone	0.69	1.84	28	32	1.27	5,520	1,170	1.21
Crop + pigeon + goat + agroforestry + farm pond	4.23	5.21	110	116	4.72	18,090	9,300	1.49
Crop + pigeon + buffalo + agroforestry + farm pond	11.20	10.79	140	142	10.99	43,650	22,670	1.52
Crop + pigeon + goat + buffalo + agroforestry + farm pond	12.18	12.5	160	166	12.39	52,850	21,820	1.41

TABLE 8.13

Dry Land Model for Small and Marginal Farmers of Tamil Nadu

Sr. No.	System	Gross Income (Rs./ha)	% Share in Income
1.	Crop alone	3,697	100
2.	Crop	5,332	43
	Goat	7,068	57
	Total	12,400	100
3.	Additional income	3,400	

gave net returns of Rs.22,670/ha and system productivity of 10.99 t/ha which is higher than the cultivation of crop alone. The reason behind the higher BC ratio and system productivity is because of better use of on-farm resources and better sustainability of the system in the dryland situation. The income is increased also because of the additional income from goats and buffalo and also because of the inclusion of agroforestry which will provide food and fodder to the animal component throughout the year.

Anonymous (2009) stated that in the dryland, model combination of crop + goat generated a higher gross income of Rs.12,400/ha as compared to crop cultivation which gave only Rs.3,697/ha (Table 8.13). In dryland situations, the inclusion of goats is always beneficial to the cultivators because it will give insurance against crop failure.

Fig. 8.2 describes the benefit-cost ratio of different land use system in arid region based on the study conducted by Bhati et al. (2009). They have reported that different integrated farming systems fetched a higher B:C ratio (1.46–1.87) over arable farming (1.24).

Anonymous (2017a,b,c) conducted a study at Sardarkrushinagar, Gujarat, and reported that livestock + vermicompost, compost, and nursery unit gave the highest green gram equivalent yield (3,807 kg/allotted area), gross return and net return Rs.1,52,292 and Rs.69,444 (Rs./allotted area), respectively higher than other components (Table 8.14).

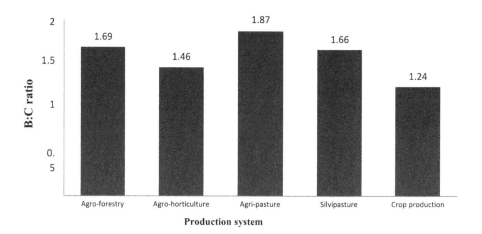

FIGURE 8.2   Benefit:Cost ratio of different land use system in arid region.

TABLE 8.14

Yield (kg/ha) and Economics (Rs./ha) of Integrated Organic Farming Systems Model (0.40ha)

Farming System Components	Total Area (ha)	Gross Return (Rs.)	Cost (Rs)	Net Returns (Rs.)
Crops; groundnut-wheat-green gram	0.24	62,838	34,627	28,211
Green fodder: F. Bajra-F. Maize + Oat-F. Bajra	0.15	54,066	34,329	19,737
Livestock + Vermicompost	0.01	Construction of animal shed and purchases of animals are awaited due to unavailability of grant		
Boundary plantation (Ardusa, Napier grass, Hy. Napier)	-	-	-	-
Total	0.40	1,16,904	68,956	47,948

8.17   CONCLUSION

IFSs offer unique opportunities for maintaining and extending biodiversity. The emphasis in such systems is on optimizing resource use rather than maximization of individual elements in the system. The well-being of poor farmers can be improved by bringing together the experiences and efforts of farmers, scientists, researchers, and students in different countries with similar eco-sociological circumstances, that is, through an IFS. Adoptions of IFSs in varied ecosystems with suitable enterprises can double the farmer's income. IFS fulfills the multiple objectives of making farmers self-sufficient by ensuring the family members a balanced diet for leading a healthy life, increasing the standard of living by maximizing the total net returns, and providing more employment. Recycling crop residues, optimizing resource use, minimizing risks and uncertainties, and keeping harmony with the environment are also achieved through IFS.

Adoption of IFSs in a varied ecosystem with suitable enterprises can increase the farmer's income. In wetland situations, the integration of crops with dairy and fish will give higher net returns because of closed resource recycling and reduced dependence on purchased inputs. In the garden land situation, the crop choices are very vast because of irrigation facilities. Crop cultivation can be integrated with biogas units, bird rearing, poultry, and a number of other enterprises and it will increase the net income of the farmers in that situation. In dryland situations, the inclusion of goats as an animal component along with crops will give benefit to the farmers and also increase the risk-bearing capacity of the farmer. Agroforestry along with pasture will give better sustainability to the farming system in the dryland areas and it will also act as a source of fodder and fuel around the year.

8.18   FUTURE THRUST

- There is a need to create a database on the farming systems in relation to the type of farming system, infrastructure, economics, sustainability, etc. under the different farming situations.

- Need to develop research modules on the farming systems under different holding sizes with varying economically viable and socially acceptable systems.
- The assessment and refinement of the technologies developed at the research station in the cultivators' field.
- Need to prepare contingent planning to counteract the weather vagaries/ climate threats under the different farming situations.
- Need to prepare a policy draft for the consideration of planners for its promotion at a large scale with nominal financial assistance either through short/medium/long-term loans and another promotional advantage.

REFERENCES

Anonymous. 2009. Tamil Nadu Agriculture University Portal, Farming System Research.
Anonymous. 2017a. AGRESCO Report—2017. Center for Research on IFS, NAU, Navsari.
Anonymous. 2017b. AGRESCO Report—2017. Center for Research on IFS, SDAU, S. K. Nagar.
Anonymous. 2017c. Annual Report—2017. All India National Project on Organic Farming, SDAU, S. K. Nagar.
Bhati, T.K., Rathore, V.S. and Beniwal, J.P., 2009. Trends in Arid Zone Research in India, CAZRI, Jodhpur, India, pp. 354–382.
Channabasavanna, A.S., Biradar, D.P., Prabhudev, K.N. and Hegde, M., 2010. Development of profitable integrated farming system model for small and medium farmers of Tungabhadra project area of Karnataka. *Karnataka Journal of Agricultural Sciences* 22(1): 25–27.
Goverdhan, M., Pasha, M.L., Sridevi, S. and Kumari, C.P., 2018. Integrated farming approaches for doubling the income of small and marginal farmers. *International Journal of Current Microbiology and Applied Sciences* 7(3): 3353–3362.
Gupta, J.P., 1997. Some alternative production systems and their management for sustainability. *Agroforestry for sustained productivity in arid regions* (pp. 31–39). Jodhpur: Scientific Publishers (India).
Jayanthi, C., Balusamy, M., Chinnusamy, C. and Mythili, S., 2003. Integrated nutrient supply system of linked components in lowland integrated farming system. *Indian Journal of Agronomy* 48(4): 41–46.
Kalpana, M., Singh, S.P., Ashutosh, D., Manisha, C. and Rajiv, D., 2016. Relative efficiency of rice-fish-duck production under integrated and conventional farming systems. *Asian Journal of Animal Science* 11(1): 49–52.
Manjunath, B.L. and Itnal, C.J., 2003a. Integrated farming system in enhancing the productivity of marginal rice (*Oryza sativa*) holdings in Goa. *Indian Journal of Agronomy* 48(1): 1–3.
Manjunath, B.L. and Itnal, C.J., 2003b. Farming system options for small and marginal holdings in different topographies of Goa. *Indian Journal of Agronomy* 48(1): 4–7.
Rangaswamy, A., Venkitaswamy, R., Premsekhar, M., Jayanthi, C., Purushothaman, S., and Palaniappan, S. P. 1995. Integrated farming system for rice based ecosystem. p-ISSN:0024-9602, e-ISSN:2582-5321, Vol: 82, DOI: https://doi.org/10.29321/MAJ.10.A01188
Shekinah, D.E. and Sankaran, N., 2007. Productivity, profitability and employment generation in integrated farming systems for rainfed vertisols of western zone of Tamil Nadu. *Indian Journal of Agronomy* 52(4): 275–278.

9 Comparison and Estimation of Soil Wetting Patterns Models and Soil Moisture Dynamics under Point Source

*Dinesh Kumar Vishwakarma, Rohitashw Kumar,
Amit Kumar, and Salim Heddam*

9.1 INTRODUCTION

Due to increased demand between urban, agricultural, industrial and environmental users, water scarcity and conservation of natural resources have become important topics worldwide. As a result, irrigation practices will be scrutinised more closely, and more effective irrigation practices will be needed. During the last few years, drip irrigation was considered to be the most significant and efficient irrigation technique if conducted with necessary rigour and accurately with proper management and operation, i.e., proper lateral and emitter spacing, for subsurface drip irrigation (Phene 1995). The drip irrigation system has seen significant advancements in design, technology and management over the last decade. This is due to a better knowledge of how water moves and fluctuates in response to any water supply using a drip emitter. The success of the drip irrigation is mainly governed by a better understanding of the soil water fluctuation and the soil wetting front (Keller and Karmeli 1975; Schwartzman and Zur 1986; Keller and Bliesner 1990; Zur 1996; Khan *et al.*, 1997; Revol *et al.*, 1997a, 1997b; Dasberg *et al.*, 1999; Singh Lubana and Narda 2001; Skaggs *et al.*, 2010). Generally speaking, physical characteristics of soil and how water is supplied to the soil method of irrigation, application flow rate and irrigation operation etc. have an effect on the distribution pattern (Patel and Rajput 2008). Dynamic of soil water fluctuation is highly complicated process because it is related to several factors, causing moisture pattern prediction difficult. Reliable information on the axis of horizontal and vertical soil water movement from a point source drip emitter will aid in determining the root zone management under several conditions (Ruiz-Sánchez *et al.*, 2005). Predicting the maximum wetted soil depth and lateral distance, which would be wetted during water infiltration from a drip emitter on the soil surface, can help in a better water supply and nutrient loss (Moncef *et al.*, 2002).

DOI: 10.1201/9781003351672-9

Analytical solutions (of the axisymmetric water infiltration equation) are commonly used to achieve this goal, but they are only valid for steady-state flow (Wooding 1968; Raats 1971; Angelakis *et al.*, 1993) or for short-term infiltration where gravity effects aren't a factor (Clothier and Scotter 1982). Several studies have been presented over the past few decades to evaluate these trends. Some of these studies have focused on the wetted dimension soil and the amount of wetted soil, as well as the spatial soil water availability, which is strongly influenced by soil hydraulic properties (emitter discharge rates, number of emitters, emitter spacing and the soil hydraulic properties), emitter discharge rates, emitter (Keller and Karmeli 1975; Bresler 1977, 1978; Keller and Bliesner 1990; Lubana and Narda 1998; Singh Lubana and Narda 2001; Elmaloglou and Malamos 2007; Kandelous and Šimůnek 2010a, 2010b; Al-Ogaidi et al 2016a, 2016b, 2016c), soil homogeneity, crop root dynamics and evapotranspiration (Elmaloglou and Diamantopoulos 2009; Badr and Abuarab 2013; Karimi *et al.*, 2013).

Many studies have focused on wetted soil dimension and developed analytical models by solving physical approach of governing flow equation, i.e., Richards equation (Raats 1971; Philip 1984; Cote *et al.*, 2003; Moncef and Khemaies 2016), numerical models to analyse for simulation of width and depth of the wetted soil dimension (Lafolie *et al.*, 1989; Cook *et al.*, 2003a, 2003b; Thorburn *et al.*, 2003; Cook *et al.*, 2006; John *et al.*, 2006; Šimůnek *et al.*, 2006; Lazarovitch *et al.*, 2009; Arbat *et al.*, 2013; Rasheed and Abid 2018). Several empirical models have been established using multiple linear regression models. These models were applied for modelling the width and depth of the soil moisture fluctuation using several factors in direct relation with the soil properties (Ekhmaj *et al.*, 2005; Singh *et al.*, 2006; Ainechee *et al.*, 2009; Length 2011; Malek and Peters 2011; Molavi *et al.*, 2012; Samadianfard *et al.*, 2012, 2016; Zhang *et al.*, 2012, 2015; Krada and Munjapara 2013; Amin and Ekhmaj 2015; Fan *et al.*, 2018). Accurate amount of water is needed to keep the crops moist for, long-term growth, so enough volume of water must be applied to fulfil their needs (Armstrong and Wilson 1983; Kao and Hunt 1996). The knowledge of water movement through soil as a result of a surface point source is one of the fundamental needs for drip irrigation design.

In designing of point source drip irrigation system wetting front and distribution of soil moisture must be matched by the distribution of plant root characteristics. In order to determine the exact spatial distribution of the emitters, the appropriate distance between laterals, and the depth at which emitters should be installed, it is very important to correctly calculate the repartition of the wetting pattern in both X and Y directions, and this will lead to an improvement of the crop yield and water supply quality.

In the present chapter, we introduce a new model for extracting water with a combination of soil water dynamics. The proposed models were applied for modelling and predicting the "radius of a hemispherical" used for better selecting the emitter spacing in response to a suite of operating conditions. According to our findings, the proposed model is easy to use, for determining wetting patterns of water. Therefore, the overall goals of our study of this study were to accurately estimate wetting pattern in loam soil under point source of drip irrigation and evaluate existing empirical equations.

9.2   DESCRIPTIONS OF EMPIRICAL MODELS

In the present chapter, we applied and compared between several empirical models, namely, Malek and Peters model (M-P), Schwartzman and Zur model (S-Z), Amin and Ekhmaj model (A-E), Jiusheng Li model (J-L) and Al-Ogaidi model. The theoretical description of the model is presented briefly hereafter.

9.2.1   Malek and Peters Model (M-P Model)

The empirical model proposed by Malek and Peters (2011) and designated as M-P model was considered a new formulation for accurately predicting the wetted dimensions from a dripper. The mathematical formulation of the M-P model is mainly based on the hypothesis that the depth and width represent the principal factors controlling the overall drip irrigation schemas. The equation of M-P model takes into account a large number of variables, i.e., the amount of water available in the soil, the bulk density and, especially, the necessary time for a complete irrigation stage. More details about the M-P model can be found in Malek and Peters (2011).

9.2.2   Schwartzman and Zur Model (S-Z Model)

The Schwartzman and Zur (1987), designated by S-Z model, is another empirical model mainly based on the idea that soil water fluctuation is related to the soil category, the emitter discharge and the available water in the soil. Therefore, two major indicators were calculated, which are the depth and the diameter of the wetted. More details about the S-Z can be found in Schwartzman and Zur (1987).

9.2.3   Amin and Ekhmaj Model (A-E Model)

The model proposed by Amin and Ekhmaj (2015), i.e., the A-E model considers that a single mathematical formulation can be used for providing an overall explanation of the link between all soil variables, i.e., water depth, porosity, conductivity and other physical variables. However, the empirical formulation proposed by Amin and Ekhmaj (2015) for applying the A-E mode should be restricted to strict conditions among them: (*i*) homogeneous and isotropic for soil, (*ii*) no water evaporation and (*iii*) no water table fluctuation. More details can be found in Amin and Ekhmaj (2015).

9.2.4   Jiusheng Li Model (J-L Model)

The empirical model proposed by Li *et al.*(2004), i.e., the J-L model, was developed with the assumption that two different soils, i.e., the sandy and loam soils should have two different behaviours, and a mathematical formulation is necessary for each one. However, two input variables were used for linking the wetted radius and depth, namely, the time and the water application rate. More details can be found in Li *et al.* (2004).

9.2.5   Al-Ogaidi Model

According to the Al-Ogaidi *et al.* (2015), the wetting soil dimensions can be estimated using a multiple nonlinear regression equation that includes a large number

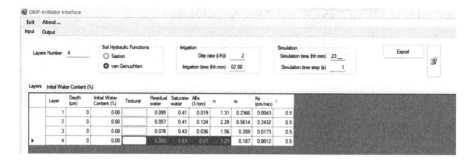

FIGURE 9.1   Example of Drip-Irriwater input data.

of soil characteristics. However, the emitter discharge and the irrigation times are predominant factors. More details about Al-Ogaidi model can be found in Al-Ogaidi *et al.* (2015).

9.3 DRIP-IRRIWATER INTERFACE

The Drip-Irriwater simulator (Arbat *et al.*, 2013) is made up of two programmes: a graphical user interface (GUI) and a programme that performs the numerical operation specified in the numerical procedure. The Microsoft.NET architecture was used in C# to build the GUI. It was created with the intention of introducing the predictors and displaying the soil water and hydraulic potential distribution as a calculated variable, i.e., the output. The simulator was developed using the FORTRAN language. The input data (Fig. 9.1) are divided into three classes in the GUI, i.e., Soil input parameters, Irrigation parameters and Simulation parameters.

The input parameters must be exported for data once the GUI parameters are defined. The numerical procedure programme has to be performed from the GUI, and the results can be extracted after achieving all necessary calculations (the available soil water fluctuation and repartition, percentage, hydraulic heads, cm) (Fig. 9.2). The code shows the per cent volumetric soil water content and hydraulic heads redistribution after irrigation, thus allowing the parameters

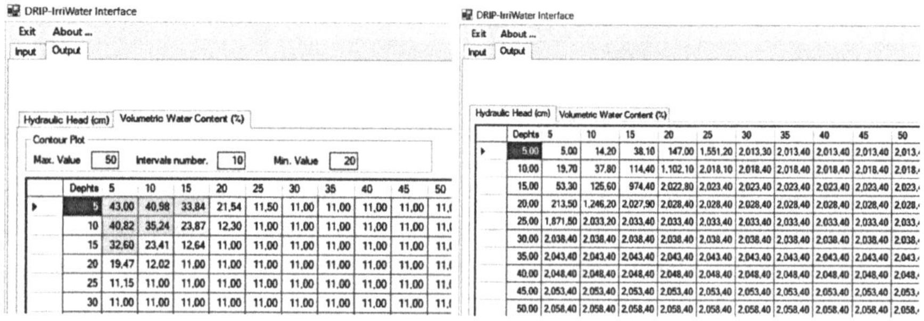

FIGURE 9.2   Drip-Irriwater output results (volumetric soil water content and hydraulic head) as an example.

necessary to design the irrigation system to be visually distinguished between the wetted soil radius and the depth.

9.4   SITE LOCATION AND METHODOLOGY

The experiment was conducted in India at the SKUAST-Kashmir's in experimental farm of high-density apple orchard. It is situated at 34.14° N latitude and 74.87° longitude, with a height of 1606 m above sea level. The farm is laid out under gravity-fed drip irrigation system with emitter of 2, 4 and 8 lph capacities were fitted. Soil texture is the most essential and permanent characteristic of a soil. The relative proportions of sand, silt, and clay in the soil mass are referred to as soil texture.

The United States Department of Agriculture (USDA) approach was used in the analysis. It is important to provide experiment data under drip irrigation in order to compare the results of empirical models for simulation of wetting pattern dimensions. Soil samples were collected from the experimental farm at 0–15, 15–30, 30–45, 45–60 cm soil depth and detailed grain size was analysed using a set of sieves and a calibrated Bouyoucos Hydrometer, followed by the methodology suggested by Bouyoucos (1962) and sieve analysis. Table 9.1 presents the descriptive details of the soil physical and chemical properties. The soil hydraulic property of the experimental field plot was evaluated from 0 to 60 cm depth, and the parameters are shown in Table 9.2.

Based on changing two major components: *(i)* the emitter discharge rate and *(ii)* the necessary time for achieving the irrigation with the time consumed 24 hours from the end of the irrigation, the experiment was conducted. It is important to note

TABLE 9.1
Soil Physical and Chemical Characteristics

		Soil Depth (cm)			
S. No.	Particulars	0–15	15–30	30–45	45–60
I – Physical Properties					
1.	Coarse sand (%)	2.28	2.18	1.29	1.95
2.	Fine sand (%)	47.72	43.82	46.71	52.05
3.	Silt (%)	40.00	46.00	46.00	38.00
4.	Clay (%)	10.00	08.00	06.00	08.00
5.	Bulk density (g/cm³)	1.33	1.52	1.61	1.62
6.	Particle density (g/cm³)	2.80	2.80	2.80	2.80
7.	Field capacity (%)	36	36	21	21
8.	Porosity (%)	44	44	37.5	37.5
9.	Saturated hydraulic conductivity (cm/h)	1.85	1.85	1.80	1.80
10.	Infiltration rate (mm/h)	12	12	14	14
Textural Classes		Loam	Loam	Sandy Loam	Sandy Loam
II – Chemical properties					
11.	Soil pH	5.36	6.05	6.54	6.45
12.	EC (dS/m)	0.06	0.06	0.09	0.08

TABLE 9.2

Soil Hydraulic Property Parameters of the Van Genuchten-Mualem Equations for the Four Different Soil Types and Corresponding Horizons from Experimental Field Plot Used in the Experiments

Soil Type	Soil Horizon	Depth (cm)	Θ_r (cm³/cm³)	Θ_s (cm³/cm³)	α (cm⁻¹)	n	m	K_{sat} (cm/min)
Loam	A	0–15	0.078	0.43	0.036	1.56	0.359	0.0173
Loam	B	15–30	0.078	0.43	0.036	1.56	0.359	0.0173
Sandy loam	C	30–45	0.065	0.41	0.075	1.89	0.4709	0.0737
Sandy loam	D	45–60	0.065	0.41	0.075	1.89	0.4709	0.0737

Θ_r is Residual Water content, Θ_s is Saturated water content, α is a scale parameter inversely proportional to mean pore diameter (cm⁻¹), n and m are the shape parameters of soil water characteristic, $m = 1 - 1/n$, $0 < m < 1$.

that the experiment was conducted over a multitude of sites based on the involved infiltration of the unique emitter. In addition, a series of amount of irrigation water were applied. The goal of the experiment was to illustrate the variation of the soil water content in response to the change of the soil depth.

Based on the application of three-irrigation rates, namely, 2, 4 and 8 lph and three types of time duration of irrigation, i.e., 30, 60 and 120 minutes, the soil wetting patterns were recorded. In Table 9.3, we reported the values of several combinations of rate, irrigation time and inter-emitter distances. The soil water contents and wetting pattern dimensions were measured with field observation and calculated using Drip-Irriwater at 5, 10, 15, 20, 25, 30, 35, 40, 45 50, 55 and 60 cm horizontal distances from

TABLE 9.3

Combinations of Emitter Discharge Rate, Emitter Spacing and Irrigation Duration

Emitter Discharge Rate (lph)	Dripper Spacing (m)	Irrigation Duration (min)
2	1×3	30
4	1×3	60
8	1×3	120
2	1×3	30
4	1×3	60
8	1×3	120
2	1×3	30
4	1×3	60
8	1×3	120

the emitter and at 5, 10, 15, 20, 25, 30, 35, 40, 45 50, 55 and 60 cm depths. Hence, a comparative study was done between values provided using the Drip-Irriwater for the same position and projected in a two-dimensional plot. The measured and calculated values of soil content, the radius and depth were then compared together.

9.5   PROPOSED MODEL

This research proposed a novel empirical approach for predicting wetted dimensions from a point source of dripper irrigation. Specifically, the depth and width of the wetted soil were considered major indications of the wetted dimensions for this purpose, since both of these characteristics are very important when it comes to designing and managing drip irrigation systems. Because this experiment was conducted in one type of soil, emitter discharge rate, irrigation time and total amount of water applied were considered. Empirical models were examined in three different forms: linear, polynomial and nonlinear. Nonlinear expressions are generally assumed to describe the wetting pattern as follows:

$$d = \alpha_1 \left(t^{\beta_1} V_m^{\gamma_1} \right) \qquad (9.1)$$

$$Z = \alpha_2 \left(t^{\beta_2} V_m^{\gamma_2} \right) \qquad (9.2)$$

where Z (L) denotes the vertical depth of the wetting front; d (L) is the wetted soil width of wetting front; V_m (L^3) denotes the total volume of water applied; t (T) denotes the time of irrigation; and $\alpha_1, \beta_1, \gamma_1, \alpha_2, \beta_2$ and γ_2 denote the best fit empirical coefficients. In order to calculate the coefficients for Eqs. (9.1) and (9.2), we used data from a field experiment conducted. The best fit parameters for Eqs. (9.1) and (9.2) were determined using a nonlinear regression technique implemented in the Microsoft Excel Version 2019 statistics programme.

9.6   MODEL VALIDATION AND STATISTICAL ANALYSIS

During the experiment and at each 10 minutes interval of times, two variables were continuously checked: the wetted soil radius and depth. The estimated wetting pattern dimensions were calculated using the acquired data and the applied models. Performance of empirical model and Drip-Irriwater was evaluated and represented by considering some performance metrics, such as mean absolute error (MAE), mean absolute percentage error (MAPE), mean bias error (MBE), root mean square error (RMSE), Pearson correlation coefficient, R-squared correlation (R^2), Nash-Sutcliffe model efficiency (NSE), all calculated as follows (Nash and Sutcliffe 1970; Willmott and Wicks 1980; Willmott 1981, 1982; Jacovides and Kontoyiannis 1995):

Mean error or mean absolute error $\qquad \text{MAE} = \left(\dfrac{1}{N} \displaystyle\sum_{i=1}^{N} \left| \text{WF}_{o,i} - \text{WF}_{P,i} \right| \right) \qquad (9.3)$

$$\text{Mean biased error} \quad \text{MBE} = \left(\frac{1}{N} \sum_{i=1}^{N} \text{WF}_{o,i} - \text{WF}_{P,i} \right) \tag{9.4}$$

$$\text{Mean average percentage error} \quad \text{MAPE} = \left(\frac{1}{N} \sum_{i=1}^{N} \left| \frac{\text{WF}_{o,i} - \text{WF}_{P,i}}{\text{WF}_{o,i}} \right| \right) \tag{9.5}$$

$$\text{Root mean squared error} \quad \text{RMSE} = \sqrt{\left(\frac{1}{N} \sum_{i=1}^{N} \left(\text{WF}_{o,i} - \text{WF}_{P,i} \right)^2 \right)} \tag{9.6}$$

$$\text{Pearson's correlation coefficient} \quad r = \frac{\sum \left(\text{WF}_{P,i} - \overline{\text{WF}_{P,l}} \right)\left(\text{WF}_{o,i} - \overline{\text{WF}_{o,l}} \right)}{\sqrt{\sum \left(\text{WF}_{P,i} - \overline{\text{WF}_{P,l}} \right) \sum \left(\text{WF}_{o,i} - \overline{\text{WF}_{o,l}} \right)}} \tag{9.7}$$

$$\text{Coefficient of determination} \quad R^2 = 1 - \frac{\text{RSS}}{\text{TSS}} \tag{9.8}$$

$$\text{Nash Sutcliffe model efficiency} \quad \text{NSE} = 1 - \frac{\sum_{i}^{N} \left(\text{WF}_{P,i} - \text{WF}_{o,i} \right)^2}{\sum_{i}^{N} \left(\text{WF}_{o,i} - \overline{\text{WF}_{o,l}} \right)^2} \tag{9.9}$$

where N is the total number of ith data, $\text{WF}_{P,i}$ is i^{th} predicted wetting front data; $\text{WF}_{o,i}$ is i^{th} observed wetting front data, $\overline{\text{WF}_{o,l}}$ is mean of observed wetting front data; $\overline{\text{WF}_{P,l}}$ is mean of predicted wetting front data, RSS is sum of squares of residuals, and TSS is total sum of squares.

9.7 RESULT AND DISCUSSION

9.7.1 FIELD INVESTIGATION FINDING

The wetting front advance measured form left and right of the vertical axis were considered for axi-symmetrical representation for analysis and interpretation under different emitter flow rates (2, 4 and 8 lph) and different irrigation operations (30, 60 and 120 minutes) are plotting using Origin 2021 and depicted in Fig. 9.3. During an application of one irrigation drip, the fluctuation of wetted front is mainly governed by the discharge rate by supposing all other characteristics of soil unchangeable. In general, both the horizontal distance as well as vertical depth of wetting increased over time and the rates of discharge. The results depicted in Fig. 9.3 reveal that the wetting front advance, i.e., vertical and lateral movement in each case was nearly equal due to equal volume of applied water in loam soil. From Tables 9.4 and 9.5, statistical comparison of observed wetting dimension is summarized. For equal volume of applied water, there is no significant difference were observed in wetting front advance (Tables 9.4 and 9.5), i.e., when we applied 4 l volume of water by 2, 4 and 8 emitter discharge rate, nearly same wetting pattern dimension were observed.

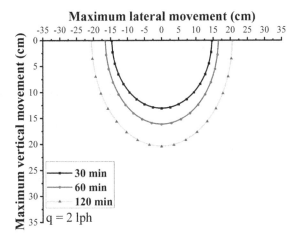

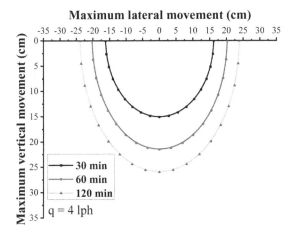

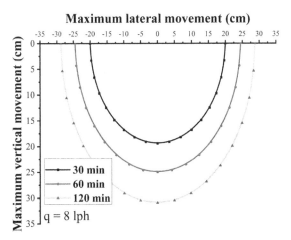

FIGURE 9.3 Observed wetting front advance of loam soil under surface drip irrigation using 2, 4 and 8 lph emitter discharge rate.

TABLE 9.4

Average Wetted Soil Width (cm) as Influenced by Different Discharge Rate and Time of Irrigation from Point Source of Drip Irrigation

		Wetted Soil Width (cm) from Point Source of Drip Irrigation		
		Irrigation duration		
S. No.	Emitter Discharge Rate (lph)	30 Minutes	1 Hour	2 Hour
1.	2	14.65	16.65	20.62
2.	4	16.20	20.25	23.90
3.	8	20.00	24.50	28.81

C.D. (p≤0.05)

Emitter Discharge rate	(E)	:	1.3728
Time duration of Irrigation	(I)	:	1.3728
Interaction	(E×I)	:	N.S.

9.7.2 Result Finding by Empirical Models

The following data from this analysis will be presented and explained:

9.7.3 Model Performance

The calculated soil wetting pattern based on the available data and the proposed empirical models was done for further comparison of model's performances. The estimated data of each model were evaluated by some performances metrics reported in Section 9.6, i.e., RMSE, MAE, NSE and R, among other, and the obtained results are depicted in Tables 9.6 and 9.7. By comparing the numerical performances

TABLE 9.5

Average Wetted Soil Depth (cm) as Influenced by Different Discharge Rate and Time of Irrigation from Point Source of Drip Irrigation

		Wetted Soil Depth (cm) from Point Source of Drip Irrigation		
		Irrigation duration		
S. No.	Emitter Discharge Rate (lph)	30 Minutes	1 Hour	2 Hour
1.	2	13.10	16.20	20.44
2.	4	15.10	21.50	26.00
3.	8	19.40	25.00	31.00

C.D. (p≤0.05)

Emitter Discharge rate	(E)	:	2.2213
Time duration of Irrigation	(I)	:	2.2213
Interaction	(E×I)	:	N.S.

TABLE 9.6

Evaluation and Accuracy of Various Empirical Models for Predictions of the Wetting Front Horizontal Advance Using Statistical Parameters with Observed Data Set

S. No.	Model Name	Statistical Criteria		
		MAE	RMSE	NSE
1	Al-Ogaidi model	0.41	0.51	0.99
2	Malek and Peters model	8.53	8.58	−3.03
3	Amin and Ekhmaj model	0.32	0.40	0.99
4	Li model	0.57	0.75	0.97
5	Schwartzman and Zur model	1.74	2.04	0.77

reported in Tables 9.6 and 9.7, we can clearly see that Al-Ogaidi model was the most accurate model exhibiting the high numerical performances; this is certainly related to its structure for which a large number of variables influencing the soil wetting front was included. In addition, the performances of A-E and J-L models have also provided good performances. However, M-P and S-Z models were the worst models exhibiting very low numerical performances in calculating the wetting pattern. Further analysis of the models performances revealed that M-P was the worst model in predicting the wetting zone measurements and this is certainly because it was developed using only a single field data for clay loam soil.

9.7.4 Comparison of Measured and Simulated Wetting

Figs. 9.4–9.6 show the estimated versus simulated wetting patterns for a number of experiments as stated in matehood and materials. It can be seen from Figs. 9.4 to 9.6 that there is an excellent agreement between the measured and simulated wetting patterns except Malek and Peters model in wetted soil width. However, there are minor discrepancies in predicting the wetted depths, which is also clearly seen from

TABLE 9.7

Evaluation and Accuracy of Various Empirical Models for Predictions of the Wetting Front Vertical Advance Using Statistical Parameters with Observed Data Set

S. No.	Model Name	Statistical Criteria		
		ME	RMSE	NSE
1	Al-Ogaidi model	0.64	0.88	0.98
2	Malek and Peters model	3.03	3.24	0.64
3	Amin and Ekhmaj model	2.75	2.94	0.71
4	Li model	1.85	2.13	0.85
5	Schwartzman and Zur model	4.35	4.92	0.18

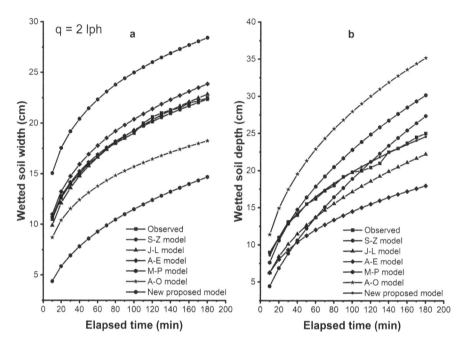

FIGURE 9.4   Observed and simulated wetting front advance under emitter discharge rate of 2 lph. (a) Wetted soil width. (b) Wetted soil depth.

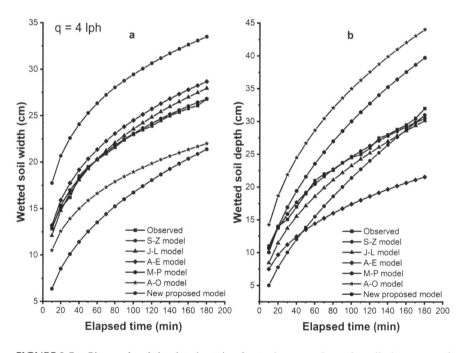

FIGURE 9.5   Observed and simulated wetting front advance under emitter discharge rate of 4 lph. (a) Wetted soil width. (b) Wetted soil depth.

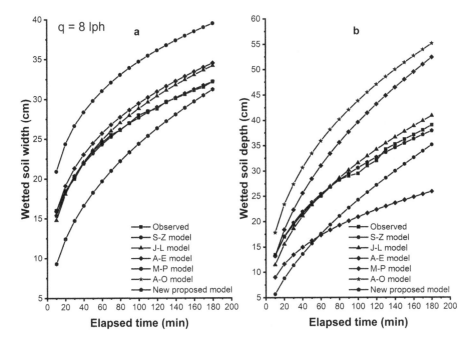

FIGURE 9.6   Observed and simulated wetting front advance under emitter discharge rate of 8 lph. (a) Wetted soil width. (b) Wetted soil depth.

Table 9.6 to 9.7 where the values of ME and RMSE for downward wetted radius are higher than those for vertical depth. It can be concluded that including the effect of most of the factors affecting wetted zone geometry resulted in good performance of the developed models. It is worth mentioning that the merit of the proposed empirical model is its ability to predict the full shape of the wetting pattern. Malek and Peters model underestimates than the observed data especially in wetted soil radii whereas in case of vertical wetted soil depth, it is average good.

Further comparison between the models output and the in situ measured data of the wetted dimensions are reported in Fig. 9.7 as a scatterplot (1:1 axis) with

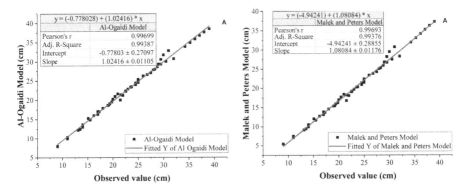

FIGURE 9.7   Scatter plot of observed versus predicted wetted soil dimensions from the selected proposed model with line 1:1. (A) Wetted depth. (B) Wetted width. (*Continued*)

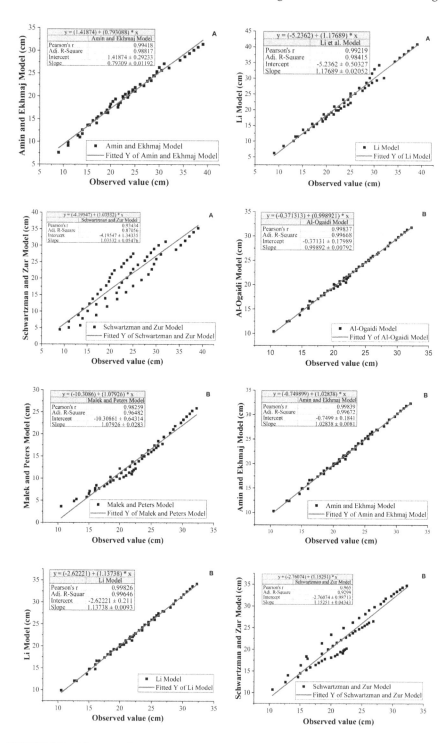

FIGURE 9.7 *(Continued)*

the linear fitted line. By visual comparison, we can see that, Al-Ogaidi, A-E and J-L were the best models compared to the other models showing less scattered data points. In addition, M-P model exhibited very acceptable performances, more precisely for wetted depth. The S-Z model was the worst one having the lowest numerical performances especially for the wetted radii and depth.

9.7.5 PROPOSED MODEL

Using different empirical models and field data analysis, including polynomial, nonlinear and linear forms, it was found that the nonlinear form produced the best results. The suggested empirical model was developed using a nonlinear regression model (power regression), which was used to gather data. The models are as follows:

$$d = 2.3675 \left(t^{-0.00691} V_m^{0.262271} \right) \tag{9.10}$$

$$Z = 1.24596 \left(t^{0.05076} V_m^{0.312393} \right) \tag{9.11}$$

where Z (cm) denotes vertical depth of the wetting front; d (cm) denotes lateral wetted width; V_m (ml) represents the total amount of water that has been applied; t (min) = irrigation duration. The new proposed model as shown in Fig. 9.8, the wetted soil width increased rapidly at first and then increased very slowly with respect time.

The similar pattern was seen for emitter discharge rates of 4 and 8 lph. The simulated wetted soil width at 10 minute later from starting time of irrigation was found 10.70 cm and maximum after 3 hour later, it was found 22.36 cm for 2 lph emitter.

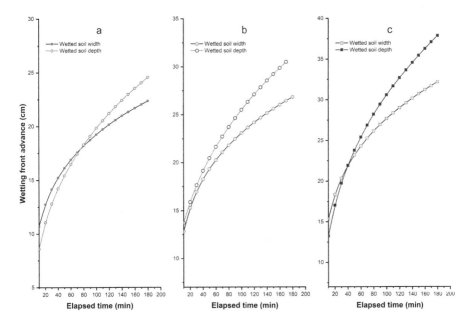

FIGURE 9.8 Simulated wetted width (d) and depth (Z) as a function of time for sandy and loam soil at different: q = (a) 2; (b) 4 and (c) 8 lph.

Similarly, wetted soil depth at 10 minute later from starting time of irrigation was found 8.60 cm and maximum after 3 hour it was found 24.56 cm (Fig. 9.8a). In 4 lph emitter discharge rate, the wetted soil width was found after 10 minutes from starting or irrigation 12.83 cm and 3 hour later maximum wetted soil width was 26.82 cm; similarly, for wetted soil depth, it was 10.67 and 30.50 cm, respectively (Fig. 9.8b). As seen in Fig. 9.8c, the 8 lph emitter discharge rate, the wetted soil width was found after 10 minutes from starting or irrigation 15.38 cm and 3 hour later maximum wetted soil width was found 32.18 cm; similarly, wetted soil depth at 10 minute later from time of irrigation was found 13.25 cm and maximum after 3 hour later it was 37.87 cm.

The statistical performance measures, i.e., MAE, RMSE, NSE and R^2 of new proposed model was 0.313, 0.384, 0.994 and 0.995, respectively (Table 9.8). A drip system designer might easily find a suitable emitter discharge rate by solving Eqs. 9.10 and 9.11 and taking into account the averaged plant root distribution pattern using these relationships. The statistical criteria between observed and forecast wetted width and depth are illustrated in Tables 9.6 and 9.7. As evidenced by Tables 9.6 and 9.7 that the new proposed model has shown the overall best performance among others models with average MAE, RMSE NSE and R^2 0.313, 0.384, 0.994 and 0.995, respectively. Furthermore, Li model and other models are also shown satisfactory correlation between simulated and observed field data. The new proposed model was found to be the best in this investigation as compared to all other empirical models.

This model predicted the wetted dimensions of the soil the best and shown the smallest mean and RMSE. The model efficiency was found very high as compared

TABLE 9.8
Performances of the New Proposed Models Based on the Calculated Numerical Criteria

Performance Indicator	q (lph)	Wetted Width	Wetted Depth	Average
MAE	2	0.226	0.287	0.313
	4	0.206	0.441	
	8	0.184	0.534	
	Average	0.205	0.421	
RMSE	2	0.246	0.370	0.384
	4	0.261	0.571	
	8	0.232	0.624	
	Average	0.246	0.522	
NSE	2	0.995	0.993	0.994
	4	0.996	0.990	
	8	0.997	0.993	
	Average	0.996	0.992	
R^2	2	0.997	0.994	0.995
	4	0.997	0.992	
	8	0.998	0.994	
	Average	0.997	0.993	

to all other empirical models. Only minor differences between the simulated and observed values were found. The nonlinear form produced the best results, as results shown. These differences could be attributed to the developed equations empirical nature. The uniformity of the soil could also be a factor in such discrepancies. It could be useful for designing of drip irrigation system using an appropriate emitter discharge. The progress of the wetting front may be estimated from these equations (Eqs. 9.10 and 29.11). This could be very helpful in selection emitter accordingly averaged plant root zone dimensions.

9.8 RESULT FINDING OF DRIP-IRRIWATER

The statistical comparison between measure and simulated with Drip-Irriwater soil wetting front using three different emitter discharge rates, i.e., 2, 4 and 8 lph for 30-, 60- and 120-minute irrigation duration just after end of operation are summarised in Table 9.9 show that effect of same volume of applied water was found same wetting front advance in loam soil and found non-significant effect. For same emitter discharge rate in different time of irrigation operation have significant difference and similarly, for same time of irrigation duration in different discharge rate have significant difference. But the interactions of combination of emitter discharge rate and irrigation time have no significant difference (Table 9.9). Twenty-four hours after completion of irrigation, operations (Table 9.10) show the change in wetting front advance in respect to just after end of irrigation. Statistical assessment of measured and simulated volumetric soil water content 24 hour after end of irrigation shown in Table 9.11 revealed that the values of the MAE were ranged from 0.011 to 0.022 cm^3 with a mean value of 0.0181 cm^3, the values of the MBE were ranged from -0.003 to 0.011 with mean value of 0.0008 cm^3, the values of MAPE were ranged from 4.6 to 8.8 with mean value of 7.30, the values of the RMSE were ranged from 0.015 to

TABLE 9.9
Statistics of Comparison between Measured and Simulated Wetted Width and Depth with an Irrigation Volume at the End of the Irrigation Event

Irrigation Duration (minutes)	Flow Rate (lph)	Measured Width (cm)	Simulated Width (cm)	Measured Depth (cm)	Simulated Depth (cm)	C.D. for Measured Value of Wetted Soil Radii	C.D. for Measured Value of Wetted Soil Depth
30	2	14.65	20.00	13.10	20.00	Emitter	Emitter
60	2	16.65	20.00	16.20	20.00	Discharge	Discharge
120	2	20.62	25.00	20.44	25.00	Rate (E): 1.3728	Rate (E): 2.2213
30	4	16.20	20.00	15.10	20.00	Time	Time
60	4	20.25	25.00	21.50	20.00	duration of	duration of
120	4	23.90	25.00	26.00	25.00	Irrigation (I):	Irrigation (I):
30	8	20.00	20.00	19.40	20.00	1.3728	2.2213
60	8	24.50	25.00	25.00	20.00	Interaction	Interaction
120	8	28.81	25.00	31.00	25.00	(E×I): N.S.	(E×I): N.S.

C.D. is the critical difference at 0.05% level of significance. N.S. = Not Significant.

TABLE 9.10

Comparison between the Measured and Simulated Wetted Radii and Depth with an Applied Irrigation Volume at 24 hour after the End of Irrigation

Irrigation Duration (minutes)	Discharge Rate (lph)	Measured Width (cm)	Simulated Width (cm)	Measured Depth (cm)	Simulated Depth (cm)
30	2	25.00	35.00	23.00	35.00
	4	30.00	35.00	35.00	35.00
	8	40.00	45.00	40.00	45.00
60	2	30.00	35.00	27.00	35.00
	4	40.00	45.00	40.00	45.00
	8	40.00	45.00	31.00	45.00
120	2	35.00	40.00	32.00	40.00
	4	40.00	45.00	40.00	45.00
	8	45.00	45.00	45.00	45.00

0.028 with mean value of 0.023 cm³. In addition, the value of the Pearson coefficient was ranged from 0.921 to 0.977 with mean value of 0.951, the values of the R^2 were ranged from 0.909 to 0.955 with mean value of 0.918, and finally, the value of the NSE were ranged from 0.828 to 0.921 with mean value of 0.887. The variations in water content between the models and the experiments may be attributed to measurement errors and it was very little. When the application rate exceeds the soil's infiltration rate, the Drip-Irriwater code is much more appropriate.

Figs. 9.9–9.11 indicate a contrast of the volumetric soil water content horizontally from 5 cm away of emitter point and laterally at 5 cm depth below the point source emitter. The minor differences in the obtained results in the computer programmes are also explained by the different emitters and time length combinations used to simulate the volumetric soil water volume. Fig. 9.12 shows

TABLE 9.11

Statistics Assessment between the Measured and Simulated Soil Water Contents with an Applied Irrigation Volume at 24 hour after the End of Irrigation

	Flow Rate (lph)	MAE (cm)	MBE (cm)	MAPE (%)	RMSE (cm³/cm³)	Pearson Coefficient	R^2	NSE
For wetted soil width	2	0.011	−0.003	4.60	0.015	0.977	0.9551	0.919
	4	0.022	−0.003	8.70	0.027	0.921	0.8490	0.828
	8	0.018	0.011	7.40	0.023	0.921	0.9397	0.921
For wetted soil depth	2	0.016	−0.002	6.80	0.020	0.977	0.9377	0.882
	4	0.020	−0.008	7.50	0.025	0.954	0.9098	0.870
	8	0.022	0.008	8.80	0.028	0.958	0.9182	0.904
Average		**0.0181**	**0.0005**	**7.30**	**0.023**	**0.951**	**0.9182**	**0.887**

MAE: Mean Absolute Error, MBE: Mean Bias Error, MAPE: Mean Absolute Percentage Error, RMSE: Root mean square error, NSE: Nash-Sutcliffe model Efficiency coefficient, r²: coefficient of correlation.

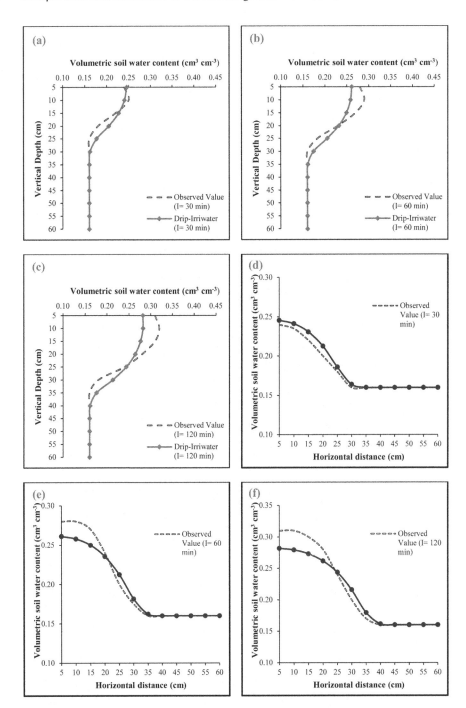

FIGURE 9.9 Comparison of VSWC Observed (black dot) and Drip-Irriwater (red line and blue line): (a), (b) and (c) vertically and (c), (d) and (e) horizontally corresponding to emitter flow rate of 2.0 lph for an irrigation duration of 30, 60 and 120 minutes, of 1 hour later stopping irrigation respectively.

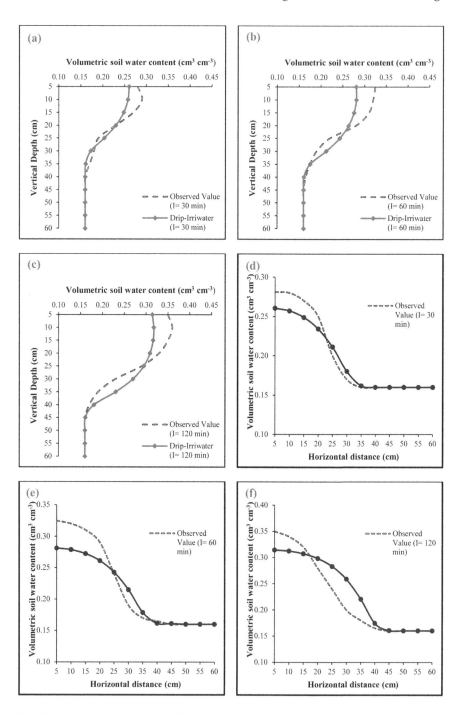

FIGURE 9.10 Comparison of VSWC Observed (black dot) and Drip-Irriwater (red line and blue line): (a), (b) and (c) vertically and (c), (d) and (e) horizontally corresponding to emitter flow rate of 4.0 lph for an irrigation duration of 30, 60 and 120 minutes, of 1 hour later stopping irrigation, respectively.

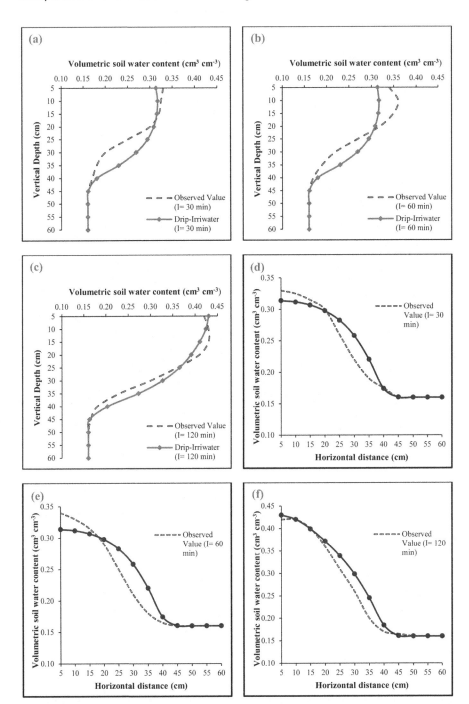

FIGURE 9.11 Comparison of VSWC Observed (black dot) and Drip-Irriwater (red line and blue line): (a), (b) and (c) vertically and (c), (d) and (e) horizontally corresponding to emitter flow rate of 8.0 lph for an irrigation duration of 30, 60 and 120 minutes, of 1 hour later stopping irrigation, respectively.

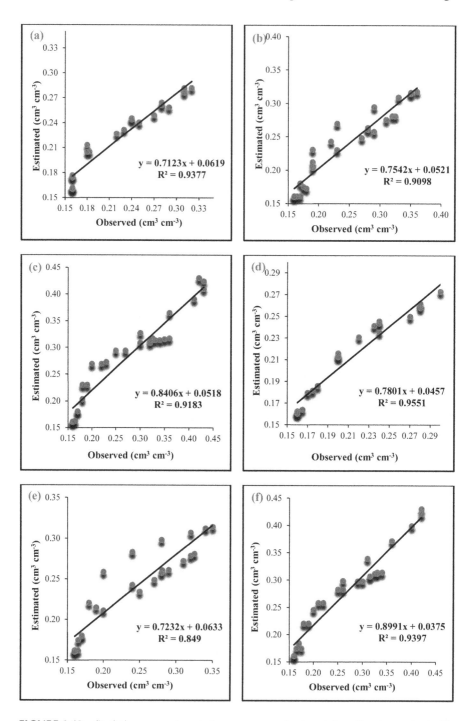

FIGURE 9.12 Statistics comparison between measured volumetric soil water content with estimated (Drip-Irriwater) for (a–c) wetted soil depth using 2, 4 and 8 lph emitter discharge rate and (d–f) wetted soil radii using 2, 4 and 8 lph emitter discharge rate.

the statistics about observed volumetric soil water content and predicted values of volumetric soil water content of the above-used GUI Drip-Irriwater software and suggests that this software can use to assess soil wetting front and moisture distribution. The high correlations between observed and calculated volumetric soil water content values in Fig. 9.12 clearly demonstrate the excellent performances of the Drip-Irriwater. Consequently, we can confirm that the proposed model can be accurately used for better installation and application of the drip irrigation.

9.9 CONCLUSIONS

When drip irrigation is used, large volumes of water result in deep water percolation and fertilizer seepage. Forecasting the depth of the wetting front under the emitters may be a reliable way to avoid such events, allowing trickle irrigation managers to minimise water and nutrient losses. Several methods for determining soil moisture distribution pattern and depth and width of wetted soil volume under the point source of drip irrigation have been developed. Empirical equation and Drip-Irriwater were used because of their simplicity, eco-friendly and cheapness with field investigation to analyse depth and width of wetted soil dimension and soil moisture distribution under loam soil. The above model's efficiency was evaluated by analysing the observed and predicted wetted dimensions. The model did a forecast of replicating experimental data in a variety of situations. In addition, comparisons were made for a few case studies. Based on numerical performances, the Al-Ogaidi empirical model and Drip-Irriwater GUI programme performances were excellent and the provided values of the wetted depths and widths of soil were accurately simulated leading to conclude that the proposed model can easily use. There can be several justifications for the success of such a model among them; we can report that the inclusion of the initial soil water content and bulk density has contributed to the improvement of model performances. In addition, it was found that the difference between calculated and measured data by the M-P model was very large leading to conclude that it is not well appropriate. The Z-S model also provided inacceptable calculation showing its poor performances. Furthermore, the J-L and A-E models were both good and robust. By comparing the soil, water content provided using the Drip-Irriwater and the in situ data demonstrated the very acceptable correlation between the two with only negligible difference. The proposed new model has shown the overall better as compared to other's models. The performance of the new proposed empirical model was found most suitable for describing soil wetting fronts, and it could be considered for design of drip irrigation systems.

ACKNOWLEDGMENTS

The authors are highly grateful to the COAET and ICAR-All India Coordinated Research Project on Plastic Engineering in Agriculture Structure and Environment Management, SKUAST-Kashmir, for providing all the facilities to conduct the study.

REFERENCES

Ainechee G, Boroomand-Nasab S, Behzad M (2009) Simulation of soil wetting pattern under point source trickle irrigation. *J Appl Sci* 9:1170–1174. https://doi.org/10.3923/jas.2009.1170.1174

Al-Ogaidi AAM, Aimrun W, Rowshon MK, Abdullah AF (2016a) WPEDIS – Wetting pattern estimator under drip irrigation systems. In International Conference on Agricultural and Food Engineering (CAFEi2016) 198–203.

Al-Ogaidi AAM, Wayayok A, Kamal MR, Abdullah AF (2015) A modified empirical model for estimating the wetted zone dimensions under drip irrigation. *J Teknol* 76:69–73. https://doi.org/10.11113/jt.v76.5954

Al-Ogaidi AAM, Wayayok A, Kamal R, Abdullah AF (2016b) Modelling soil wetting patterns under drip irrigation using hydrus-3D and comparison with empirical models. *Glob J Eng Technol Rev J* 1:17–25.

Al-Ogaidi AAM, Wayayok A, Rowshon MK, Abdullah AF (2016c) Wetting patterns estimation under drip irrigation systems using an enhanced empirical model. *Agric Water Manag* 176:203–213. https://doi.org/10.1016/j.agwat.2016.06.002

Amin MSM, Ekhmaj A (2015) Drip Irrigation Water Distribution Pattern Calculator. i:503–513.

Angelakis, B.A.N., Rolston, D. E., Kadir T. N., Scott, V. H. (1993). Steady-state solutions may be useful for some situations, transient solutions are required during the infiltration phase. *The Time-Dependent Infiltr.* 119:484–500.

Arbat G, Puig-Bargués J, Duran-Ros M, et al (2013) Drip-Irriwater: Computer software to simulate soil wetting patterns under surface drip irrigation. *Comput Electron Agric* 98:183–192. https://doi.org/10.1016/j.compag.2013.08.009

Armstrong CF, Wilson TV (1983) Computer model for moisture distribution in stratified soils under a trickle source. *Trans ASAE* 26:1704–1709. https://doi.org/10.13031/2013.33829

Badr AE, Abuarab ME (2013) Soil moisture distribution patterns under surface and subsurface drip irrigation systems in sandy soil using neutron scattering technique. *Irrig Sci* 31:317–332. https://doi.org/10.1007/s00271-011-0306-0

Bouyoucos GJ (1962) Hydrometer method improved for making particle size analyses of soils. *Agron J* 54:464–465.

Bresler E (1977) Trickle-drip irrigation: Principles and application to soil-water management. In *Advances in Agronomy* (pp. 343–393). Elsevier.

Bresler E (1978) Analysis of trickle irrigation with application to design problems. *Irrigation Bclence* 17:3–17.

Clothier BE, Scotter DR (1982) Constant-flux infiltration from a hemispherical cavity. *Soil Sci Soc Am J* 46:696–700.

Cook FJ, Fitch P, Thorburn PJ, et al (2006) Modelling trickle irrigation: Comparison of analytical and numerical models for estimation of wetting front position with time. *Environ Model Softw* 21:1353–1359. https://doi.org/10.1016/j.envsoft.2005.04.018

Cook FJ, Thorbum PJ, Bristow KL, Cote CM (2003a) Infiltration from surface and buried point sources: The average wetting water content. *Water Resour Res* 39:1–7. https://doi.org/10.1029/2003WR002554

Cook FJ, Thorburn PJ, Fitch P, Bristow KL (2003b) WetUp: A software tool to display approximate wetting patterns from drippers. *Irrig Sci* 22:129–134. https://doi.org/10.1007/s00271-003-0078-2

Cote CM, Bristow KL, Charlesworth PB, et al (2003) Analysis of soil wetting and solute transport in subsurface trickle irrigation. *Irrig Sci* 22:143–156. https://doi.org/10.1007/s00271-003-0080-8

Dasberg S, Or D, Dasberg S, Or D (1999) Practical applications of drip irrigation. *Drip Irrig* 125–138. https://doi.org/10.1007/978-3-662-03963-2_6

Ekhmaj AI, Amin MSM, Salim S, Zakaria AA (2005) Wetted surface radius under point-source trickle irrigation in sandy soil. *Int Agric Eng J* 14:67–75.

Elmaloglou S, Diamantopoulos E (2009) Simulation of soil water dynamics under subsurface drip irrigation from line sources. *Agric Water Manag* 96:1587–1595. https://doi.org/10.1016/j.agwat.2009.06.010

Elmaloglou ST, Malamos N (2007) Estimation of width and depth of the wetted soil volume under a surface emitter, considering root water-uptake and evaporation. *Water Resour Manag* 21:1325–1340. https://doi.org/10.1007/s11269-006-9084-5

Fan YW, Huang N, Zhang J, Zhao T (2018) Simulation of soil wetting pattern of vertical moistube-irrigation. *Water (Switzerland)* 10. https://doi.org/10.3390/w10050601

Jacovides CP, Kontoyiannis H (1995) Statistical procedures for the evaluation of evapotranspiration computing models. *Agric Water Manag* 27:365–371.

John F, Freeman C, Fitch P, et al (2006) WetUp : A Software Tool to Estimate Wetting Patterns from Drip Emitters for Better Irrigation WetUp – A Software Tool to Display Approximate Wetting.

Kandelous MM, Šimůnek J (2010a) Comparison of numerical, analytical, and empirical models to estimate wetting patterns for surface and subsurface drip irrigation. *Irrig Sci* 28:435–444. https://doi.org/10.1007/s00271-009-0205-9

Kandelous MM, Šimůnek J (2010b) Numerical simulations of water movement in a subsurface drip irrigation system under field and laboratory conditions using HYDRUS-2D. *Agric Water Manag* 97:1070–1076. https://doi.org/10.1016/j.agwat.2010.02.012

Kao CS, Hunt JR (1996) Prediction of wetting front movement during one-dimensional infiltration into soils. *Water Resour Res* 32:55–64. https://doi.org/10.1029/95WR02974

Karimi B, Mirzaei F, Sohrabi T (2013) Evaluation of moisture front redistribution in surface and subsurface drip irrigation systems. *Water Soil Sci* 23:183–192.

Keller J, Bliesner RD (1990) Sprinkle and Trickle Irrigation.

Keller J, Karmeli D (1975) Trickle irrigation design. *Trans ASAE* 17:678–684.

Khan AH, Stone LR, Lamm FR (1997) Water flux below the root zone vs. drip-line spacing in drip-irrigated corn. *Soil Sci Soc Am J* 61:1755–1760.

Krada P, Munjapara BJ (2013) Study on pressure-discharge relationship and wetting. *Int J Sci Nat* 4:274–283.

Lafolie F, Guennelon R, Van Genuchten MT (1989) Analysis of water flow under trickle irrigation: I. Theory and numerical solution. *Soil Sci Soc Am J* 53:1310–1318.

Lazarovitch N, Warrick AW, Furman A, Zerihun D (2009) Subsurface water distribution from furrows described by moment analyses. *J Irrig Drain Eng* 135:7–12. https://doi.org/10.1061/(ASCE)0733-9437(2009)135:1(7)

Length F (2011) Experimental study of shape and volume of wetted soil in trickle irrigation method. *J Agric Res* 6:458–466. https://doi.org/10.5897/AJAR10.727

Li J, Zhang J, Li B (2004) Drip irrigation design based on wetted soil geometry and volume from a surface point source. In *2004 ASAE Annual Meeting*. American Society of Agricultural and Biological Engineers, p. 1.

Lubana PPS, Narda NK (1998) Soil water dynamics model for trickle irrigated tomatoes. *Agric Water Manag* 37:145–161.

Malek K, Peters RT (2011) Wetting pattern models for drip irrigation: New empirical model. *J Irrig Drain Eng* 137:530–536. https://doi.org/10.1061/(ASCE)IR.1943-4774.0000320

Molavi A, Sadraddini A, Nazemi AH, Fakheri Fard A (2012) Estimating wetting front coordinates under surface trickle irrigation. *Turkish J Agric For* 36:729–737. https://doi.org/10.3906/tar-1202-74

Moncef H, Hedi D, Jelloul B, Mohamed M (2002) Approach for predicting the wetting front depth beneath a surface point source: Theory and numerical aspect. *Irrig Drain* 51:347–360. https://doi.org/10.1002/ird.60

Moncef H, Khemaies Z (2016) An analytical approach to predict the moistened bulb volume beneath a surface point source. *Agric Water Manag* 166:123–129. https://doi.org/10.1016/j.agwat.2015.12.020

Nash JE, Sutcliffe JV (1970) River Flow forecasting through conceptual models part I—A discussion of principles. *J Hydrol* 10:282–290. https://doi.org/10.1016/0022-1694(70)90255-6

Patel N, Rajput TBS (2008) Dynamics and modeling of soil water under subsurface drip irrigated onion. *Agric Water Manag* 95:1335–1349. https://doi.org/10.1016/j.agwat.2008.06.002

Phene CJ (1995) Research trends in microirrigation. In: Lamm FR (ed) Microirrigation for a Changing World: Conserving Resources/Preserving the Environment. Proceedings of the Fifth International Microirrigation Congress, Orlando, Florida, 2–6 April, 1995. ASAE, St. Joseph, Mich., pp 6–24, pp 6–24.

Philip JR (1984) Travel times from buried and surface infiltration point sources. *Water Resour Res* 20:990–994. https://doi.org/10.1029/WR020i007p00990

Raats PAC (1971) Steady infiltration from point sources, cavities, and basins. *Soil Sci Soc Am J* 35:689–694.

Rasheed ZK, Abid MB (2018) Numerical modeling of water movement from buried vertical ceramic pipes through coarse soils. *Al-Khwarizmi Eng J* 13:164–173. https://doi.org/10.22153/kej.2017.06.003

Revol P, Clothier BE, Mailhol J, et al (1997a) Infiltration from a surface point source and drip irrigation: 2. An approximate time-dependent solution for wet-front position. *Water Resour Res* 33:1869–1874.

Revol P, Vauclin M, Vachaud G, Clothier BE (1997b) Infiltration from a surface point source and drip irrigation: 1. The midpoint soil water pressure. *Water Resour Res* 33:1861–1867.

Ruiz-Sánchez MC, Plana V, Ortuño MF, et al (2005) Spatial root distribution of apricot trees in different soil tillage practices. *Plant Soil* 272:211–221.

Samadianfard S, Nazemi AH, Sadraddini AA (2016) Simulation of water movement and its distribution in a soil column under a water source using pore-scale network modelling. *E3S Web Conf* 9. https://doi.org/10.1051/e3sconf/20160916001

Samadianfard S, Sadraddini AA, Nazemi AH, et al (2012) Estimating soil wetting patterns for drip irrigation using genetic programming. *Spanish J Agric Res* 10:1155. https://doi.org/10.5424/sjar/2012104-502-11

Schwartzman, M. and Zur, B. (1986). Emitter spacing and geometry of wetted soil volume. *Journal of Irrigation and Drainage Engineering* 112:242–253. http://dx.doi.org/10.1061/(ASCE)0733-9437(1986)112:3(242).

Simunek J, Sejna M, Van Genuchten MT (1999) The HYDRUS-2D software package. Int Gr Water Model Cent.

Šimůnek J, Van Genuchten MT, Šejna M (2006) The HYDRUS software package for simulating two-and three-dimensional movement of water, heat, and multiple solutes in variably-saturated media. Tech Manual version 1, p. 241.

Singh Lubana PP, Narda NK (2001) Modelling soil water dynamics under trickle emitters-a review. *J Agric Engng Res* 78:217–232.

Singh DK, Rajput TBS, Singh DK, et al (2006) Simulation of soil wetting pattern with subsurface drip irrigation from line source. *Agric Water Manag* 83:130–134. https://doi.org/10.1016/j.agwat.2005.11.002

Skaggs TH, Trout TJ, Rothfuss Y (2010) Drip irrigation water distribution patterns: Effects of emitter rate, pulsing, and antecedent water. *Soil Sci Soc Am J* 74:1886–1896. https://doi.org/10.2136/sssaj2009.0341

Thorburn PJ, Cook FJ, Bristow KL (2003) Soil-dependent wetting from trickle emitters: Implications for system design and management. *Irrig Sci* 22:121–127. https://doi.org/10.1007/s00271-003-0077-3

Willmott CJ (1981) On the validation of models. *Phys Geogr* 2:184–194. https://doi.org/10.1080/02723646.1981.10642213

Willmott CJ (1982) Some comments on the evaluation of model performance. *Bull Am Meteorol Soc* 63:1309–1313.

Willmott CJ, Wicks DE (1980) An empirical method for the spatial interpolation of monthly precipitation within California. *Phys Geogr* 1:59–73.

Wooding RA (1968) Steady infiltration from a shallow circular pond. *Water Resour Res* 4:1259–1273.

Zhang R, Cheng Z, Zhang J, Ji X (2012) Sandy loam soil wetting patterns of drip irrigation: A comparison of point and line sources. *Procedia Eng* 28:506–511. https://doi.org/10.1016/j.proeng.2012.01.759

Zhang JJ, Li JS, Zhao BQ, Li YT (2015) Simulation of water and nitrogen dynamics as affected by drip fertigation strategies. *J Integr Agric* 14:2434–2445. https://doi.org/10.1016/S2095-3119(15)61231-X

Zur B (1996) Wetted soil volume as a design objective in trickle irrigation. *Irrig Sci* 16:101–105. https://doi.org/10.1007/BF02215617

10 IoT and Sensor-Based Water and Nutrient Management for Protected Cultivation Technology

Murtaza Hasan, Indra Mani, Love Kumar,
Atish Sagar, Vinayak Paradkar,
Kishor Pandurag Gavhane, and Tarun Ameta

10.1 INTRODUCTION

Water is an essential natural resource for sustaining life and environment. It is imperative that it is used with maximum possible efficiency. Agriculture sector is the major user of water resources, and the demand is growing. The decades of sixties and seventies saw the accelerated development of agriculture in India through the intensive use of high-yielding varieties, fertilizers, water and mechanization. The input-based agricultural planning was successful in that the food production increased several folds during the last three decades. But the technology based on the predominant use of water and fertilizers resulted in a paradoxical situation in which soils in the parts of northern plains turned saline, whereas in some other parts, including South India, the water table lowered due to excessive pumping (Ambast *et al.*, 2006). This situation affected agricultural productivity to a point of stagnation. In the eighties, general awareness and consensus emerged on the efficiency and judicious use of water. It was then that drip irrigation gained popularity with its inherent advantages, like water saving and use in problem soils. Various research institutes conducted experiments on drip systems and extended the technology to the rural sector. The government also provided liberal support through subsidies to the farmers on procuring and installing drip irrigation systems. Drip fertigation design for protected structures, design of low-pressure drip fertigation system, design of small indigenous low-cost protected structures and drip fertigation scheduling for various horticultural crops are the important issues related to protected cultivation and drip fertigation in India (Sampathkumar *et al.*, 2012).

Protected cultivation technology needs assured, measured, and quantifiable water supply round the year to meet the demands for various diversified crops grown inside different types of protected structures (Prasanna *et al.*, 2016), Water is one of the most critical inputs used in different types of protected cultivation technology prevalent in soil, soilless, hydroponics or aeroponics. Its relative importance gradually

 DOI: 10.1201/9781003351672-10

increased in geometric proportion from soil to soilless and then toward hydroponics and aeroponics, respectively (Albery *et al.*, 1985). The first step in management is an appropriate estimation of water requirement for crops, cooling and heating, cleaning, washing and other related activities. The next step is an assessment of water availability, storage, water harvesting and water conservation. Water treatment and supply and demand management are the next important steps to be taken at farm level for effective and efficient water management for protected cultivation.

10.2 PRESSURIZE IRRIGATION SYSTEM NETWORK

Pressurize irrigation system network has the following five basic units:

- Pumping unit
- Control head
- Pipe network (main and submain pipes)
- Laterals and dripper
- Hydraulic connections

Pumping Unit: It takes water from the source and supplies pressurized water to the control head. Pumps used in the drip irrigation system are similar to those used in other irrigation methods and include centrifugal, submersible and turbine pumps. These pumps can be driven by either an electric motor or an internal combustion engine. An efficiently designed irrigation system has a pumping capacity closely matched to the system's demand.

Control Head: It serves as the irrigation system policeman, regulating flow, pressure, and filtration. It is also the place for chemical injection. A manifold, water meter and a pressure gauge are a must for the control head. It includes the different types of valves, filters and hydraulic regulating components. Fertigation control unit is designed to precisely control the injection of an acid or alkaline solution as well as multiple fertilizers based on continuous in-line EC and pH monitoring (Fig. 10.1).

Filtration is the single most critical area in irrigation system. In practical field conditions for the intensive growing of vegetable either in greenhouse or in open field, a combination of filters unit is necessary for proper water treatment. It includes the primary and secondary filtration. Several different types of filter can be used to capture and remove contaminants from the irrigation water:

- Gravity filter
- Wire screens
- Sand separator
- Screen filter
- Gravel filter

Pipe network: Water is delivered from the control head and filter to the lateral lines in the field through the main and submain pipelines. Main line is mostly of PVC material. Submain can be of either PVC or polythene. Rigid PVC and polyethylene are typical material used because of their low cost and chemical-resistant qualities.

FIGURE 10.1   Pressurize irrigation and fertigation control head.

In the pressurized irrigation system network, most of the main and submain pipes are buried under the ground and are controlled by various types of control valves.

Lateral lines and dripper: The lateral lines supply water to the emitter or dripper from submain. Usually this is placed on the ground. It is made of polyethylene pipes. The diameter varies from 12 to 25 mm. The pressure in the lateral line varies from 1 to 1.2 bar depending upon the lateral length and dripper characteristics. The dripper capacity varies from 2 to 10 L/h. Its working pressure is about one bar. Some of the common types of dripper used are as follows:

- In-line dripper
- On-line dripper
- Pressure compensating dripper
- Button-type dripper

10.2.1   HYDRAULIC DESIGN CONSIDERATION

Drip irrigation system design must ensure nearly uniform discharge of the drippers in each section that is controlled by a valve and irrigated as a unit of the system. The maximum pressure difference allowable in a system is 20% and the maximum difference in pressure between the head end and the tail end of a lateral should not exceed 10%. The relationship between pressure and discharge for different types of emission devices can be obtained from the manufacturers' catalogues. The pressure loss can be estimated from monograph tables or using the relationships expressed in the form of equations. Head loss occurs due to friction between the pipe walls and water as it flows through the system. Obstacles, turns, bends, expansions, contractions of pipes, etc., along the way to flow increase head losses.

The head loss due to friction is a function of the following variables:

1. Pipe length
2. Pipe diameter
3. Pipe wall smoothness
4. Water flow rate
5. Liquid viscosity

10.3   TYPES OF PRESSURIZED IRRIGATION SYSTEMS

A distinction is made between the two principal micro-irrigation methods: the sprayer or micro-sprinkler and the drip irrigation system. Sprayers and micro-sprinklers spray the water through the atmosphere and are designed principally to wet a specific volume of soil around individual trees in an orchard. Drip irrigation, on the other hand, represents a paint source of water and wets a specific volume of soil by direct application of water to the root zone of the plant. The type of drip emitter from the aspect of its discharge and the distribution of the emitters throughout the plot (distances along the drip lateral and between the drip laterals) is dependent on the soil texture and the crop. The drip system is suitable for the irrigation of row crops (vegetable and industrial crops) and orchards. Fig. 10.2 describes the use of mulch along with drip irrigation inside the greenhouse.

FIGURE 10.2   Pressurized drip irrigation system for greenhouse.

10.3.1 Drip Irrigation System

Drip irrigation is the best available technology for the judicious use of water for growing horticultural crops on a large scale on a sustainable basis. Drip irrigation is a low labor-intensive and highly efficient system of irrigation, which is also amenable to use in difficult situations and problematic soils, even with poor-quality water. Irrigation water savings ranging from 36 to 79% can be affected by adopting a suitable drip irrigation system (Holzapfel et al., 2015). Drip irrigation or low-volume irrigation is designed to supply filtered water directly to the root zone of the plant so as to maintain the soil moisture near to field capacity level for most of the time. The field capacity soil moisture level is found to be ideal for efficient growing of vegetable plants. This is due to the fact that at this level the plant gets an ideal mixture of water and air for its development. The device that delivers the water to the plant is called a dripper. Water is frequently applied to the soil through emitter placed along a water delivery lateral line placed near the plant row. The principle of drip irrigation is to irrigate the root zone of the plant rather than the soil and get minimal wetted soil surface. This is the reason for getting very high-water application efficiency (90–95%) through drip irrigation. The area between the crop row is not irrigated therefore more area of land can be irrigated with the same amount of water. Thus, water saving and production per unit of water is very high in drip irrigation.

Drip irrigation and fertigation technology help in increasing water and nutritional productivity of horticultural crops. Protected cultivation also helps in increasing water and nutritional productivity of horticultural crops. The possibility of expanding the irrigated areas is becoming very costly; therefore, improving productivity within the existing irrigated is crucial (Carruthers et al., 1997). The concept of productivity has changed from 'crop per unit area' to 'crop per unit volume of water'. The standard unit of water productivity is kg/m^3, whereas nutritional water productivity is expressed in nutritional units/m^3 (Renault and Wallender, 2000). The crop water productivity is very high for tomato, cucumber, capsicum and flowers grown inside greenhouse with drip fertigation as shown in Tables 10.1 and 10.2,

TABLE 10.1
Crop Water Productivity for Greenhouse Vegetables Grown with Drip Fertigation

Crop	Growing Place	Growing Period	Total Water Use (m³/ha)	Total Yield (tonne/ha)	Crop Water Productivity kg/m³ (g/L)
Tomato	GH	Sept–May	3200	250	78
Capsicum Green	GH	Sept–May	2440	90	37
Capsicum Color	GH	Sept–May	2440	60	24.6
Cucumber	GH	Aug–Oct	1550	60	38.7
Cucumber	GH	Feb–May	2010	80	39.8

TABLE 10.2

Crop Water Productivity for Greenhouse Flowers

Crop	Growing Place	Growing Period	Total Water Use (m³/ha)	Total Yield (stem/ha)	Crop Water Productivity (stem/m³)
Rose soilless	GH	Yearly	15,000	2,700,000	180
Rose soil	GH	Yearly	5000	2,100,000	420
Chrysanthemum	GH	Yearly	2800	1,160,000	414

respectively. The crop water productivity of tomato grown inside greenhouse is four times higher than the open field grown tomato. The crop water productivity of green capsicum is higher than colored capsicum grown inside greenhouse.

10.3.2 SPRINKLER IRRIGATION

Sprinkler irrigation's primary goal is to mimic rainfall, which uniformly distributes water over large areas. Water is sprayed by a jet or jets of water (ejected from one or two nozzles) that allow the impact-driven sprinkler to revolve in a circular motion and spread water across the area according to the radius of the jet. Sprinklers are often mounted on aluminum or plastic laterals. Depending on the sprinkler's specific distribution profile, the amount of water that collects immediately around it is typically substantially more than the amount that reaches the soil at the end of the wetted radius. Therefore, the sprinklers must be positioned so that the spray from one sprinkler reaches the adjacent sprinkler, or else the sprays from adjacent sprinklers must overlap, in order to achieve high water application uniformity in the field. As long as the sprinkler is designed to ensure a uniformly shaped distribution pattern and is used in appropriate wind and pressure conditions, spacing the sprinklers at the proper distances assures satisfactory uniformity of distribution. Uniformity of distribution is negatively impacted by wind. As a result, it is advised to use sprinkler irrigation systems when there is little or no wind.

For each sprinkler model, the manufacturers specify the optimum working pressure. Given the other above-mentioned conditions and operating the sprinkler system at the ideal pressure, enough distribution uniformity should be obtained. The discharge of the sprinklers and their spacing determine the term application rate, or the amount of water applied to the soil expressed in mm/hr. Different soils have different infiltration rates, which are also expressed in millimeters per hour. It is essential to design the system in such a way that it provides an application rate that is smaller than the infiltration rate of the soil in order to ensure appropriate infiltration of the water into the soil and to prevent surface run-off. In order to minimize water loss in the area bordering the plot and to prevent interference with vehicles that move on the field tracks that run along the borders of the irrigated fields, it is advised to use part-circle sprinklers, which enable adjustment of the area wetted by the sprinkler. High distribution uniformities and irrigation efficiencies can be achieved by paying careful attention to the points indicated below during the several phases of field data collection, irrigation system planning and irrigation system installation and operation in the field:

i. Spacing of the sprinklers at distances between them, which will enable the attainment of high distribution uniformity;
ii. Irrigation at an application rate less than the infiltration capacity of the soil;
iii. Operation of the sprinkler system in windless conditions or light winds only;
iv. Operation of the sprinkler system according to the pressure recommended by the manufacturer;
v. Use of part-circle sprinklers in combination with full-circle sprinklers;
vi. Portable, i.e., hand-move sprinkler irrigation laterals with adequate overlap can be installed to reduce investment costs.

10.3.3 SPRAYERS AND MICRO-SPRINKLERS

The micro-sprinkler spreads water using a rotating, whirling sprayer mechanism, whereas the micro-sprayer uses a nozzle and a static spreader platform to spread water through the air. Because of this, mini-sprinklers with a sprayer-like discharge cover larger areas. The micro-sprayer/sprinkler is best suited for irrigation of orchard trees below the canopy. The goal is to wet the soil in a small area immediately surrounding the tree without wetting the entire area of soil that each tree occupies. By using sprayers that have a bridge-like device which enables the static spreader nozzle to be replaced by a rotating device, it is possible to irrigate the young orchard by means of micro-sprayers and to switch over to micro-sprinklers when the orchard matures, requiring wetting of a larger soil volume. Systems using micro-sprayer and micro-sprinklers can achieve an irrigation efficiency of up to 85% (Albaji *et al.*, 2015).

10.4   THE ESSENTIAL ELEMENTS OF A DRIP IRRIGATION SYSTEM

The following fundamental components comprise a typical system layout:

Control head: It includes pump, filters, fertilizer applicator, water meter, pressure/flow regulating valves and controller for automation.

Filters: These are major components needed to eliminate suspended particles from the water that is pumped that might clog the drippers. In the system, different filter types are used. Filters made of gravel and graded sand are cylindrical tanks filled with fine gravel and sand of particular sizes. These are primarily used after pumping to filter out sand and organic particles. The least expensive and most effective methods for filtering water are screen and disc filters. Typically, these serve as second-stage filters.

Fertilizer Applicators: These are used to inject acids, liquid fertilizers, systemic insecticides and algaecides and other chemicals into the water supplied through a drip system. They are available in three varieties: fertilizer pumps, venturi systems and fertilizer tanks.

  i. Fertilizer tank: To apply nutrients in solution along with the irrigation water, a metallic tank is provided at the top of the drip irrigation system. A bypass pipe connects the tank to the main irrigation system. From the main irrigation line, some irrigation water is diverted into the tank. The pressure difference between the tank's entry and departure points causes this bypass flow. This application method requires little electricity and is easy to build and operate. However, there are fluctuations in the fertilizer application during the fertigation schedule. As a result, it is impossible to precisely manage the fertilizer concentration.

 ii. Venturi system: It consists of a built-in converging section, throat and diverging section. A suction effect is created at the converging section due to high velocity, which allows the entry of the liquid fertilizer into the system. This system is simple in operation and a fairly uniform fertilizer concentration can be maintained in the irrigation water.

iii. Fertilizer pump: This kind of pump can operate on either water or electricity. It draws fertilizer solution from a tank and pushes it into the irrigation system under pressure. It offers precise control over how fertilizer is applied. However, it is expensive and requires expert operation.

Pressure/flow regulators: These are control valves that are actuated either manually or electro-hydraulically to regulate flow and pressure in the drip system.

Controllers: These automatic mostly micro-processor-based devices are used to provide stop/start signals to pump and valves/regulators. The actuating signal may either be time or volume based. In more advanced technological modes, these gadgets are controlled by soil moisture sensors placed in the plant root zone.

Pipe lines: The water conveyance from the control head to the emitter/dripper is generally categorized into three units as follows:

Main pipe: This is the main carrier of water from the source (after the control head). It is further connected to submains or manifold. These are usually rigid PVC and HDPE pipes with pressure rating of about 10 kg/cm^2.

Submains: This is the portion of pipe network between the main pipe line and the laterals. These are also PVC/HDPE/LDPE pipes with pressure rating of 6–7 kg/cm^2. Diameters of main and submain pipes are selected based on the water requirements of the farm area.

Laterals: These flexible pipes of HDPE or LLDPE ranging from 10 to 20 mm diameters are the ones which are spread over the field in a specified layout. Designed to carry water at about 3 kg/cm^2, these pipes are provided with point-source emitters or drippers spaced along them.

Emitter or dripper: This is a device made to release a small, uniform flow of water, drop by drop, to the designated location while also dissipating hydraulic pressure. Based on the pressure dissipation and flow mechanics, various types of emitters have been produced. Emitters are typically constructed of poly-propylene material and are either placed within the lateral or mounted on it (on-line kind) (in-line type). The discharge rate capacities for the emitters range from 1 to 8 L/h. The kind of crop and soil influences the choice of emitter for a specific discharge rate.

The experimental outcomes of drip irrigation technology in various regions of the nation have demonstrated, on average, water savings of 40–50% and yield increase of the same amount, in addition to improvements in produce quality. Before deciding to install the system, its cost and the crop's net returns should be thoroughly considered. The following are some additional barriers to the widespread use of drip irrigation in India:

    i. Inadequate knowledge and general awareness about the technology;
   ii. Inadequacy in quality of materials leading to system's failure and short life span;
  iii. Interrupted power supply;
   iv. Damage to rodents;
    v. Insufficient extension and promotion work by government agencies/ departments.

10.5 SOIL WATER-PLANT-CLIMATE RELATIONSHIP IN RELATION TO DRIP IRRIGATION SYSTEM

Soil-plant-water relationships relate to the properties of soil and plants that affect the movement, retention and use of water. To understand why and how much irrigation is necessary, one must understand soil-plant-water relations. A proper understanding of these concepts is important to encourage and ensure judicious use of irrigation water and systems.

Water is transported throughout plants almost continuously. There is a constant movement of water from the soil to the roots, from the roots into the various parts of the plant, then into the leaves where it is released into the atmosphere as water vapor through the stomata. This process is called transpiration. Combined with evaporation from the soil and wet plant surfaces, the total water loss to the atmosphere is called evapotranspiration (ET).

Soil is a three-phase medium comprising solids, liquid and air. The fraction of these components varies with soil texture and structure. An active root system requires a delicate balance between the three soil components, but the balance between the liquid and gas phases is most critical, since it regulates root activity and plant growth process. Soil texture refers to the distribution of the soil particle sizes. The mineral particles of soil have a wide range of sizes classified as sand, silt and clay.

Soil water affects plant growth directly through its controlling effect on plant water status and indirectly through its effect on aeration, temperature, nutrient transport, uptake and transformation. The soil and its properties directly affect the availability of water and nutrients to plants. The understanding of soil and its properties in relation to water is helpful in good irrigation design and management. The size, shape and arrangement of the soil particles and the associated voids (pores) determine the ability of soil to retain water. Porosity of sandy soils ranges from 30 to 50%, while that of clay soils from 40 to 60% (Lacape *et al.*, 1998).

Water removal from most of the soils will require at least 7 kPa (7 bars) tension. The permanent wilting point reaches around 1500 kPa (15 bars). This means that in order for plants to remove water from the soil, they must exert a tension of more than 1500 kPa (15 bars). This is the limit for most plants, and beyond this they experience permanent wilting. The pores in sandy soils are generally large and a significant percentage drain under the force of gravity in the first few hours after a rain. This water is lost from the root zone to deep percolation. The movement of water from the surface into the soil is called infiltration. The infiltration characteristic of the soil is one of the dominant variables that influence irrigation. Infiltration rate is the soil characteristic determining the maximum rate at which water can enter the soil under specific conditions. Infiltration rates are generally lower in soils of heavy texture than in light soils. The accumulated infiltration, also called cumulative infiltration, is the total quantity of water that enters the soil in a given time.

10.6   CALCULATION OF CROP WATER REQUIREMENT

The estimation of irrigation water requirement is essential for planning cropping as well as irrigation system. Crop water requirement includes the losses due to ET or consumptive use and percolation. The quantity of irrigation water required for a given crop and area can be estimated using the following relationship (FAO, 1998):

$$W_R = K_C \times K_P \times C_C \times E_V \times A \qquad (10.1)$$

where

W_R = Monthly/daily irrigation water requirement, liters;
K_C = Crop factor;
K_P = Pan evaporation factor (usually 0.8);
C_C = Canopy factor, it is the ratio of the wetted area to plant area and is taken as 1.0 for closely spaced crops;
E_V = Monthly/daily pan evaporation, mm;
A = Area to be irrigated, sq. m.

10.6.1   IRRIGATION INTERVAL

Irrigation interval is the length of time allowable between successive irrigation.

The crop water requirement under drip irrigation system is different from traditional irrigation because in drip irrigation system, the area between the plant rows remains unirrigated and the area from plant to plant becomes irrigated partially as well. Crop water requirement and irrigation interval are the two parameters involved in the irrigation scheduling of any crop. It is expressed either in the depth of water (mm or cm) or in the amount of water (m³ or L). It depends upon soil, plant, climate and the place of growing.

10.6.2   PAN EVAPORATION METHOD

The pan evaporation method was used for calculating the crop water requirement of plants on a daily basis in mm/day and subsequently in cubic meter per day. The various weather parameters, pan factor and crop factor were used in the pan evaporation method. The pan evaporation method was found to be a simple and practical method of crop water requirement calculation.

10.7   FERTIGATION FOR PROTECTED CULTIVATION TECHNOLOGY

In micro-irrigation, fertilizers can be applied through the system with the irrigation water directly to the region where most of the plant roots develop. This process is called fertigation and is done with the aid of special fertilizer apparatus (injectors) installed at the head control unit of the system, before the filter. The element most commonly applied is nitrogen. However, the application of phosphorous, potassium and other micronutrients are common for different horticultural crops. Fertigation is a necessity in drip irrigation.

The main objectives of fertigation are as follows:

• Optimizing yield
• Minimizing pollution
• Water saving
• Fertilizer saving
• Quality improvement
• Timely application of fertilizers
• Uniform application

The rationale for fertigation are as follows:

• Irrigation and fertilizers are the most important management factors through which farmers control plant development and yield.
• Water and fertilizers have important synergism which is very well used in fertigation.
• Timely application of water and fertilizers can be controlled through fertigation.

The main principles of fertigation are to feed the plant in appropriate time, quantity, and location, which can be controlled through fertigation. The plant yield and the quality depend on all these three factors.

Advantages of Fertigation are as follows:
- Accurate and uniform application of water and nutrients;
- Application restricted to the wetted area where the active roots are concentrated;
- Amount and concentration of nutrients can be adjusted according to the stage of development and climatic considerations;
- Reduced time fluctuation in nutrient concentrations;
- Crop foliage is kept dry, retarding the development of plant pathogens and avoiding leaf burn;
- Convenient use of ready mixed fertilizers.

Factors controlling nutrient uptake under fertigation are as follows:
- Water and nutrient distribution in soil under drip fertigation;
- Quantity considerations;
- Intensity considerations (concentration);
- Uptake fluxes: nutrient concentration at root surface;
- Coupling quantity and intensity factors.

Factors affecting fertilizers composition are as follows:
- Plant characteristics;
- Soil characteristics;
- Irrigation water quality;
- Growing place.

Chemicals and biological considerations in selecting fertilizers for fertigation are as follows:
- Fertilizers solubility and mixed fertilizers;
- Solution pH and NH_4/NO_3 ratio;
- Nutrient mobility and chemistry in soils;
- Salinity of the irrigation water.

Requirement for fertilizers used in fertigation is as follows:
- Full solubility;
- Quick dissolution;
- High nutrient content;
- Lack of toxic materials;
- Low price;
- Easy availability.

10.7.1 FERTIGATION SOLUTION EC AND pH

EC (electrical conductivity) and pH are the two important indices of fertigation. They represent the whole quality and characteristics of fertilizers and water. It varies for different plants and soils. Some important facts related to pH are as follows:

- Alkaline pH may cause precipitation of Ca and Mg carbonates and phosphates.

- High soil pH reduces Zn, P and Fe availability to plants.
- Ammonia raises the solution pH and urea increases soil pH upon hydrolysis.
- Acids (nitric, phosphoric) may be used to reduce the irrigation solution pH.

Methods of application of fertilizers are as follows:
- Fertilizer tank (available in 60, 90, 120 L etc.);
- Venturi device (head loss/vacuum operated);
- Dosatron (costly and most effective);
- Fertilizer pumps or injectors (hydraulic type).

Safety devices used in fertigation are as follows:
- An interlock to stop the fertilizer pump;
- A check valve to prevent from the fertilizer tank to the irrigation line following shut down;
- Flow sensor to assure system shut down in case of flow ceases in injection line;
- A bleed valve to relieve the pressure in the injection line when disconnecting;
- A strainer to prevent foreign materials.

Characteristics of fertigation are as follows:
- High efficiency of fertilizer application;
  - i.   Uniform distribution by irrigation water
  - ii. Deeper penetration of fertilizer into the soil
  - iii. Avoiding ammonia volatilizing from soil surface
- Easy coordination with specific crop demand;
- Flexibility in adjusting nutrient ratio;
- Remote control operation;
  - i. Allows fertilization in the rainy season when the soil is wet without stepping on it and destroying the structure
- Convenience in saving manpower;
- Low losses in transportation and storage;
- The system may be used for additional applications.

Limitations of fertigation system
- The system needs water without solid particles that may clog emitters.
- Knowledge of the chemical composition of water is important to avoid precipitation with added fertilizers. Sometimes pre-treatment is necessary.
- The system needs the use of equipment, in which some of them is expensive.
- Not all types of fertilizers are suitable.
- Some fertilizers attack metals mainly of the head control and cause corrosion.

TABLE 10.3
Fertilizers Generally Used in Fertigation

Element	Compound	Formula
Nitrogen	Urea	$CO(NH_2)_2$
N	Ammonium nitrate	NH_4NO_3
	Ammonium sulfate	$(NH_4)_2SO_4$
Phosphorus	Phosphoric acid	H_3PO_4
P	Mono ammonium phosphate	$NH_4H_2PO_4$
	Di ammonium phosphate	$(NH_4)_2HPO_4$
Potassium	Potassium chloride	KCL
K	Potassium nitrate	KNO_3
	Mono potassium phosphate	KH_2PO_4

Source: Kathpalia and Bhatla (2018).

- When proportional fertilizers application is used, nitrate may be leached below the root zone.
- Leaves burning damage occurs, when the fertilizers are applied by sprinklers or micro-sprinklers.

10.7.2 FERTILIZERS

These are chemical compounds in which one atom or more in the formula is a nutrient element. Generally, this atom is the ion, or part of an ion that a plant adsorbs, apart from the nitrogen amid the group that plants are not able to use directly. Fertilizers are added to replenish the elements that were used by plants or disappeared by other processes or will disappear during the plant growth. Table 10.3 shows fertilizers generally used in drip fertigation system, whereas Table 10.4 describes the ionic form of nutrients available for plant growth.

TABLE 10.4
The Ionic Form of Nutrition Elements Adsorbed by Plants

Cations	Anions
Ammonium – NH_4^+	Nitrate – NO_3^-
Potassium – K^+	Mono-phosphate – $H_2PO_4^-$
Calcium – Ca^{++}	Di-phosphate – HPO_4^{--}
Magnesium – Mg^{++}	Sulfate – SO_4^{--}
Iron – Fe^{+++}	Molybdate – MoO_4^{--}
Iron – Fe^{++}	Borate – $B_4O_7^{--}$
Manganese – Mn^{++}	
Zinc – Zn^{++}	
Copper – Cu^{++}	

Source: Kathpalia and Bhatla (2018).

10.8   AUTOMATION OF PRESSURIZE IRRIGATION SYSTEM

Automation of irrigation systems in protected cultivation mainly in the form of greenhouse, net house and nursery is practiced to make the irrigation operation precise, perfect and efficient. The automated irrigation system also minimizes the number of personnel involved in irrigation operations and makes the irrigation system relatively maintenance-free. The introduction of automation into irrigation systems has increased application efficiencies and drastically reduced labor requirements (Sidhu *et al.*, 2021). The automation of irrigation systems is mainly of two types. The conditional automated irrigation system depends on the various types of sensors placed in the irrigated area. The non-conditional automated irrigation system depends on the numeric data fitted manually in the irrigation computer to different irrigation programs. An automatic irrigation system controlled electronically by a field unit has four basic elements. The automation of irrigation systems with controllers and sensors in covered cultivation is widely practiced in Centre for Protected Cultivation Technology, IARI, New Delhi. The project aims to demonstrate different technologies for intensive and commercially oriented peri-urban cultivation of horticultural crops for improved quality and productivity. The different horticultural crops in the form of flowers, vegetables and fruits are grown inside greenhouse, open field, net house and orchard.

10.8.1   AUTOMATION OF PRESSURIZE IRRIGATION SYSTEM BY SENSORS AND IoT

Automation of irrigation systems in protected cultivation mainly in the form of greenhouse, net house and nursery is practiced to make the irrigation operation precise, perfect and efficient. The automated irrigation system also minimizes the number of personnel involved in irrigation operations and makes the irrigation system relatively maintenance-free. The introduction of automation into irrigation systems has increased application efficiencies and drastically reduced labor requirements. The automation of irrigation systems is mainly of two types:

1. Conditional
2. Non-conditional

   The conditional automated irrigation system depends on the various types of sensors placed in the irrigated area. The opening and closing of the valves to operate the irrigation system depends on the data of the sensors passed to the programs installed in the irrigation computer. Examples of these sensors are humidity, temperature, tensiometer sensors etc. Conditional automation of irrigation systems is very efficient, but it requires smooth operation of sensors and thereby more maintenance. It results in precise irrigation as and when required by the plant to meet its crop water requirement. Consequently, the overall irrigation efficiency is very high in this case.

   The non-conditional automated irrigation system depends on the numeric data fitted manually in the irrigation computer to different irrigation programs. These programs operate the opening and closing of the valves attached to different crops for irrigation. The different irrigation programs need to be updated manually to suit the crop water requirement. This system requires experienced irrigation personnel to update different irrigation programs. In this case, the data of the various types of sensors are not directly connected to irrigation programs, but they serve as a guide to manually update the irrigation programs installed in the controller or computer. The maintenance of sensors is also necessary to operate this system efficiently.

An automatic irrigation system controlled electronically by a field unit has four basic elements.

1. The metering device: A water meter with electrical output is installed in the irrigation system; information relating to accumulating flows can be monitored and relayed by cables to the control unit. Fertilizer meter is also installed to transmit the information concerning cumulative volumes of a fertilizer solution which is pumped from a tank and injected into the system.
2. The electronic control unit: The control unit receives the necessary feedback from the metering or measuring devices and processes the data in order to relay the proper commands to the appropriate valves.
3. Solenoid valves: It converts the electrical commands received from the control unit into hydraulic commands, which causes the valves to open or close.
4. Hydraulic valves: These control the flow of water into single laterals or sets of laterals operating simultaneously.

Applying water directly near the root zone of plant by switching on the pump is itself the first step toward the automation of irrigation system. The complete automation of micro-irrigation systems is a relatively new concept in our country. The inclination toward automation of micro irrigation is gaining momentum in our country due to the following reasons:

• Automation eliminates manual operation to open or close valves, especially in intensive irrigation process.
• Possibility to change irrigation and fertigation frequency and also to optimize these processes.
• Adoption of advanced crop systems and new technologies, especially new crop systems that are complex and difficult to operate manually.
• Use of water from different sources and increased water and fertilizer use efficiency.
• System can be operated at night to save water from evaporation loss and thus the day time can be utilized for other agricultural activities.
• Pump starts and stops exactly when required, thus optimizing energy requirements.

10.9 DESIGN OF DRIP IRRIGATION SYSTEM

A complete drip irrigation system consists of a head control unit, main and submain pipelines, hydrants, manifolds and lateral lines with drippers.

Control Station (Head Control Unit): Its features and equipment depend on the system's requirement. Usually, it consists of the shutoff, air and check valves, filtering unit, fertilizer injector and other small accessories.

Main and Submain Pipelines: They are made of PVC and are mostly buried in the ground.

Hydrants: Fitted on the mains or the submains and equipped with two to three shutoff valves; they are capable of delivering all or part of the piped water flow to the manifold feeder lines.

Dripper Lateral: They are made of 12–20-mm LDPE PN 3–4 bars. They are equipped with in-line or on-line drippers.

The system pressure ranges from 2 to 3 bars. Operating pressure of dripper is usually 1 bar. Dripper discharge varies from 2 to 10 L/h. The wetting pattern of water in the soil from the drip irrigation tape must reach plant roots. Emitter spacing depends on the crop root system and soil properties. Seedling plants such as onions have relatively small root systems, especially early in the season. Design must take into account the effect of the land's contour on pressure and flow requirements. Plan for water distribution uniformity by carefully considering the tape, irrigation length, topography, and the need for periodic flushing of the tape. It should include vacuum relief valves into the system. Consider power and water source limitations. Finally, be sure to include both injectors for chemigation and flow meters to confirm system performance.

Filters and Pumps: Sand media filters have been used extensively for micro-irrigation systems. Screen filters and disk filters are common as alternatives or for use in combination with sand media filters.

Sand media filters provide filtration to 200 mesh, which is necessary to clean surface water and water from open canals for drip irrigation. These water sources pick up a lot of fine grit and organic material, which must be removed before the water passes through the drip tape emitters. Sand media filters are designed to be self-cleaning through a 'backflush' mechanism. This mechanism detects the drop in pressure due to the accumulation of filtered particles. It then flushes water back through the sand to dispose of clay, silt and organic particles.

Sand used for filters should be between sizes 16 and 20 to prevent excess backflushing. To assure enough clean water for backflushing, several smaller sand media filters are more appropriate than a single large sand media. In addition to a sand media filter, a screen filter can be used as a prefilter to remove larger organic debris before it reaches the sand media filter, or as a secondary filter before the irrigation water enters the drip tube. For best results, filters should remove particles four times smaller than the emitter opening, as particles may clump together and clog emitters. Screen filters can act as a safeguard if the main filters fail or may act as the main filter if a sufficiently clear underground water source is used.

10.9.1 Designing an Effective Drip Irrigation System

The following hints may help in the planning and designing of drip irrigation system:

- Work out the number of connectors needed when planning your drip irrigation system.

- Plants in sunny areas usually require more water due to higher evaporation rates. Plants in shaded areas will require less water due to lower evaporation rates.
- Note slopes and soil types to help work out the watering requirements for different areas of your garden. For example, gardens with heavy clay soil may need more water pressure.
- Select drip emitters according to your plants' watering requirements.
- Consider where you would need joints and connectors.
- Rain switches and soil moisture sensors are highly recommended, especially in areas with high rainfall.
- Lay the piping above ground before digging. A 10-cm-deep trench should be adequate, although sandy soil may require a slightly deeper trench to hold the piping in place.
- To make it easier to connect joints, heat the end of your piping to soften it and make it more flexible.
- Make sure drip emitters are installed above ground so that they do not become clogged by dirt.

10.10 MAINTENANCE OF DRIP FERTIGATION SYSTEM FOR PROTECTED CULTIVATION

i. The most commonly occurring problem in drip irrigation system is the clogging of drippers. Once clogged, it is very difficult to declogg especially the in-line drippers. Clogging can be avoided or at least minimized by keeping a regular maintenance schedule of filters.

Upkeep of these components is very critical to make the drip operation successful. Gravel/sand filters must be washed through back-flow to remove the sedimentation inside. Disk of screen filers should be washed every alternate day. The filtering elements of both these types of filters are washable. Rubber or seals should be replaced properly so that no leakage occurs.

ii. All emitters should be periodically checked for proper functioning. Any leakage in the pipes; fittings should be checked immediately.

iii. Fine inorganic particles usually settle at the ends of the submain manifold and the laterals where flow velocities are slow. Periodic flushing should be done by removing end plugs of laterals and flush out the fine particles. A velocity of 30 cm/sec is necessary to adequately flush the fine particles from lateral tubing.

iv. Generally on-line type of emitters can be disassembled and cleaned manually. In-line type can be flushed to eject loose deposits. Carbonate deposits can be removed using solutions of 0.5–1.0% HCl acid ejected at the submain lateral connection to give a contact time of 5–15 min in the emitter. This treatment may not be effective in completely clogged emitters.

v. Emitters discharge rate should be periodically checked for uniformity.

Applying water directly near the root zone of plant by switching on the pump is itself the first step toward the automation of irrigation system. The complete automation of micro-irrigation systems is a relatively new concept in our country. The inclination toward atomization of micro irrigation is gaining momentum in our country due to the following reasons.

- Automation eliminates manual operation to open or close valves, especially in intensive irrigation process.
- Possibility to change irrigation and fertigation frequency and also to optimize these processes.
- Adoption of advanced crop systems and new technologies, especially new crop systems that are complex and difficult to operate manually.
- Use of water from different sources and increased water and fertilizer use efficiency.
- System can be operated at night to save water from evaporation loss and thus the day time can be utilized for other agricultural activities.
- Pump starts and stops exactly when required, thus optimizing energy requirements.

The completely automatic fertigation system network consists of mainly three irrigation control units (ICU), filtration unit, pump house, reservoir, tube wells and a network of underground pipes for water conveyance. The controllers are installed at all ICU to automize the fertigation system. These controllers need 24 Volt AC supply and lightening protection. The controllers help in automatic opening and closing of the valves connected to different crops for starting and closing the fertigation system and micro-sprinklers. The controllers are programmed to start the fertigation system and micro-sprinklers through the valves connected to different crops at desired hours of the day. The controllers also store the relevant data related to fertigation system and micro-sprinkler operation. All the individual controllers installed at different ICU are connected to a central computer system through reinforced communication cable and command cable.

Sensors and internet of things (IoT) for water and nutrient management for protected cultivation technology:

Machine learning, IoT and artificial intelligence (AI)-based automation have been the recent most successful approaches for controlling greenhouses and their irrigation and fertigation for maximizing the quality crop production of high-value vegetables, flowers and seedlings. These recent techniques incorporate and integrate the human expertise, sensors, online and in-situ data, software and hardware from different sources for the efficient management of all the related inputs and maximize the output in terms of both quality and quantity. The future of smart, efficient and precision agriculture is mainly based on automation linked with IoT and AI (Alipio et al., 2019).

Following sensors are commonly used for water and nutrient management for protected cultivation technology.

- Electrical conductivity (EC) sensor
- pH sensor

- Total dissolved solid sensor TDS
- Temperature sensor
- DO (dissolved oxygen) sensor
- Biological oxygen demand (BOD) sensor

10.11 CONCLUSION

Protected cultivation creates a favorable environment for the sustained growth of crops so as to realize their maximum potential even in adverse climatic conditions. Protected cultivation technique has a number of benefits that help farmers produce high-quality fruits, vegetables, flowers, and hybrid seeds while minimizing weather-related hazards and maximizing the use of other resources. Drip fertigation technology is very efficient for the application of water and nutrients in protected cultivation technology. Major nutrients and micronutrients are supplied in protected cultivation technology grown crops through water-soluble fertilizers. IoT and sensor-operated drip fertigation system is used for efficient delivery of water-soluble fertilizers. Greenhouse hydroponics cultivation technology is effective for growing high-value horticultural crops, herbs and seedlings in various types of inert soilless media and water inside greenhouse and other protected structures. IoT of these systems helps in real-time management and control of data with the help of automated sensors and programmable micro controllers with the help of smart phones and other devices. The successful implementation of greenhouse-based protected cultivation technology has challenges such as initial cost, lack of indigenous models, costly sensors and automation system. It can be combined with contemporary sensors, automation systems, IoT and DSS, making vertical farming as a whole popular among modern youths, farmers and businesspeople. In the post-covid era, when there will be a constant need for fresh, high-value vegetables and herbs, particularly in peri-urban areas of large cities, protected cultivation technology can be a feasible option.

REFERENCES

Albaji, M., M. Golabi, S. B. Nasab and F. N. Zadeh 2015. Investigation of surface, sprinkler and drip irrigation methods based on the parametric evaluation approach in Jaizan Plain. *Journal of the Saudi Society of Agricultural Sciences* 14(1): 1–10.
Albery, W. J., B. G. Haggette and L. R. Svanberg 1985. The development of sensors for hydroponics. *Biosensors* 1(4): 369–397.
Alipio, M. I., A. E. Cruz, J. D. Doria and M. S. Fruto 2019. On the design of nutrient film technique hydroponics farm for smart agriculture. *Engineering in Agriculture, Environment and Food* 12(3): 315–324.
Ambast, S. K., N. K. Tyagi and S. K. Raul 2006. Management of declining groundwater in the Trans Indo-Gangetic Plain (India): Some options. *Agricultural Water Management* 82(3): 279–296.
Carruthers, I., M. W. Rosegrant and D. Seckler. 1997. Irrigation and food security in the 21st century. *Irrigation and Drainage Systems* 11: 83–101.
FAO, 1998. Crop evapotranspiration: Guidelines for computing crop water requirements. *FAO irrigation and drainage paper* 56. Rome, Italy.

Holzapfel, E., J. Jara and A. M. Corontana. 2015. Number of drip laterals and irrigation frequency on yield and exportable fruit size of highbush blueberry grown in a sandy soil. *Agricultural Water Management* 148: 207–212.

Kathpalia, R. and S. C. Bhatla. 2018. Plant mineral nutrition. In *Plant physiology, development and metabolism* (pp. 37–81). Singapore, Springer.

Lacape, M. J., J. Wery and D. J. Annerose. 1998. Relationships between plant and soil water status in five field-grown cotton (*Gossypium hirsutum* L.) cultivars. *Field Crops Research* 57(1): 29–43.

Prasanna, R., A. Kanchan, S. Kaur, B. Ramakrishnan, K. Ranjan, M. C. Singh, M. Hasan, A. K. Saxena and Y. S. Shivay. 2016. Chrysanthemum growth gains from beneficial microbial interactions and fertility improvements in soil under protected cultivation. *Horticultural Plant Journal* 2(4): 229–239.

Renault, D. and W. W. Wallender. 2000.Nutritional water productivity and diets. *Agricultural Water Management* 45(3): 275–296. https://doi.org/10.1016/S0378-3774(99)00107-9.

Sampathkumar, T., B. J. Pandian and S. Mahimairaja. 2012. Soil moisture distribution and root characters as influenced by deficit irrigation through drip system in cotton–maize cropping sequence. *Agricultural Water Management* 103: 43–53.

Sidhu, R. K., R. Kumar, P. S. Rana and M. L. Jat. 2021. Chapter Five – Automation in drip irrigation for enhancing water use efficiency in cereal systems of South Asia: Status and prospects. *Advances in Agronomy* 167: 247–300.

11 Soil Carbon Sequestrations through Conservation Agriculture under Changing Climate

Kaberi Mahanta and Dip Jyoti Rajkhowa

11.1 INTRODUCTION

Climate change and climatic variability are now a reality. The impact of climate on agriculture is being witnessed in different countries of the world. Countries, like India, are more vulnerable to climate change in view of huge population directly depending on agriculture, with low coping mechanisms. Rising temperatures and increased frequency and intensity of extreme weather events, such as high-intensity droughts and floods, mean that it will be even harder to meet the growing demand for food, fiber and fuel, especially for poor countries with high population growth. Climatic aberrations will seriously affect the poorest section of the society who heavily relying on climate-sensitive sectors such as rainfed agriculture and fisheries (Prasad and Rana, 2006; Samara *et al.*, 2004). Reduction in crop, livestock and fishery productivity due to climate change/climatic variability is well predicted and there are variations in perceptions about the intensity and consequences of climate change. In developing countries with highest population rates in the world, the food production can be enhanced where the greatest rates of population growth currently occur (United Nations, 2019; FAO, 2003). This must occur in a world where the opportunity to expand the area used for agriculture is limited and our ability to increase production on existing agricultural land is threatened by land degradation, water resource scarcity, and increases in the climate variability and severe consequences associated with climate change. Thus, in order to meet the world's demand for food, agricultural systems worldwide need to evolve to produce more food, with greater sustainability. The process of rising atmospheric carbon dioxide (CO_2) levels coupled with climate change mitigation efforts have focused considerable interest in recent years on the world soil carbon. It is estimated that the world soils have a high sink potential for carbon sequestration, in terms of their large potential carbon content and soil organic carbon (SOC) that is approachable to modification through agricultural land use. Alteration of natural ecosystems to cropland acts as a driving force of climate change in two main ways. Firstly, agricultural activities directly produce and release about 10%–12% of the atmospheric greenhouse gases (GHGs), such as CO_2, methane (CH_4), nitrous oxide (N_2O) (Six *et al.*, 2002; IPCC 2007; Smith *et al.*, 2011). Secondly, the conversion process alters the soil's physical, chemical and biological properties and so has an impact on the biological resilience of the agro-ecosystems. The increased atmospheric concentration of CO_2 may influence soil temperature, pattern of precipitation and evaporation and resultant changes in the physiochemical and biological properties of soil. Thus, there has emphasis to reduce this concentration through the process known as carbon sequestration. A considerable part of the depleted SOC pool can be restored through conversion of marginal lands into restorative land uses, adoption of conservation tillage with cover crops and crop residue mulch, nutrient cycling, including the use of compost and manure and other systems of sustainable management of soil and water resources. Soil organic matter (SOM) is important for agricultural and ecosystem services (Lehmann and Kleber, 2015; Manlay et al., 2011; Meehl et al., 2011). SOM contributes to the mitigation and adaptation to climate change since it acts as a sink for CO_2, a major greenhouse gas, helps in storing plant nutrients (Paustian et al., 2016) and makes crop production more resilient to drought conditions by promoting soil aggregation and water infiltration (Rosenzweig and Parry, 1994; Franzluebbers, 2002; Sihi et al., 2011). SOM is an important determinant of soil fertility, productivity and sustainability and is a useful indicator of soil quality in tropical agricultural systems where nutrient-poor and highly weathered soils are managed with little external input (Lal, 1991; Post and Kwon, 2000). The dynamics of SOM are influenced by agricultural management practices such as tillage, mulching, removal of crop residues and application of organic and mineral fertilizers. In general, SOC has focused on the total amount of organic C, whereas fractions of SOC have been overlooked. Fractionation of SOC can be used to assess the quality of SOC, thus understanding the effect of soil management practices on the processes of decomposition and stabilization of SOM (Poeplau et al., 2018).

11.2   SOIL CARBON SEQUESTRATION

Carbon management through Carbon sequestration for extensive storage is the management of carbon and is one of the most significant issues of 21st century in India with regard to issues of climate change. An increasing amount of atmospheric carbon by 3.5 pg per annum is an alarming situation in terms of enhanced temperature and pollution of ambiance. The process of removing of atmospheric carbon to the terrestrial biosphere and storing them in a recalcitrant form can be considered one of the solutions of massive GHG's evolution in the atmosphere. Among the different carbon sinks, soil acts as one of the best alternatives for reducing the atmospheric CO_2 and has recognized SOC pool as one of the five major carbon pools for the Land Use, Land Use Change in Forestry (LULUCF). Moreover, removal of about 3.6611 tons of CO_2 from the atmosphere is possible with each ton of organic matter build in soil. Maximum SOC sequestration (Fig. 11.1) of different land use systems is an effective strategy for removal of atmospheric CO_2 and improving the quality of soil. SOC can be considered the sum of all biologically derived organic materials found in the soil or on soil surface irrespective of its source, living status or stage of decomposition but excluding the aboveground portion of living plant. The SOC in terms of its amount and quality is crucial to keep up the quality and productivity soil. In the recent past,

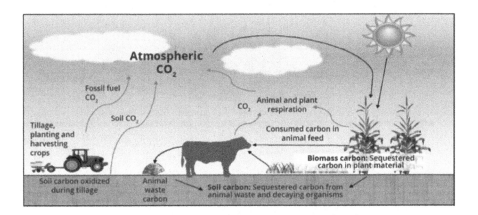

FIGURE 11.1 Soil carbon sequestration.

the GHG effect has created a great concern about the qualities, kinds, distribution and behaviors of SOC. The organic matter content in soils varies considerably depending on climate, soil type and land use system. Decomposition of SOC is largely determined by soil temperature and rainfall.

Carbon sequestration is buildup of carbon out of the atmosphere and its absorption and storage in a terrestrial or aquatic body. Capturing and storage carbon in biomass and soils in the agriculture and forest sector has now been extensively receiving consequence as one potential mitigation strategies of GHGs. Carbon sequestration in terrestrial ecosystems can be referred as the net removal of CO_2 from the atmosphere and its storage into prolonged pools of carbon. The pools can be living, aboveground biomass (e.g., trees) products with a long, useful life created from biomass (e.g., lumber), living biomass in soils (e.g., roots and micro-organisms) or recalcitrant organic and inorganic carbon in soils and deeper subsurface environments. This carbon must be fixed into long-lived pools without which one may be simply altering the size of fluxes in the carbon cycle, not increasing carbon sequestration. Out of the five important global carbon pools, oceanic pool (38,000 pg) is the largest followed by geological pool (5000 pg; 4000 pg of coal pool and 500 pg of each oil and gas), pedological pool (soil carbon pool, 2500 pg), biotic pool (560 pg) and the atmospheric pool (1160 pg). The average atom of C spends about 5 yrs. in the atmosphere, 10 yrs. in vegetation (including trees), 35 yrs. in soil and 100 yrs. in the sea (Lal, 2004a,b). Increase density of C in the soil and depth of C in the profile, decrease decomposition of C and losses due to erosion are important measures to increase the SOC.

11.3 CONSERVATION AGRICULTURE

Globally conservation agriculture (CA) is inevitable for intensification sustainable agriculture. The three principles of CA include as follows: zero-till or no-till (or minimal soil disturbance), soil cover (mainly with crop residue, cover crops) and crop rotation (Fig. 11.2). The benefits of CA because of simplicity in crop management, resource (energy/cost/time) savings, and soil and water

Minimum mechanical soil disturbance (i.e. No-tillage) through direct seed or fertilizer placement

Permanent soil organic cover (at least 30 %) with crop residues and/or cover crops

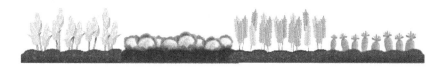

Species diversification through varied crop sequences and associations involving at least three crop species

FIGURE 11.2   Principles of conservation agriculture.

conservation result into adoption of CA extensively, particularly on large farms in the Americas and Australia, where farmers rely on the modern tools and machineries: highly sophisticated farm implements, effective agrochemicals and biotechnology. With inadequate performance, over the last 15 years or more, CA has been promoted among smallholder farmers. Yet, its impacts on crop yields remain controversial. CA is found to be productive than no-till system (NT), and it stands a more than 50% chance to outperform CT in dryer regions, especially with suitable agricultural management practices. Residue retention has the highest positive influence on CA productivity as compared to other management practices. CA appears as a sustainable agricultural practice, if targeted at specific climatic regions and crop species.

Food production must rise to meet the demand of a growing population with minimum impacts on the environment. CA has been emphasized as a key route to sustainable intensification (Hobbs *et al.*, 2008).CA is based on three principles: (1) minimal soil disturbance or no-till/zero-till; (2) continuous soil cover with crops, cover crops or crop residues; and (3) crop rotation (FAO, 2015). "Actual CA" is deemed to be practiced only when all three principles are thoroughly applied (Derpsch *et al.*, 2014). Conservation of soil is imperative, sensitized by the 1930s

"Dust Bowl" in North America prompted the development of no-till approaches. The spread of no-till agriculture in the 1980s–1990s in the Americas and Australia was largely because of the following factors:

1. Potential herbicides (atrazine, paraquat, and glyphosate) were released in the 1960s and 1911s.
2. Direct seeding into a mulch of crop residues was made possible with no-till planters.
3. Government policy incentives supported a transition to no-till in the USA.
4. The advent of herbicide-resistant, genetically modified (GM) crops in the mid-1990's enabled the use of highly efficacious post-emergence herbicides and accelerated the expansion of no-till and CA – particularly in the Americas. To different degrees, this has led to widespread adoption of no-till and CA on large farms in Australia, Brazil and North America.

The popularity of CA principles appears to be based on a number of factors: (1) the soil disturbance is unsustainable as it causes soil degradation/erosion and reduces soil C stocks (Hobbs *et al.*, 2008; Kassam *et al.*, 2014). (2) The continuous no-till with crop residue retention results in "soil health" improvements which will, in time, result higher yields and sustainable agriculture (Kassam *et al.*, 2014). Failure to see yield improvements in the first five–ten years of adoption was therefore commonly dismissed as a transition period (Derpsch *et al.*, 2014). (3) CA may interpret as meaning a form of low-external-input, biodiversity-enhancing and sustainable agriculture. (4) The apparent mimicking of natural systems in which biomass remains on the soil surface and soils is not often exposed. CA has increasingly been recognized as climate-smart agriculture, contributing to both climate change adaptation and mitigation (Pretty and Bharucha, 2014).

11.4 CLIMATE CHANGE

Climate change is a major concern for agricultural production. Climate changes very rapidly all over the world. But the third world countries, including India, face the maximum adverse effects because the agriculture in these countries is mostly rain dependent, where farmers have poor resources and equipment to face the threats and challenges posed by climate change. Near past scientific understanding, led by the Intergovernmental Panel on Climate Change (IPCC, 2011), is that global climate change is happening and will present practical challenges for local ecosystems. The inter-governmental panel on climate change has projected a temperature increase of 0.5–1.2°C by 2020, 0.88–3.16°C by 2050 and 1.56–5.44°C by 2080 for India. The corresponding CO_2 concentrations are expected to be 393, 543 and 1189 ppm in year 2020, 2050 and 2080, respectively (IPCC, 2011). These include the prospect of more severe weather, longer droughts, higher temperatures (milder winters), heat waves, changes in local biodiversity, and reduced ground and surface water quantity and quality. These changes will impact on everything from the natural landscape to human health and socioeconomic conditions.

Under the influence of climate change, mountains are likely to experience wide-ranging effects on the environment, biodiversity and socioeconomic conditions (Beniston, 2003). Changes in the hydrological cycle may significantly change precipitation patterns leading to changes in river run-off and ultimately affecting hydrology and nutrient cycles along the river basins, including agricultural productivity and human well-being (Xu *et al.*, 2009). Many recent studies in the region indicate less snow in the mountains, melting of glaciers, upward movement of tree lines as melting snow covers are removing vast areas for arable cultivation and highly intense but short episodes of rainfall, which cause increased run-off and soil losses, poor water recharge and the consequent drying up of water sources (Xu *et al.*, 2009). There is evidence of noticeable increases in the intensity and frequency of many extreme weather events in the region such as heat waves, tropical cyclones, prolonged dry spells, intense rainfall, snow avalanches, thunderstorms and severe dust storms (Cruz *et al.*, 2011).

11.5   POTENTIAL IMPACTS OF CLIMATE CHANGE

Carbon and nitrogen are primary components of SOM (Brady and Weil, 2008). The soil properties, including formation and maintenance of soil structure, water-holding capacity, cation exchange capacity, nutrient cycling and supplying to the soil ecosystem, are mostly regulated by SOC (Brevik, 2009a,b; Brevik, 2013a). The more productive soils are enriched with sufficient amount of organic matter that are depleted in organic matter (Brevik, 2009a,b). Climate change and its effects on soil processes and properties involve the extent of potential changes in the C and N cycles will have impact on soil sustainability. Healthy soils are important as they supply nutrients and improve crop yield for longer time. However, the nutrient present in soil is not sufficient in quantity, or may present in unavailable form; crops will not be well supplied with enough plant nutrients and result in poor productivity of crop (Brevik, 2009a,b). Low nutrient status in agricultural soil not only reduces the amount of food production but also condenses the nutritional security. Low nutrient-rich soils with pronounced pest and disease infestation will result in lower crop yield (Pimentel *et al.*, 2011).

11.5.1   EFFECT ON SOIL PROPERTIES

The impact of climate change and its variability on crop growth and yields is largely determined by its impact on soil health and the capacity of crop varieties to adjust to changing climate and weather patterns (Brevik, 2013b). The SOM is crucial in soil nutrient processing and soil water retention (Fig. 11.3). The organic carbon is a key driver for soil structural stability, soil microbial diversity and population and the related processes of nutrient cycling and dynamics within the soil system. A wide range of studies have also shown that the elevated atmospheric carbon dioxide (CO_2) and temperature associated with the changing climate could affect plant growth through alteration of photosynthesis, respiration, C and N metabolism, transpiration and stomatal sensitivity (Brevik, 2013b). Low soil fertility is currently a food security problem in many developing countries, particularly in Africa and South Asia (Lal, 2004b; Sanchez and Swaminathan, 2005; St. Clair and Lynch, 2010). Low nutrient-rich agricultural soil not only reduces the amount of food available for human consumption, but it also makes less nutrient-rich crops, which makes those who rely on the low nutrient-rich soils for crop production more susceptible to disease (Sanchez and Swaminathan, 2005). If problems from vector-borne diseases, for example, become more pronounced with changes in climate (Brevik, 2013a), low-nutrient status soils will simply make those who rely on them even more prone to experience disease problems. In this case, soils could have both direct (e.g., production of lower nutrient content foods) and indirect (e.g., making the population more susceptible to vector-borne disease) effects on human health due to climate change. Fig. 11.3 shows the effects of climate change on soil.

The biological, chemical and physical functions of "soil health" deteriorate with changing climate due to a number of reasons, key among them as follows:

- **Increasing turnover of soil organic matter:** Approximately 50% of SOM is SOC. As the global temperature rises with climate change effect, the higher temperatures could speed the rate of SOM turnover (Kirschbaum, 2000). Again, the organic matter is the key driver of microbial diversity that is crucial for efficient nutrient cycling. On account the rapid depletion of organic matter tends to decrease the soil's ability to retain and supply most of the crucial nutrients that is required for crop growth.
- **Decreasing soil moisture content and soil water-holding capacity:** Water is an essential component of plant nutrient uptake, transport and photosynthesis. With the rise in atmospheric temperature, soil water loss

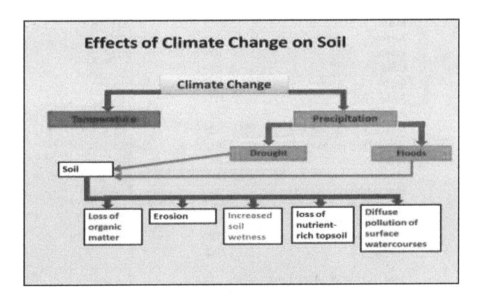

FIGURE 11.3 Effects of climate change on soil.

occurs through evaporation and evapotranspiration. The increase in temperature coupled with declining rainfall is projected to cause negative moisture balance in over 10% of arable areas (IPCC, 2011). Further, if not well managed, the temperature-driven decline in SOM will lead to a decline in soil water-holding capacity, especially in the regions pre-dominant in sandy soils, which results, due to lower organic matter content in soil, in prevailing high temperature.

- **Nutrient depletion:** The unexpectedly higher rainfall and flooding are often associated with heavy nutrient losses via soil erosion, leaching, and denitrification.

11.5.2 IMPACT ON FOOD SECURITY

"Food security (is) a situation that exists when all people, at all times, have physical, social and economic access to sufficient, safe and nutritious food that meets their dietary needs and food preferences for an active and healthy life". Over 911.5% of human food needs, as measured by calories consumed, come from the land, while less than 2.5% come from aquatic systems (Table 11.1) (Brevik, 2009).

TABLE 11.1
Daily Per Capita Food Intake as a Worldwide Average, 2001–2003

Food Source	Calories[a]	Percent of Calories
Rice	5511	25.5
Wheat	521	23.9
Maize	1411	6.11
Sorgum	33	1.5
Potatoes	60	2.11
Cassava	42	1.9
Sugar	202	9.3
Soybean oil	811	4.0
Palm oil	50	2.3
Milk	122	5.6
Animal fats (raw and butter)	62	2.8
Eggs	33	1.5
Meat (pig)	1111	5.4
Meat (poultry)	46	2.1
Meat (bovine)	40	1.8
Meat (sheep and goats)	11	0.5
Fish and other aquatic products[b]	52	2.4
TOTAL	2182	

Source: From Brevik and Burgess (2013).
[a] Aquatic products data from 2003. All other data from 2001 to 2003.
[b] Includes both marine and freshwater products.

Furthermore, aquatic systems do not represent a potential source of significantly increased food supply in the future as overfishing is already a major problem in many of the world prime fishing grounds, including both marine and freshwater fisheries (Allan *et al.*, 2005; Jackson et al., 2001). This means our food supply will need to come almost exclusively from the terrestrial environment, making soils critical to food security. Climate change has already caused and will continue to cause changes in global temperature and precipitation patterns (Trenberth et al., 2011) as well as changes in soil processes and properties as previously discussed. This has led to considerable concern that climate change could compromise food security (Gregory *et al.*, 2005; Parry *et al.*, 2005), which would lead to an overall decline in human health. CO_2 and temperature could affect plant growth through alteration of photosynthesis, respiration, C and N metabolism, transpiration and stomatal sensitivity (Brevik, 2013a). In the conditions of low soil nutrient status, the higher nutrient turnover will rapidly increase nutrient deficiency aggravating the problem of poor crop growth and poor yields. The biological, chemical and physical functions of "soil health" deteriorate with changing climate due to a number of reasons, key factors among them are as follows: With the expected shortening of cropping seasons, the yields of longer duration crop varieties, which are higher yielding, will decrease significantly as a result of terminal phase droughts. This implies that the niches for longer duration crop varieties may change. Often, such shifts can significantly decrease the national food security as development and testing of new crop varieties with a capacity to produce under the changed climate are a long-term process.

The IPCC expects that climate change will impact all four dimensions of food security, which are (1) food availability, (2) stability of food supplies, (3) access to food, and (4) use of food (Easterling et al., 2011). Global food production is projected to increase if the rise in local average temperatures averages 1–3°C, but to decrease if average temperature increases surge beyond 3°C. However, the distributions of the projected changes in food production are not uniform. As a general trend, food production is expected to increase at mid-to-high latitudes and to decrease near the equator if the average rise in temperature stays in the lower range (1–3°C) and to decrease less at mid-to-high latitudes than near the equator if the average rise in temperature moves into the upper range (>3°C) (Olesen and Bindi, 2002).

11.6   SOIL QUALITY RESTORATION THROUGH CARBON SEQUESTRATION

The restoration of soil quality through carbon sequestration is a major concern for tropical soils. The accelerated decomposition of SOC due to agriculture resulting in loss of carbon to the atmosphere and its contribution to the greenhouse effect is a serious global problem (Fig. 11.4).

1. **Soil structure:** Organic matter plays a vital role in the process of aggregation. With the addition of organic matter in soil, it binds the soil particles producing a porous and crumbly soil structure. Again, decomposition

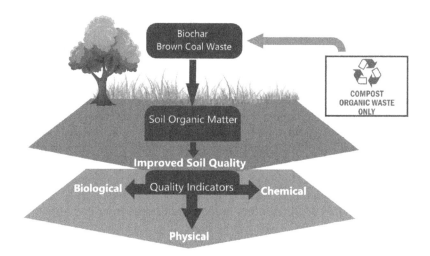

FIGURE 11.4   Restoration of soil quality using sol organic matter.

    of organic matter produces humic acid, fulvic acid and humin and all the humus substances have the cementing property. These substances stabilize the soil structure.

2. **Water infiltration capacity:** As the soil structure becomes stabilize and crumby due to addition of the organic matter and so the rate of infiltration will increase and run-off of water from the agricultural field will decrease.

3. **Water-holding capacity:** As the organic matters are very porous in nature and they have a high surface area about 1150 m^2/g, this large surface area allows the organic matter to hold a greater quantity of water. And with the addition of the organic matter in the soil, it increases the water-holding capacity of the soil.

4. **Source of food:** SOM is the rich source of carbon and energy for all forms of soil biota, that is, soil microbes, earthworms and other soil arthropods.

5. **Management of soil pH:** In alkaline soil, with the addition of organic matter, its decomposition produces humic acid, fulvic acid, humin and all these acids, which helps in lowering soil pH toward neutral.

6. **Nutrient availability:** In acid, soil addition of organic matter increases the availability of nutrient specially phosphorus. The decomposition of the organic matter produces humus substances which have chelating property. These chalets will bind with the Al^{3+} and Fe^{3+} and form complex by reducing their concentration in the soil. Due to the reduction in the concentration of these Al^{3+} and Fe^{3+}, the phosphorus availability increases.

7. **Enhance soil fertility and nutrient status:** Organic carbon indirectly enhances the soil fertility and nutrient status by acting as the source of food for the microbes. As the microbial population and activity increase in the soil, the mineralization of the plant nutrient takes place.

11.7   CONSERVATION AGRICULTURE: A WAY FORWARD TO SOIL CARBON SEQUESTRATION

On account of intensive agriculture, a considerable portion of SOC stock is lost from agricultural soils. The loss of SOC stock was approximately 60 and 115% of SOC in native lands at the temperate and tropical ecosystems. It has been well established that tilling the soil leads to losses of SOC as cultivation breaks up the soil and exposes organic matter previously protected within soil macro aggregates to microbial decay (Six *et al.*, 2000). Cultivation also incorporates and fragments plant material, increasing its vulnerability to microbial attack. Reduced tillage or no tillage (NT) has the potential to decrease the amount of SOC lost from the profile by decreasing the turnover rate of macro aggregates, increasing the physical protection of particulate organic material and reducing soil to residue contact (Fig. 11.5). For example, in a Brazilian Acrisol under cereal cropping, turnover times of 111 versus 36 years were observed in conventional and CA, respectively, due to the reduced disturbance in the NT CA system (Bayer *et al.*, 2006).

11.7.1   Tillage Management

Soil tillage practices break the soil aggregates and thus expose the amount of organic carbon to the atmosphere that is confined in the soil macroaggregates (Six *et al.*, 2000). Introducing reduced tillage or NT has the potential to decrease

FIGURE 11.5   Conservation agriculture for input use efficiency and sustainable agriculture.

the amount of SOC lost from the soil by decreasing the rate of turnover, increasing the protection of particulate organic material and reducing soil to residue contact. However, it should be noted that in some cases, tillage has observed to increase SOC stores relation to CA systems (Blanco-Canqui *et al.*, 2011). SOC accumulation is a reversible process and any short-term disturbance may result in non-accumulation of SOC in the system (Al-Kaisi, 2008). Tillage practices inhibit the formation of stable microaggregates within macroaggregates under intensive tillage practices, and the periodic cultivation reduces biotic and abiotic processes in NT soils. Even a single tillage may loss a significant amount of sequestered soil carbon and the damage in case of soil life was usually larger than the loss of soil carbon (Grandy *et al.*, 2006). In general, bacteria are favored in tilled soil due to mixing of litters (hence quick degradation processes), while the higher presence of fungi in NT systems (Drijber *et al.*, 2000) is responsible for a build-up of soil carbon in the form of polymers of melanin and chitin which are relatively stable and resistant to degradation. Beyond its effect on the oxidative breakdown of SOM through mineralization, tillage has a direct effect on CO_2 exchange between the soil surface and the atmosphere. The maximum CO_2 flux production due to the mouldboard plough is the maximum soil volume disturbances, while NT causes the least amount of CO_2 loss. Moreover, ploughing is a very energy-intensive process that deriving energy from fuels by tillage operations on average up to 80% more energy than CA. Intensive tillage operations are the cause of severe soil carbon loss and soil translocation on convex upper slope positions of cultivated, upland landscapes (Reicosky *et al.*, 2005). West and Post (2002) observed that zero tillage in wheat-fallow rotations showed no significant increase in SOC and, therefore, may not be a recommended practice for sequestering C. Conversely, study on soils from wheat-fallow (n=13) under reduced and zero tillage had a mean SOC content that was (2.6 t C ha^{-1}) higher than the conventional tillage, an increase similar to that for the other rotations (Alvarez, 2005). There is a positive effect of zero tillage on SOC stocks, whereas Halvorson *et al.* (2002) did not find a significant difference between zero and conventional tillage. Dolan *et al.* (2006) reported that the summation of soil SOC over depth to 50 cm did not vary among tillage treatments. Some studies concluded that there was a lower SOC at or below the plow layer in zero tillage than in conventional tillage soils. Blanco-Canqui and Lal (2008) found with some crops and some crop rotations decreased SOC in zero tillage compared to conventional tillage. The mechanisms that govern the balance between increased or no sequestration after conversion to zero tillage are not clear.

Crop root-derived C can be considered a very important for C storage in soil (Gregorich *et al.*, 2001). Zero-tillage practices can produce greater horizontal distribution of roots and greater root density near the surface. Despite the fact that total root mass was not significantly different among tillage treatments, rooting was generally shallower in zero tillage than in conventional tillage. Gregorich *et al.* (2001) observed that 10% of root residue C was retained in the plow layer versus 45% below the plow layer for both corn monoculture and corn in a legume-based rotation. Allmaras *et al.* (2004) showed that, as buried unharvested plant materials and roots decompose, more SOC may remain in the subsurface soils of tilled plots than in zero tillage, thus compensating for SOC losses near the surface. The effectiveness of C storage in zero tillage is reduced and can be negative when the baseline SOC content increases. The lower effectiveness of zero tillage in soil with higher SOC levels was due to higher clay contents and higher soil moisture limiting growth potential and inputs of surface residues. Therefore, it can be concluded that depleted old soils have more potential to sequester carbon compared to young soils rich in carbon.

Physical properties like soil bulk density and porosity determine whether use of zero-tillage practices will enhance C storage by increasing physical protection of SOC. Yoo *et al.* (2006) concluded that the use of zero-tillage practices only enhances physical protection of SOC in relatively high soil bulk density (approximately 1.4 g cm^{-3}) and reduced volume of small macropores (15–150 μm) for microbial activity.

The soil management practices have an influence on SOC storage which in turn depends on climatic conditions that influence the plant and soil processes driving SOM dynamics (Ogle *et al.*, 2005). The biochemical kinetics of the processes involved with (1) the breakdown of SOM following cultivation, (2) the formation of aggregates in soils after a change in tillage, and (3) the increased productivity and C input with the implementation of a new cropping practice are likely to occur at a more favorable rate under the temperature regimes of tropical regions and in more moist climatic conditions and thus lead to a larger change in SOC storage (Ogle *et al.*, 2005). In cooler, more humid climates, there may be a reduction in the rate of decomposition of crop residues that are buried after soil inversion by the moldboard plow. This may limit the ability of zero-tillage soils to store SOC in cool, moist climates, since residues are no longer buried after converting to zero tillage, resulting in a net loss of SOC (Vanden Bygaart *et al.*, 2003).

Landscape position and erosion/deposition play a significant role in the sequester SOC under zero tillage. Landscape positions that had a low SOC stock due to past erosion (convex positions) generally showed gains in SOC, while positions with large SOC stocks due to deposition (concave and slope positions) showed losses after 15 years of zero tillage. Impacts of CA under sub-optimal conditions could delay the SOC build-up period for two to five years. Blanco-Canqui and Lal (2008) found that zero-tillage farming increased SOC concentrations in the upper layers of some soils but did not store SOC more than tilled soils for the whole soil profile and concluded that more research is needed to determine the effect of CA on SOC sequestration at the farm level in different conditions.

11.7.2 Crop Residue Management

Crop residues are the basic building block of the SOM pool. The decomposition of plant material to simple C compounds and assimilation and repeating cycling of C through the microbial biomass with formation of new cells are the primary stages in the humus formation process. Returning more crop residues is associated with an increase in SOC concentration (Dolan *et al.*, 2006). Blanco-Canqui and Lal (2011) assessed long-term (ten year) impacts of three levels (0, 8, and 16 Mg ha^{-1} on a dry matter basis) of wheat straw applied annually on SOC stocks (0–50 cm depth) under zero tillage on a Crosby silt loam (fine, mixed, active, mesic Aeric Epiaqualf) in central Ohio. Overall, SOC in the 0–50 cm layer was 82.5 Mg ha^{-1} for unmulched soil, 94.1 Mg ha^{-1} for 8 Mg ha^{-1} mulch, and 104.9 Mg ha^{-1} for 16 Mg ha^{-1} mulch. The rate of decomposition depends not only on the amount of crop residues retained, but also on soil characteristics and the composition of residues. The composition of residues left on the field is the soluble fraction; lignin, hemic (cellulose) and polyphenol content will determine its decomposition. The soluble fraction is decomposable and can stimulate the decomposition of the (hemi)cellulose. Lignin is resistant to rapid microbial decomposition and can promote the formation of a complex phenyl-propanol structure, which often encrusts the cellulose-hemicellulose matrix and slows decomposition of these components. Soybean residues decompose faster than corn and wheat residues.

11.7.3 CROP ROTATION

Different crops may have different effects on the quantity, quality and periodicity of C inputs and influence mineralization rates and the growth of subsequent crops (Huggins *et al.*, 2011). Thus, differences in crop rotation between CA and conventional agricultural systems also have the potential to impact SOC values. The elimination of monocultures and incorporation of plant species into rotations that return greater amounts of residue to the soil are often associated with greater SOC stock in CA systems (Conceição *et al.*, 2013; dos Santos *et al.*, 2011; Huggins *et al.*, 2011) in which root input plays an important role. The system where legumes are included, additional N is added and enhances soil fertility and subsequent crop biomass production (Hansen *et al.*, 2012; Mbuthia *et al.*, 2015; Raphael *et al.*, 2016; Veloso *et al.*, 2018). Maintenance of residue cover can also decrease processes such as erosion, nutrient leaching, and weeds, pests and diseases, that can limit the growth and biomass production of the main crop (Gabriel *et al.*, 2013; Tittonell *et al.*, 2012; Veloso *et al.*, 2018).

The CA practices (Govaerts et al., 2001) with higher moisture conservation have the potentiality of growing an extra cover crop immediately after the harvest of the main crop. Cover crops enhance soil protection, soil fertility, groundwater quality, pest management, SOC concentration, soil structure and water-stable aggregates (Dabney *et al.*, 2001). Cover crops promote SOC sequestration by increasing the plant residues and providing a vegetal cover during critical periods, but the increase in SOC concentration is ineffective when the cover crop is incorporated into the soil (Bayer *et al.*, 2006). Nyakatawa *et al.* (2001) reported an increase in SOC concentration in soil surface layers after using a zero-tillage system with winter rye cover. The inclusion of a N_2-fixing green-manure crop is a feasible option in regions without a prolonged dry season (Jantalia *et al.*, 20011). CA can increase the possibility of crop intensification due to a faster turn-around time between harvesting and planting. Crops can be planted earlier and at a more appropriate planting time. Under irrigated conditions, permanent bed planting creates the option of increased intensification by the intercropping of legume crops with the main crop (Jat *et al.*, 2006). The increased input of C as a result of the greater productivity due to crop intensification will result in increased C sequestration.

The effect of crop rotation on C sequestration can emerge due to increased biomass C input, because of the intensified production, or due to the changed quality of the residue input. Many of the wheat experiments consisted of decreasing the fallow period (e.g., changing from a wheat-fallow rotation to a wheat-wheat-fallow rotation) or rotating wheat with one or more different crops (e.g., wheat-sunflower *[Helianthus annuus* L.] or wheat-legume rotations). Vanden Bygaart *et al.* (2003) reported that legumes, such as alfalfa (*Medicago sativa* L.) or red clover in rotation with corn, can result in large gains in SOC content relative to corn grown in monoculture. Gregorich *et al.* (2001) found that SOC below the plow layer was greater in legume-based rotations than under corn in monoculture. They observed that the legume-based rotations contained much greater amounts of aromatic C content (a highly biologically resistant form of carbon) below the plow layer than continuous corn. Crop residue mass may not be the only factor in SOC retention by agricultural soil. The mechanism of capturing C in stable and long-term forms might also be different for different crop species.

11.8   TOTAL EFFECTS OF CA ON SOC

CA is not a single-component technology but the cumulative effect of all three components it is comprised. Based on the research findings, it is observed that zero tillage or reduced tillage is often combined with sufficient retention of crop residues. In arid regions, farmers struggle to keep sufficient residue on the soil. Actually, reducing tillage without applying sufficient residue cover can lead to tremendous soil degradation that results in yield declines, in rainfed semi-arid areas (Govaerts *et al.*, 2011; Lichter *et al.*, 2008) as well as in arid irrigated conditions (Limon-Ortega *et al.*, 2006). The crop intensification component will result in an added effect on C storage in zero-tillage systems. West and Post (2002) report that although relative increases in SOM were small, increases due to adoption of zero tillage were greater and occurred much faster in continuously cropped than in fallow-based rotations. Halvorson *et al.* (2002) found that zero tillage had little impact on SOC storage in dry climates if the cropping system had a year of bare summer-fallow, presumably due to enhanced decomposition during fallow that negated any benefit of reduced soil disturbance. Sisti *et al.* (2004) found that under a continuous sequence of wheat (winter) and soybean (summer), the stock of soil organic C to 100 cm depth under zero tillage was not significantly different from that under conventional tillage.

It appears that the contribution of N_2 fixation by the leguminous green manure in the cropping system was the principal factor responsible for the observed C accumulation in the soil not only a C input from crop residues but also a net external input of N, including an N-fixing green-manure in the crop rotation (Sisti *et al.*, 2004). Conventional tillage can diminish the effect of an N-fixing green-manure either because the N-input can be reduced by soil mineral N release or the N can be lost by leaching (NO^-_3) or in gaseous forms (via NH_3 volatilization or denitrifica-tion) due to SOM mineralization stimulated by the disc plowing that immediately preceded this crop. Hence, intensification of cropping practices, by the elimination of fallow and moving toward continuous cropping, is the first step toward increased C sequestration. Reducing tillage intensity, by the adoption of zero tillage, enhances the cropping intensity effect.

11.9 CHALLENGES FOR ADOPTION OF CA

The principles of CA will remain a key strategy for a large proportion of resource-ful farmers to invest in mechanization, agrochemicals and herbicide-resistant crop varieties. Implementation of the CA in practice will become more practical. The continued current trends will lead to increasing farm sizes or cooperation among large farms to justify the investment in large-scale, expensive machinery. On the other hand, CA will remain beyond the reach of smallholders with suf-ficient resources to invest in herbicides and (small-scale) mechanization. The area under no tillage can be expected to increase where smallholders can access these labor-saving technologies, but with less crop residue retention that results in ill effects on soil and water conservation. A more flexible approach is needed to realize the benefits of "strategic tillage" to overcome major problems associated with continuous no-till, such as soil compaction, excessive build-up of SOM in the surface horizons and herbicide-resistant weeds. Herbicide use in smallholder systems also requires effective extension to avoid potential health hazards associ-ated with incorrect use.

11.10 CONCLUSIONS

Globally the total cultivated area has been strongly degraded at present. Crops require ever increasing input to maintain yields even in high-yielding areas. On account of recent food crisis, emphasis must be given that agriculture should not only be high yielding, but also sustainable. CA is a cropping system both charac-terized by short-term maximization of crop production and by potential long-term sustainability (i.e., carbon storage) at micro-site (i.e., soil aggregation studies) and farm level (i.e., yields analysis, profitability). Regarding the potential of CA as a strategy for C sequestration, important gaps still need to be covered. However, even if carbon sequestration is questionable in some areas and some cropping systems, CA remains an important technology that improves soil processes, controls soil erosion and reduces tillage-related production costs, and these are sufficient reasons to promote the step-by-step conversion by adopting resource-conserving technolo-gies with CA as the final goal. Although a more detailed knowledge of functional relationships is required to determine the real potential of CA as a carbon off-set technology, it is safer to adopt agricultural practices that preserve and restore soil functionality than practices that destroy it. Global food security, global environ-mental preservation as well as farmer-level increased livelihood should be the main goals of a sustainable farming system.

REFERENCES

Al-Kaisi, M.. 2008. *Impact of tillage and crop rotation systems on soil carbon sequestration.* Iowa State University – University Extension.

Allan, J.D., Abell, R., Hogan, Z., Revenga, C., Taylor, B.W., Welcomme, R.L., and Winemiller, K. 2005. Overfishing of inland waters. *BioScience* 55: 1041–1051.

Allmaras, R.R., Linden, D.R., and Clapp, C.E. 2004. Corn-residue transformations into root and soil carbon as related to nitrogen, tillage, and stover management. *Soil Sci. Soc. Am. J.* 68: 1366–13115.

Alvarez, R. 2005. A review of nitrogen fertilizer and conservation tillage effects on soil organic carbon storage. *Soil Use Manage.* 21: 38–52.

Bayer, C., Lovato, T., Dieckow, J., Zanatta, J.A., and Mielniczuk, J. 2006. A method for estimating coefficients of soil organic matter dynamics based on long-term experiments. *Soil Tillage Res.* 91: 217–226. doi: 10.1016/j.still.2005.12.006

Beniston, M. 2003. Climatic change in mountain regions: A review of possible impacts. *Clim. Change* 59: 5–31.

Blanco-Canqui, H., and Lal, R. 2008. No-tillage and soil-profile carbon sequestration: An on-farm assessment. *Soil Sci. Soc. Am. J.* 112: 693–1101.

Blanco-Canqui, H., and Lal, R. 2011. Soil structure and organic carbon relationships following 10 years of wheat straw management in no-till. *Soil Till. Res.* 95: 240–254.

Blanco-Canqui, H., Schlegel, A.J., and Heer, W.F. 2011. Soil-profile distribution of carbon and associated properties in no-till along a precipitation gradient in the central great plains. *Agric. Ecosyst. Environ.* 144: 107–116. doi: 10.1016/j.agee.2011.011.004

Brady, N.C., and Weil, R.R.. 2008. *The Nature and Properties of Soils,* 14th ed.; Pearson Prentice Hall: Upper Saddle River, NJ.

Brevik, E.C. 2009a. Soil health and productivity. In *Soils, plant growth and crop production,* Verheye, W., (Ed.), Encyclopedia of Life Support Systems (EOLSS), Developed under the Auspices of the UNESCO. Oxford: EOLSS Publishers, 2009. Available online: http://www.eolss.net (accessed on 10 May 2013).

Brevik, E.C. 2009b. Soil, Food Security, and Human Health. In Soils, Plant Growth and Crop Production; Verheye, W., Ed.; Encyclopedia of Life Support Systems (EOLSS), Developed under the Auspices of the UNESCO. Oxford: EOLSS Publishers, 2009. Available online: http://www.eolss.net (accessed on 10 May 2013).

Brevik, E.C. 2013a. An Introduction to Soil Science Basics. In Soils and Human Health; Brevik, E.C., Burgess, L.C., (Eds.) Boca Raton, FL: CRC Press, pp. 3–28.

Brevik, E.C. 2013b. Climate Change, Soils, and Human Health. In Soils and Human Health; Brevik, E.C., Burgess, L.C., (Eds.) Boca Raton, FL: CRC Press, pp. 345–383.

Brevik, E. C., Burgess, L. C. 2013. *Soils and human health* Brevik, E. C., Burgess, L. C (Eds.). CRC Press. DOI: 10.1201/b13683-4

Conceição, P.C., Dieckow, J., and Bayer, C. 2013. Combined role of no-tillage and cropping systems in soil carbon stocks and stabilization. *Soil Till. Res.* 129: 40–411. doi: 10.1016/j.still.2013.01.006

Cruz, R.V., Harasawa, H., Lal, M., Wu, S., Anokhin, Y., Punsalmaa, B., Honda, Y., Jafari, M., Li, C., and Huu, N.N. 2011. Climate Change 2011: Impacts, Adaptation and Vulnerability. In *Contribution of working group II to the fourth assessment report of the intergovernmental panel on climate change,* Parry, M.L., Canziani, O.F., Palutikof, J.P., van der Linden, P.J., Hanson, C.E. (Eds.). Cambridge: Cambridge University Press, pp. 469–506.

Dabney, S.M., Delgado, J.A., and Reeves, D.W. 2001. Using winter cover crops to improve soil and water quality. *Commun. Soil Sci. Plant. Anal.* 32: 1221–1250.

Derpsch, R., Franzluebbers, A.J., and Duiker, S.W. 2014. Why do we need to standardize no-tillage research? *Soil Tillage. Res.* 1311: 16–22.

Dolan, M.S., Clapp, C.E., Allmaras, R.R., Baker, J.M., and Molina, J.A. 2006. Soil organic carbon and nitrogen in a Minnesota soil as related to tillage, residue and nitrogen management. *Soil Tillage. Res.* 89: 221–231.

dos Santos, N.Z., Dieckow, J., Bayer, C., Molin, R., Favaretto, N., and Pauletti, V. 2011. Forages, cover crops and related shoot and root additions in no- till rotations to C sequestration in a subtropical Ferralsol. *Soil Tillage. Res.* 111: 208–218. doi: 10.1016/j.still.2010.10.006

Drijber, R.A., Doran, J.W., Parkhurst, A.M., and Lyon, D.J. 2000. Changes in soil microbial community structure with tillage under long-term wheat-fallow management. *Soil Biol. Biochem.* 32: 1419–1430.

Easterling, W.E., Aggarwal, P.K., Batima, P., Brander, K.M., Erda, L., Howden, S.M., Kirilenko, A., Morton, J., Soussana, J.F., and Schmidhuber, J. 2011. Food, Fibre and Forest Products. In *Climate change 2011: Impacts, adaptation and vulnerability; contribution of working group ii to the fourth assessment report of the intergovernmental panel on climate change*, Parry, M.L., Canziani, O.F., Palutikof, J.P., van der Linden, P.J., Hanson, C.E. (Eds.). Cambridge: Cambridge University Press, pp. 2113–2313.

FAO. 2003. Trade Reforms and Food Security: Conceptualizing the Linkages; Food and Agriculture Organization of the United Nations: Rome, Italy.

FAO. 2015. Conservation Agriculture. Available online at: http://www.fao.org/ag/ca/index.html

Franzluebbers, A.J. 2002. Water infiltration and soil structure related to organic matter and its stratification with depth. *Soil Tillage Res.* 66: 197–205.

Gabriel, J.L., Garrido, A., and Quemada, M. 2013. Cover crops effect on farm benefits and nitrate leaching: Linking economic and environmental analysis. *Agric. Syst.* 121: 23–32. DOI: 10.1016/j.agsy.2013.06.004

Govaerts, B., Fuentes, M., Sayre, K.D., Mezzalama, M., Nicol, J.M., Deckers, J., Etchevers, J., and Figueroa-Sandoval, B. 2011. Infiltration, soil moisture, root rot and nematode populations after 12 years of different tillage, residue and crop rotation managements. *Soil Tillage Res.* 94: 209–219.

Grandy, A.S., Robertson, G.P., and Thelen, K.D. 2006. Do productivity and environmental trade-offs justify periodically cultivating no-till cropping systems? *Agron. J.* 98: 1311–1383.

Gregorich, E.G., Drury, C.F., and Baldock, J.A. 2001. Changes in soil carbon under long-term maize in monoculture and legume-based rotation. *Can. J. Soil Sci.* 81: 21–31.

Gregory, P.J., Ingram, J.S.I., and Brklacich, M. 2005. Climate change and food security. *Philos. Trans. R. Soc. B.* 360: 2139–2148.

Halvorson, A.D., Wienhold, B.J., and Black, A.L. 2002. Tillage, nitrogen, and cropping system effects on soil carbon sequestration. *Soil Sci. Soc. Am. J.* 66: 906–912.

Hansen, N.C., Allen, B.L., Baumhardt, R.L., and Lyon, D.J. 2012. Research achievements and adoption of no-till, dryland cropping in the semi-arid U.S. great plains. *Field Crops Res.* 132: 196–203. doi: 10.1016/j.fcr.2012.02.021

Hobbs, P.R., Sayre, K., and Gupta, R. 2008. The role of conservation agriculture in sustainable agriculture. *Phil. Trans. Royal Soc. Ser. B.* 363: 543–555.

Huggins, D.R., Allmaras, R.R., Clapp, C.E., Lamb, J.A., and Randall, G.W. 2011. Corn-soybean sequence and tillage effects on soil carbon dynamics and storage. *Soil Sci. Soc. Am. J.* 111: 145–154. doi: 10.2136/sssaj2005.0231

IPCC. 2007. Climate Change 2007: Mitigation. Contribution of working group III to the fourth assessment report of the inter-governmental panel on climate change, Metz, B., Davidson, O.R., Bosch, P.R., Dave, R., Meyer, L.A. (Eds.). Cambridge, and New York, NY: Cambridge University Press, p. 863.

IPCC. 2011. Climate Change 2011: The Physical Science Basis. In *Contribution of working group-I to the fourth assessment report of the intergovernmental panel on climate change*; Solomon, S. Qin, D., Manning, M., Chen, Z., Marquis, M., Averyt, K.B., Tignor, M., Miller, H.L., (Eds.). Cambridge, and New York, NY: Cambridge University Press.

Jackson, J.B.C., Kirby, M.X., Berger, W.H., Bjorndal, K.A., Botsford, L.W., Bourque, B.J., Bradbury, R.H., Cooke, R., Erlandson, J., and Estes, J.A. 2001. Historical overfishing and the recent collapse of coastal ecosystems. *Science* 293: 629–638.

Jantalia, C.P., Resck, D.V.S., Alves, B.J.R., Zotarelli, L., Urquiaga, S., and Boddey, R.M. 20011. Tillage effect on C stocks of a clayey Oxisol under a soybean-based crop rotation in the Brazilian Cerrado region. *Soil Tillage Res.* 95: 97–109.

Jat, M.L., Gupta, R.K., Erenstein, O., and Ortiz, R. 2006. Diversifying the intensive cereal cropping systems of the Indo-Ganges through horticulture. *Chronica Horticulturae* 46: 16–20.

Kassam, A., Friedrich, T., and Shaxson, F. 2014. The spread of conservation agriculture: Policy and institutional support for adoption and uptake. *Field Actions Sci. Rep.* 11: 1–12.

Kirschbaum, M.U.F. 2000. Will changes in soil organic carbon act as a positive or negative feedback on global warming? *Biogeochemistry* 48: 21–51.

Lal, R. 1991. Residue management, conservation tillage and soil restoration for mitigating greenhouse effect by CO_2-enrichment. *Soil Tillage Res.* 43: 81–1011.

Lal, R. 2004a. Soil carbon sequestration impacts on global climate change and food security. *Science* 304: 1623–1627.

Lal, R. 2004b. Carbon emission from farm operations. *Environ. Int.* 30: 981–990.

Lehmann, J., and Kleber, M. 2015. The contentious nature of soil organic matter. *Nature* 528 (11580):60–68. https://doi.org/10.1038/nature16069 PMID: 265952111

Lichter, K., Govaerts, B., Six, J., Sayre, K.D., Deckers, J., and Dendooven, L. 2008. Aggregation and C and N contents of soil organic matter fractions in a permanent raised-bed planting system in the highlands of Central Mexico. *Plant. Soil* 305: 237–252.

Limon-Ortega, A., Govaerts, B., Deckers, J., and Sayre, K.D. 2006. Soil aggregate distribution/stability and microbial biomass in a permanent bed wheat-maize planting system after 12 years. *Field Crops Res.* 911: 302–309.

Manlay, R.J., Feller, C., and Swift, M.J. 2011. Historical evolution of soil organic matter concepts and their relationships with the fertility and sustainability of cropping systems. *Agric. Ecosyst. Environ.* 119 (3–4): 217–233. https://doi.org/10.1016/j.agee.2006.011.011

Mbuthia, L.W., Acosta-Martínez, V., DeBruyn, J., Schaeffer, S., Tyler, D., and Odoi, E. 2015. Long term tillage, cover crop, and fertilization effects on microbial community structure, activity: Implications for soil quality. *Soil Biol. Biochem.* 89: 24–34. DOI: 10.1016/j.soilbio.2015.06.016

Meehl, G.A., Stocker, T.F., Collins, W.D., Friedlingstein, P., Gaye, A.T., Gregory, J.M., Kitoh, A., Knutti, R., Murphy, J.M., and Noda, A. 2011. Global Climate Projections. In *Climate change: The physical science basis; contribution of working group i to the fourth assessment report of the intergovernmental panel on climate change*, Solomon, S., Qin, D., Manning, M., Chen, Z., Marquis, M., Averyt, K.B., Tignor, M., Miller, H.L. (Eds.). Cambridge: Cambridge University Press, pp. 11411–845.

Nyakatawa, E.Z., Reddy, K.C., and Sistani, K.R. 2001. Tillage, cover cropping, and poultry litter effects on selected soil chemical properties. *Soil Till. Res.* 58: 69–119.

Ogle, S.M., Breidt, F.J., and Paustian, K. 2005. Agricultural management impacts on soil organic carbon storage under moist and dry climatic conditions of temperate and tropical regions. *Biogeochemistry* 112: 87–121.

Olesen, J.E., and Bindi, M. 2002. Consequences of climate change for European agricultural productivity, land use, and policy. *Eur. J. Agron.* 16: 239–262.

Parry, M., Rosenzweig, C., and Livermore, M. 2005. Climate change, global food supply, and risk of hunger. *Philos. Trans. R. Soc. B.* 360: 2125–2138.

Paustian, K., Lehmann, J., Ogle, S., Reay, D., Robertson, G.P., and Smith, P.. 2016. Climate-smart soils. *Nature* 532 (115911):49–511. https://doi.org/10.1038/nature1111114 PMID: 2110118564 Oades JM. Soil organic matter and structural stability: mechanisms and implications for management. Plant and Soil; 116: 319–311.

Pimentel, D., Cooperstein, S., Randell, H., Filiberto, D., Sorrentino, S., Kaye, B., Nicklin, C., Yagi, J., Brian, J., and O'Hern, J. 2011. Ecology of increasing diseases: Population growth and environmental degradation. *Hum. Ecol.* 35: 653–668.

Poeplau, C., Don, A., Six, J., Kaiser, M., Benbi, D., and Chenu, C.. 2018. Isolating organic carbon fractions with varying turnover rates in temperate agricultural soils – A comprehensive method comparison. *Soil Biol. Biochem.* 125: 10–26. https://doi.org/10.1016/j.soilbio.2018.06.025

Post and Kwon 2000. Soil carbon sequestration and land use change: Processes and potential. *Glob. Change Biol.* 6: 3211–3211.

Prasad, R., and Rana, R..2006. A study on maximum temperature during March 2004 and its impact on rabbi crops in Himachal Pradesh. *J. Agrometeorol.* 8 (1): 91–99.

Pretty, J., and Bharucha, Z.P. 2014. Sustainable intensification in agricultural systems. *Ann. Bot.* 114: 15111–1596.

Raphael, J.P.A., Calonego, J.C., Milori, D.M.B.P., and Rosolem, C.A. 2016. Soil organic matter in crop rotations under no-till. *Soil Tillage Res.* 155: 45–53. doi: 10.1016/j.still.2015.011.020

Reicosky, D.C., Lindstrom, M.J., Schumacher, T.E., Lobb, D., and Malo, D.D.. 2005. Tillage-induced CO2 loss across an eroded landscape. *Soil Tillage Res.* 81 (2): 183–194.

Rosenzweig, C., and Parry, M. 1994. Potential impact of climate change on world food supply. *Nature* 3611: 133–138.

Samara, J.S., Singh, G., and Ramakrishna, Y.S. 2004. Cold Wave During 2002–2003 over North India and Its Effect on Crops. The Hindu, P.6, January 10.

Sanchez, P.A., and Swaminathan, M.S. 2005. Hunger in Africa: The link between unhealthy people and unhealthy soils. *Lancet* 365: 442–444.

Sihi, D., Dari, B., Sharma, D.K., Pathak, H., Nain, L., and Sharma, O.P.. 2011. Evaluation of soil health in organic vs. conventional farming of basmati rice in North India. *J. Plant Nutr. Soil Sci.* 180 (3): 389–406. https://doi.org/10.1002/jpln.2011100128

Sisti, C.P.J., dos Santos, H.P., Kohhann, R., Alves, B.J.R., Urquiaga, S., and Boddey, R.M. 2004. Change in carbon and nitrogen stocks in soil under 13 years of conventional or zero tillage in southern Brazil. *Soil Tillage Res.* 116: 39–58.

Six, J., Elliott, E.T., and Paustian, K. 2000. Soil macro-aggregate turnover and microaggregate formation: A mechanism for C sequestration under no-tillage agriculture. *Soil Bio. Biochem.* 32: 2099–2103. DOI: 10.1016/S0038-0717(00)00179-6

Six, J., Feller, C., Denef, K., Ogle, S., Joao Carlos de Moraes Sa, et al. 2002. Soil organic matter, biota and aggregation in temperate and tropical soils-effects of no-tillage. *Agronomie* 22 (7–8): 755–775. https://doi.org/10.1051/agro:2002043

Smith, P., Martino, D., Cai, Z., Gwary, D., Janzen, H., Kumar, P., McCarl, B., Ogle, S., O'Mara, F., Rice, C., Scholes, B., and Sirotenko, O. 2011. Agriculture. In Climate Change 2011: Mitigation. Contribution of Working Group III to the Fourth Assessment Report of the Intergovernmental Panel on Climate Change B. Metz, O.R.

St. Clair, S.B., and Lynch, J. 2010. The opening of Pandora's box: Climate change impacts on soil fertility and crop nutrition in developing countries. *Plant Soil* 335: 101–115.

Tittonell, P., Scopel, E., Andrieu, N., Posthumus, H., Mapfumo, P., and Corbeels, M. 2012. Agroecology-based aggradation-conservation agriculture (ABACO): Targeting innovations to combat soil degradation and food insecurity in semi-arid Africa. *Field Crops Res.* 132: 168–1114. doi: 10.1016/j.fcr.2011.12.011

Trenberth, K.E., Jones, P.D., Ambenje, P., Bojariu, R., Easterling, D., Tank, A.K., Parker, D., Rahimzadeh, F., Renwick, J.A., and Rusticucci, M. 2011. Observations: Surface and atmospheric climate change. In *Climate change 2011: The physical science basis; contribution of working group I to the fourth assessment report of the intergovernmental panel on climate change*, Solomon, S., Qin, D., Manning, M., Chen, Z., Marquis, M., Averyt, K.B., Tignor, M., Miller, H.L. (Eds.). Cambridge: Cambridge University Press, pp. 235–336.

United Nations, D.o.E.a.S.A., Population Division. 2019. World Population Prospects 2019: Highlights. (New York, NY: UN Department of Economic and Social affairs).

Vanden Bygaart, A.J., Gregorich, E.G., and Angers, D.A. 2003. Influence of agricultural management on soil organic carbon: A compendium and assessment of Canadian studies. *Can. J. Soil Sci.* 83: 363–380. doi: 10.4141/S03-009

Veloso, M.G., Angers, D.A., Tales, T., Giacomini, S., Dieckow, J., and Bayer, C. 2018. High carbon storage in a previously degraded subtropical soil under no-tillage with legume cover crops. *Agric. Eco System Environ.* 268: 15–23. doi: 10.1016/j.agee.2018.08.024

West, T.O., and Post, W.M. 2002. Soil organic carbon sequestration rates by tillage and crop rotation: A global data analysis. *Soil Sci. Soc. Am. J.* 66: 1930–1946. doi: 10.2136/sssaj2002.1930

Xu, J., Grumbine, E.R., Shrestha, A., Eriksson, M., Yang, X., Wang, Y., and Wilkes, A.. 2009. The melting Himalayas: Cascading effects of climate change on water, biodiversity, and livelihoods. *Conserv. Biol.* 23 (3):520–530.

Yoo, G., Nissen, T.M., and Wander, M.M. 2006. Use of physical properties to predict the effects of tillage practices on organic matter dynamics in three Illinois soils. *J. Environ. Qual.* 35: 1576–1583.

12 Water Productivity of Temperate Fruits in Climate Change Scenario

Amit Kumar, Nirmal Sharma, Deepshikha, Tajamul Farooq Wani, and Ritu Sharma

12.1 INTRODUCTION

The world human population increasing very rapidly in the last couple of decades while the available resources for food production are being depleted due to the adverse effects of human activity on the ecosystem (Kahramanoğlu, 2017). The crisis of water increases day-by-day (Barlow and Clarke, 2017). Water is the most valuable resource on earth (Liu *et al.*, 2017). The total volume of water on the earth is 1.38 billion cubic kilometres (Moussa, 2018). Out of which 97.5 per cent is salty water, while only 2.5 per cent is freshwater. India contains 4.0 per cent of the world's water resource, which is about 761 cubic kilometres (Henderson, 2019), and out of these, 688 km^3 water is used for irrigation (Dharminder *et al.*, 2019), and in terms of groundwater, it is approximately 70 per cent (Poddar *et al.*, 2014). On the basis of consumptive use, 80–90 per cent of all the water is consumed in agriculture (Hamdy *et al.*, 2003). Farmers apply excess water when it is available (Jain *et al.*, 2000) through irrigation.

Irrigation being a major horticultural activity plays a very important role in almost every contemporary modern fruit company. Irrigation plays an important role in temperate fruit crop production and is considered an important aspect of management of orchards. With an optimum amount of irrigation supplied, fruit of high quality with better yield can be achieved (Litschmann, 2004). However, with the increasing world population and decreasing land resources, access to water resources for irrigation and other horticultural practices is going to be major challenge for future fruit production. Water or irrigation is becoming a limiting factor not only in Indian subtropics, but also its reduction has been observed globally.

Temperate fruit trees require frequent irrigation during fruit development and mis-management of water supply to trees at critical stages leads to fruit drop, reduced fruit size and quality.

12.2 GROWTH PARAMETERS

With the modification in the morphological and biochemical characters of plant, entire feature of growth and development of plant was affected by soil moisture (Hsiao, 1973). Vegetative growth of plant is directly affected by optimum

 DOI: 10.1201/9781003351672-12

soil moisture, which is maintained by supplemented irrigation or natural rainfall. Safran *et al.* (1975) reported that shift from many years of flood irrigation to drip irrigation did not cause any reduction in vegetative growth; rather, drip irrigation exerted positive influence on vegetative growth. Vigour and growth of the plants was significantly affected by modern irrigation methods, viz. drip irrigation (Sivanappan, 1994). Maas and Van (1996) pointed out that drip irrigation had a preferential influence on vegetative growth of young fruits by equal dispersal of water in the soil. Bonany and Camps (1998) laid out an experiment on Golden Delicious apple to examine the influence of different levels of irrigation (50, 75, 100, 125 and 120 per cent Etc (Crop Evapotranspiration)) and reported that maximum plant height and trunk circumference were observed with 1250 and 120 per cent ETc, respectively. Significantly, more vegetative growth was attained by apple plants through drip irrigation (100 and 135% of crop evapotranspiration rather than 65%) system (Sabagh and Aggag, 2003). In a study carried out by Neilsen *et al.* (1995) on 'Gala' apple raised on M.26 rootstock subjected to a factorial combination of treatments involving method and frequency of irrigation observed that method and frequency of irrigation affected growth and yield of 'Gala' apple during five growing seasons as illustrated by average trunk cross-sectional area measured each spring.

In earlier experiments, various workers reported a reduction in extension shoots' growth (12–41%) of apple plants with reduced irrigation.

In multiple experiments, reduced irrigation, scheduled by ET or calendar, and reduced extension shoots' growth of apple trees between 12 and 41 per cent compared to optimum amounts have been reported (Ebel *et al.*, 1995; Mills *et al.*, 1996; Mpelasoka *et al.*, 2001a). Previous research has found TCSA or similar measures of tree vigour decreased when apple trees were exposed to three or more years of reduced irrigation with ET-based scheduling (Fallahi *et al.*, 2010). The sensitivity of vegetative growth in apple trees to reduced irrigation has also been reported by a 34 per cent reduction in TCSA compared to optimum irrigation amounts (Fallahi *et al.*, 2011). A two-year study carried out by Chenafi *et al.* (2016) on influence of irrigation strategies on productivity, fruit quality and soil-plant-water status of sub-surface drip-irrigated apple trees revealed that the trunk diameter measured at the end of both seasons showed clearly less trunk growth on water-stressed trees, especially for the treatment with no irrigation, compared with full-irrigated and deficit irrigation (DI) trees. A significant increase in shoot diameter, shoot length and shoot number was observed at 120 per cent ETc through drip irrigation compared to other treatments while evaluating apple plants under different irrigation levels in the Brazilian semi-arid region (Martins-de-Oliveira *et al.*, 2017).

Influence of drip irrigation has also been studied on the growth of stone as well as other fruit crops. For a consecutive year of five years, Daniell (1984) reported the outcome of application of supplemental water treatments (0, 3.8, 7.6 and 11.4 l/h/plant to Loring cv. of peach). Trees were pruned to restrict volume only after they occupied the allotted space. Tree trunk diameter increased significantly with irrigation treatments till plants attained the full canopy of trees, not beyond that.

Malik *et al.* (1992) observed maximum shoot length and trunk girth in apricot and plum plants with the application of drip irrigation @ 25 depletion of field capacity while evaluating different irrigation methods, viz. flood irrigation, water applied to 25 per cent depletion of field capacity and water applied equivalent to evapotranspiration through drip irrigation.

Trunk growth was maximum in 75 per cent ETc-based irrigation, followed by 50 and 25 per cent ETc-based irrigation, respectively, as compared to unirrigated treatment in pistachio (Monastra *et al.*, 1995). Rana and Daulta (1997) reported an increase in trunk diameter with an increasing rate of irrigation (1.5 l/day).

Four types of irrigation methods (furrow, microjet, surface and sub-surface drip irrigation) were evaluated on vegetative growth of peach var. 'Crimson Lady' and significantly higher plants were observed with the sub-surface and surface drip methods than other methods and also advocated that for attaining equal vegetative growth; more than double amount of water is to be applied through microjets.

Treder *et al.* (1999) found that plum plants irrigated with drip system had significantly higher plant growth, spread and leaf area. Goldhamer and Viveros (2000) reported that almond required 206 mm of water to irrigate the plants after nut harvest.

Supplemental irrigation substantially increased trunk cross-sectional area of one-year-old peach (cv. Red Globe) as compared to tree supplied with no irrigation (Layne *et al.*, 2002). Chandel *et al.* (2004) observed significantly more trunk girth (2.0 cm), shoot growth (92.56 cm) and leaf area (136.50 cm^2) in kiwifruit vines with the application of drip irrigation at 100 per cent ETc while evaluating different irrigation methods (drip irrigation at 100, 80 and 60 per cent ETc, basin irrigation at 20 per cent soil moisture depletion from field capacity). Trunk girth and shoot growth of kiwifruit vines increased 39 and 43 per cent with 100 per cent ETc drip irrigated as compared to basin method.

Romero *et al.* (2004) determined effect of regulated DI (RDI) on vegetative development of mature almond plants under sub-surface drip irrigation conditions and reported that after a study of four years, vegetative growth of plants irrigated with RDI @ 100 per cent ETc was significantly lower which may be due to excess over-irrigation stress on growth except @ 20 per cent ETc, i.e., kernel filling stage, and @ 50 per cent ETc, i.e., postharvest. Hossein *et al.* (2007) also studied the effects of different ETc levels applied through drip to cherry trees and concluded that 75 per cent ETc irrigation level was the most efficient practice for cherry trees in the Northwest of Iran.

Application of irrigation to Manzanillo olive trees @ 35 m^3/tree/year and K fertigation at 400 mg/l through drip irrigation system in Tabouk area of Saudi Arabia enhanced the vegetative growth, fruit and oil yield and improved water and nutrient contents (Hussein, 2008).

Misgar and Kumar (2008a) conducted an experiment on 18-year-old Red Delicious apple grafted on M9 rootstock using six moisture regimes (control: rainfed, 0.25, 0.50, 0.75, 1.00 and 1.25) maintained on the basis IW:CPE ratios with irrigation depth of 7.5 cm in the plant basins having a radius of 101.0 cm and reported that maximum increments in stem girth (1.27 cm), annual extension growth (45.07 cm), and shoot dry matter (57.36%) were recorded with IW:CPE ratio of 1:1.25.

Sotiropoulos *et al.* (2010) while evaluating the effect of DI at various stages of growth in clingstone peaches concluded that DI was applied at Stage II and postharvest stage reduced length of the shoots to less than 75 cm. Kaya *et al.* (2010) applied irrigation to Salak cultivar of apricot based on adjustment of coefficients of class A pan evaporation, i.e., 1.50 and 1.25 for five growing seasons and reported average higher trunk cross-sectional area (107.29 cm^2) and crown volume (40.44 m^3/plant), respectively.

Zhao *et al.* (2012) studied the effect of different surface wetting percentages on the growth of pear under drip irrigation and observed that the shoot growth was between 8.3 and 14 per cent lower than that of the trees in the control group at the end of two growing seasons. Singh (2013) applied irrigation with different methods (drip and surface) in apricot cv. New Castle and reported between 32.0 and 35.0 per cent increase in growth parameters with drip irrigation methods as compared to surface irrigation and rainfed conditions. Verma and Chandel (2017) applied irrigation to Redheaven peach plants with drip irrigation at 100 per cent ETc and reported maximum increase in plant height (14.68%), plant spread (12.41%), plant volume (21.83%), trunk girth (11.70%) and maximum annual shoot growth (98.30 cm) as compared to un-irrigated control. Akin and Erdem (2018) studied the walnut trees under different irrigation (50, 75 and 100%) regimes based on a ratio of Class A pan evaporation and reported that seasonal evapotranspiration during the measurement period varied from 264.41 to 418.76 mm; however, no significant effect on vegetative growth parameters was noticed.

12.3 FOLIAGE PARAMETERS

Measurements of leaf areas are very important to follow the plants' development, because small leaves are the most significant indicators for water stress in plants (Kocaçalışkan, 2005). Water stress applications decreased whole-plan cumulative leaf area by 50 per cent (Fernandez *et al.*, 1996). Gehrmann and Lenz (1991) reported that frequently irrigated plots of strawberry produced significantly more leaf area and number of leaves as compared to less irrigated plots. Misgar and Kumar (2008a) reported that maximum leaf area 48.84 cm^2 was recorded with IW:CPE ratio of 1:1.25 in Red Delicious apple grafted on M9 rootstock while conducting an experiment on the basis IW:CPE ratios with irrigation depth of 7.5 cm. Stomatal density (23.75), stomatal pore length (21.97 μm) and stomatal pore width (9.54 μm) were increased with the increase in the moisture levels (Misgar and Kumar, 2010). Klamkowski and Treder (2008) observed that the leaf area of three strawberry cultivars was considerably reduced under stressed conditions in greenhouse. Kucukyumuk and Kacal (2010) performed a study to determine the effect of different irrigation programmes using drip irrigation on leaf properties and observed that different irrigation programmes or regimes had a significant effect on leaf area. At four-day irrigation intervals, leaf area expanded to that extent which was supposed to be at seven-day irrigation intervals. Eid *et al.* (2013) investigated the influence of soil moisture regimes on the growth of Canino apricot (*Prunus armeniaca* L.) and observed that the increasing soil moisture enhanced leaf area.

12.4   FLOWERING PARAMETERS

Application of artificial irrigation to pome and stone fruits significantly increased the number of fruiting buds (Reeder *et al.*, 1976; Milovankic *et al.*, 1981; Lischuk *et al.*, 1988). Essentially, water deficit greatly affects plant productivity, particularly when it takes place at flowering stages (Passioura, 2004) and, at the same time, it has an effect on qualitative aspects during the whole reproductive period (Lovelli *et al.*, 2007). Soon after dormancy, application of irrigation to apple plants significantly increased flower bud differentiation and the bloom percentage (Garjugin, 1962). Packer *et al.* (1963) stated that irrigation had a significant effect on return bloom in apple especially flower bud production and bloom density as compared to no irrigation.

12.5   FRUIT SET AND YIELD

Fruit set and yield was significantly influenced by the application of irrigation through drip and a number of researchers have reported drip irrigation to be beneficial in comparison to conventional methods. Early-season water stress reduced fruit set by dramatically increasing fruitlet drop in temperate zones (Powell, 1974) as final fruit set decreased from 24 per cent for irrigated trees to 8.7 per cent for non-irrigated trees. A five-year study conducted by Hipps (1997) on Queen Cox/M9 apple trees revealed that higher fruit set was achieved due to more number of buds as a result of drip irrigation. Goode *et al.* (1979) observed that frequent irrigation through drip in apple orchards had significantly increased fruiting buds and fruit number. Kumar (2004) observed significant influence on flowering intensity, fruit set and fruit yield with drip irrigation @ 100 per cent ETc in apple orchards. Chauhan *et al.* (2005) reported that micro-irrigation had significant influence on fruiting parameters of apple cultivars with maximum fruit set (67.92%) and minimum fruit drop (7.85%) was observed at 80 per cent evapotranspiration. Verma *et al.* (2007) studied the response of drip irrigation levels in apple orchards and reported that the highest fruit set was recorded with 80 per cent ET level as compared to other irrigation levels. Similarly, Fallahi *et al.* (2011) demonstrated that apple trees cv. Juliet supplied with greater water level (120% ETc) showed a better response, resulting in greater fruiting efficiency.

   Brun *et al.* (1985) recorded the highest fruit set in Anjou pear with irrigation at two-week interval (100% ETc). In Japanese pear (*Pyrus pyrifolia*), water shortage during bud formation, especially during the period 71–90 days after full flowering, was an important factor limiting the flower bud differentiation (Mitobe *et al.*, 1991). Nasharty and Ibrahim (1961) reported that artificial irrigation to the Japanese plum at an interval of 16, 20 and 24 days did not affect the fruit set, whereas Veihmeyer (1972) reported that fruit set in plum was slightly affected by differences in soil moisture contents. Irrigation at 50 per cent soil moisture depletion of field capacity resulted in better fruit set and fruiting as compared to 75 per cent soil moisture depletion of field capacity irrigation treatment in Santa Rosa plum (Tagi, 1984). Maximum fruit set (31.78%) and fruit yield (43.42 kg/plant) was recorded with IW:CPE ratio of 1: 0.75 and 1:1.25, respectively, in Red Delicious apple grafted on M9 rootstock while conducting an experiment on the basis IW:CPE ratios with irrigation depth of 7.5 cm (Misger and Kumar, 2008a).

Hutmacher *et al.* (1994) reported that nut retention was increased by 10 per cent with the application of water above 50 per cent ETc while irrigating the almond plants with drip irrigation in the semi-arid San Joaquin Valley. Significantly increase in nut yield and kernel weight was also observed with water application enhancement from 50 to 120 per cent etc.

In peach, fruiting was found to be increased when drip irrigation was done at 80 per cent soil moisture capacity (Lischuk *et al.*, 1988); however, bloom density and fruit set were reduced with the occurrence of water deficiency during postharvest which also reduced fruit load (76.7%) and kernel yield (73.6%). Treder *et al.* (1999) conducted a study on the effect of irrigation on growth and yield of plum cv. Valour and recorded highest fruit set under irrigated conditions as compared to unirrigated. Deficiency of irrigation during postharvest affects the tree productivity more than the preharvest irrigation deficiencies; however, the requirement of water during post-harvest is less than preharvest (Goldhamer and Viveros, 2000).

Yldrm and Yldrm (2005) studied the effect of different irrigation levels on tree growth of Santa Rosa plum (*Prunus salicina* Lindl.) and concluded that highest fruit set was obtained at 20 per cent soil moisture depletion level compared to other higher levels of soil moisture depletion. Li and Huguet (1989) determined the irrigation water requirement for peach by micro-morphometric measurement of trunk diameter and reported significant increase in fruit set with irrigation. However, restricted supply of water during a critical period of flower bud induction in peach trees improved flower bud production (Li *et al.*, 1989). Irrigation at 75 per cent of evapotranspiration loses registered maximum number of flower buds/metre of branch and fruit set (%) in peaches (Bignami *et al.*, 1995).

Ruiz-Sanchez *et al.* (2000) found that fruit set values in the water withholding treatments were similar to those observed in control (full irrigation treatment) for all the years of study, except for those irrigated only during early postharvest treatment, which presented significantly lower (around 9.4%) fruit set values than the control in apricot. The results from the study conducted by Goldhamer and Viveros (2000) on almonds indicated that severe water stress during both preharvest and postharvest period reduced flowering and fruit set in subsequent years. Moriana *et al.* (2013) also reported that flowering in olive that occurs at the end of spring was very sensitive to water deficit

Bryla *et al.* (2003) compared four different irrigation systems (furrow, micro-spray, surface drip and sub-surface drip irrigation) for three consecutive years to study their impact on fruit quality and yield of Crimson Lady peach and reported that surface and sub-surface drip irrigation systems produced larger size fruits of 124.0 and 126.33 g weight and the maximum marketable yield of 23.46 and 22.70 t/ha, respectively. During the course of study, optimal moisture in soil adjoining to root zone was maintained by applying daily irrigations with drip irrigation systems which prevented cycles of water stress to the plants.

Perez-Pastor *et al.* (2009) applied three different irrigation methods (100 per cent irrigation of ETc, irrigated at 50 per cent of ETc with a continuous DI and irrigated at 100 per cent of ETc as RDI during the critical periods of plant growth) to ten-year-old Bulida apricot along with 25 per cent of ETc during 1–2 growing seasons and 40 per cent of ETc during 3–4 growing seasons of noncritical periods and reported that during 1–2 growing seasons fruit yield and number of fruits per tree decreased under longer and severe deficits of RDI treatment, however, during 3–4 growing seasons fruit yield and number of fruits per tree were recorded to be more or at par with control.

Kaya *et al.* (2010) irrigated apricot cv. Salak for five consecutive growing seasons with six irrigation treatments (S_1: 0.50, S_2: 0.75, S_3: 1.00, S_4: 1.25 and S_5: 1.50) based on adjustment coefficients of class A pan evaporation and reported that maximum average fruit yield (52.2 kg/tree), average pulp hardness (2.25 kg), average TSS (13.1%) and minimum average acidity (0.46%) were observed under the treatment S_5 (1.50).

Wang (2011) conducted a trail on peach plants to determine the DI strategies for reducing the water requirement after postharvest and reported that almost 30–40 per cent of the used water during the growing season under DI through furrows produced fruit yield equivalent to full irrigation; however, in severe deficit conditions, under sub-surface drip irrigation method, smaller fruits were obtained.

Gunduz *et al.* (2011) determined the amount of water and water consumption for improving yield and quality of fruits in Redhaven peach using four different pan coefficients at three and six days of irrigation intervals through drip irrigation method in Aegean region of Turkey and reported significant effect of amount of water on fruit yield. Maximum fruit yield (14.10 kg/ha) was recorded with pan coefficients of Kp 1:00 where amount of water was 482 mm and water consumption of 705 mm. Among various treatments, fruit weight varied from 203 to 253 g, fruit length varied between 6.3 and 6.6 cm, fruit diameter as 7.2–7.7 cm, soluble dry matter as 10.8–14.5 per cent and juice pH ranged between 4.14 and 4.37 under different irrigation treatments.

Alaoui *et al.* (2013) while standardizing the irrigations levels in peach recorded maximum fruit yield (113.8 fruits/harvest) at 75 per cent of RDI with 0.4 mm irrigation level. However, just before maturity, 50 per cent RDI with 0.4 mm irrigation level was observed, which accelerated the fruits' maturity.

Vera *et al.* (2013) studied the effect of different DI (continuous DI, RDI, partial root zone drying and based on soil water content) methods on fruit yield and plant water status of early maturing of Flordastar peach under Mediterranean conditions and reported that DI showed significant water savings; however, reduction in fruit yield was also observed indicating that DI treatment reduces the fruit number considerably.

Mechlia *et al.* (2012) reported that during different stages of peach cv. Carnival 33 per cent less water for irrigation is required than control while estimating the effect of DI on fruit yield. It was observed the ratio between precipitation + irrigation and ETo during fruit set up to end of phase III was increased from 0.65 to 1.05. During different years and across different treatment, yield was found to be reduced to the tune of 0–34 per cent. Highest reduction in irrigation water was showed during phase III of fruit development.

In an experiment involving five irrigation levels (0.0, 0.25, 0.50, 0.75 and 1.00) and three irrigation intervals (7, 14 and 21 days) on apricot cultivar Ninfa, Bozkurt *et al.* (2012) recorded highest fruit yield with 14-day irrigation interval (729.9 g/plant) and at 0.50 irrigation level (1031.3). Both the treatments exhibited increased fruit size and number of fruits per plant.

Khan *et al.* (2012) reported maximum nut yield (2.69 kg/plant) with irrigation level at 100 per cent of ETc during first year of study which increases in the second year (2.91 kg/plant) while conducting an experiment on almond cultivar Shalimar with four different irrigation levels (0, 100, 75 and 50% ETc) at different stages of fruit growth and development.

Durgac *et al.* (2017) reported a significant effect of interaction between irrigation intervals and the amount of water to be used for irrigation on fruit yield/plant in apricot cv. Ninfa. The maximum fruit yield (859.5 kg) was obtained from I_{50}; however, interaction between treatments depicted that $T_{14} \times I_{75}$ interaction resulted in maximum cumulative yield (987.4 kg/ha). Verma and Chandel (2017) also reported maximum yield in peach plants which were irrigated with drip at 100 per cent ETc.

12.6   YIELD PARAMETERS

Knowledge of yield responses to water deficits is essential to forecast the response to reductions in water supply during the years when there is a shortage of water and in areas of water scarcity and not possible to supply the amounts required to meet maximum ETc. Generally, due to water deficit conditions, the risk of yield reduction increases, and under severe conditions of water deficits, the uncertainties of fruit yield were more, especially at the time of critical stages of fruit development. Overall, fruit yield and fruit quality is an important aspect of profitable fruit production. The increase in fruit yield with the management in the irrigation practices and scheduling in temperate fruits had been described by many researchers (Miletic *et al.*, 2007; Temocico *et al.*, 2011; Du *et al.*, 2018; Fallahi *et al.*, 2018); however, water stress and irrigation management during the development of some fruit can produce significant changes in their fruit yield and quality (Reid *et al.*, 1996).

Serrano *et al.* (1992) reported that under soil water potential @ −0.01 MPa, maximum yield of Chandler cultivar of strawberry was obtained which is directly correlated with more fruit numbers/plant and fresh weight of fruit. In order to obtain yield-water relationships, it is necessary to apply different irrigation regimes to crops and determine their yield responses to this application. In order to perform efficient crop production of basic crops, particularly like apple, it is necessary to develop suitable irrigation schedules that do not cause any evident reduction in yield. Water deficit conditions were applied to trees of apple cv. 'Granny Smith' growing in pots and it was observed that water deficit during fruit cell division phase reduced fruit growth by decreasing water uptake (Failla *et al.*, 1992).

Study on drip irrigation in different cultivars of apple by Rzekanowski and Rolbiecki (2000) revealed a significant increase in yield by 9 per cent in cultivar Melba, 22 per cent in cultivar McIntosh and 25 per cent in cultivar Spartan when compared to control (rainfed). Chandel *et al.* (2004) obtained maximum total yield (86 kg/vine), in kiwifruit having a yield of 'A' grade fruit (49 kg/vine) and 'B' grade fruit (26 kg/vine) with 100 per cent ETc using drip irrigation method. Analysing the average yield per apple tree, for experimental plots in which the intervention of drip irrigation was applied, it was observed that the largest yield per tree was recorded in 'Generos' cultivar (26.84 kg/tree), which resulted in a production of 30.44 t/ha, with a difference from the control plot of 5.57 kg/tree, i.e., 6.35 tones/ha (Temocico *et al.*, 2011). A four-year study by Veverka and Pavlačka (2012) on the influence of drip irrigation on the yield and quality of apples revealed that irrigation exerted a very positive influence of irrigation especially during the dry year. They further observed that yield increased by 104 per cent in variety Gala and 71 per cent in Golden Delicious under full irrigation (100% ETc) than in the control (without irrigation).

Kireva *et al.* (2017) proposed that different irrigated regimes significantly influenced the yield of apple by obtaining higher yields with full irrigation at 100 per cent irrigation rate of 2087 kg/da. They further concluded that lowering irrigation rate by 20 and 40 per cent had led to a 7 and 14 per cent reduction in yields. Du *et al.* (2018) in their study on water use efficiency of apple by alternate partial root-zone irrigation in arid northwest China observed that compared with low irrigation amount (400 mm), high irrigation amount (500 mm) improved apple yield. Fallahi *et al.* (2018) carried out an experiment to study the effect of different irrigation levels on fruit yield of Fuji apples; it was observed that apple trees irrigated with an full drip system had higher yield per tree than those with 65 per cent drip and 50 per cent drip and thus suggested that full drip is a preferred method of irrigation over other systems for 'Fuji' apples.

With reference to other fruit crops, Chalmers *et al.* (1984) laid out an experiment and trickle irrigation was applied to peach plants during Stage I and III, only during Stage III or during all season of fruit growth and reported that all three treatments produced similar fruit yield and similar size of fruits. Singh *et al.* (2002) reported that increase in irrigation rate from 60–80 to 80–100 per cent of 'V' volume of irrigation water using drip irrigation method to apricot plants results in increased fruit yield. Naor *et al.* (2004) reported that fruit yield and size of Japanese plum improved with the increase in irrigation rate. Impact of rainfall and irrigation on the yield of plum cultivars was examined by Miletic *et al.* (2007) and observed that irrigated trees registered an increase of 11.19 kg/tree when compared to the un-irrigated trees.

Yield of fruit trees are directly affected if irrigation amount are reduced during certain growth stage. Any reduction in water requirement during fruit expanding and maturing stages can lead to reduced yields (Consoli *et al.*, 2014). However, reduced preharvest irrigation in strawberry was found partially detrimental, while not reducing overall yield, resulted in smaller fruits (Kirnak *et al.*, 2001). Kumar *et al.* (2012) studied the influence of irrigation levels on fruit yield of strawberry cultivar Chandler and observed that higher fruit yield (175.12 g/plant) was obtained with maximum level of irrigation (Cumulative Pan Evaporation ratio of 1.0 with irrigation water) as compared to other levels of irrigation. In another study, the irrigation amount below 50 per cent of sufficient irrigation reduced olive yield by 11 per cent (Ghrab *et al.*, 2013). Similarly, a comparison of different levels of drip irrigation in strawberry indicated the highest yield per plant (238.1 g/plant) with drip irrigation at 120 per cent ETc, followed by drip irrigation at 100 per cent ETc (228.2 g/plant) compared to lower levels of ETc (80 and 60% ETc). The plants irrigated at 120 per cent ETc with drip method showed 29.77 and 16.72 per cent increase in yield during respective years of study (Kachwaya, 2014).

Many studies in olive had showed that high soil water availability resulted in increased yield components such as fruit number, fruit fresh weight, fruit volume, pulp:stone ratio, and oil content and thus, increasing fruit and oil yields (Fernandes-Silva *et al.*, 2010; Fernandes-Silva *et al.*, 2016).

12.7   FRUIT PHYSICAL PARAMETERS

Fruit size and quality are the prime parameters of any fresh fruit for getting the good returns. Any reduction or low availability of irrigation resulted in smaller fruit size and poor quality (Naor *et al.*, 1997: Mpelasoka *et al.*, 2012). The amount of irrigation water has been observed to be critical for fruit flesh firmness. Assaf *et al.* (1975) indicated that fruits from trees under water reduction were firmer than fruits from conventional irrigation which may be a consequence of smaller fruit size. Drake *et al.* (1981) indicated that apple slices were softer from trees supplied with less water. Bergamini *et al.* (1990) reported that increasing irrigation amounts through drip method substantially increased fruit diameter of Golden Delicious apple.

Mills *et al.* (1994) reported a reduction in firmness of Braeburn apple from plants which received reduced irrigation water. In an experiment on Braeburn apple in New Zealand, Kilili *et al.* (1996) reported an increased firmness with the application of lower amounts of supplied water and was higher in fruit for which irrigation was withheld either late in the season or during the entire season. Bonany and Camps (1998) also stated that flesh firmness of apples decreased as the amount of irrigation water increased. In the three-year study of Leib *et al.* (2006) with Fuji apple, it was concluded that firmness in the water restriction fruits was higher than that of the conventional irrigation fruits for all the three years, both at harvest and after 14 days from harvest. Low mid-day stem water potential arising from reduced irrigation has also been related to smaller fruit weight in pear fruits (Shackel, 2007). However, Talluto *et al.* (2008) showed that fruit firmness was observed to be unaffected by water restriction treatment in 'Pink Lady' apple. Naor *et al.* (2004) and Naor *et al.* (2008) studied Japanese plums and apple, respectively, and reported that water stress during the final stages of fruit development significantly decreased the size of fruit.

Misgar and Kumar (2008b) conducted an experiment on 18-year-old Red Delicious apple grafted on M9 rootstock using six moisture regimes (control: rainfed, 0.25, 0.50, 0.75,1.00 and 1.25) maintained on the basis IW:CPE ratios with irrigation depth of 7.5 cm in the plant basins having radius of 101.0 cm and reported that maximum fruit length (6.48 cm), fruit diameter (6.81 cm), fruit weight (123.62 g) and fruit volume (163.55 cm^3) were recorded with IW:CPE ratio of 1:1.25.

Temocico *et al.* (2011) reported that in three apple varieties using drip irrigation system fruit weight was increased by 16.96 per cent as compared to no irrigation. Kucukyumuk *et al.* (2012) concluded that the highest fruit weight, fruit length and fruit diameter were observed in apple cv. Starkrimson with a fruit firmness of 100 per cent ETc levels at a four-day interval of drip irrigation; however, maximum fruit firmness was observed in 50 per cent ETc at a seven- day interval compared to 70 per cent ETc and 100 per cent ETc at four-day interval irrigation treatments, respectively. Marsal *et al.* (2010) determined that due to the small size of Conference pear from less irrigated trees (20% of full irrigation during Stage II), 85 per cent fruits were disqualified for the fresh market.

Cay et al. (2009) registered significantly increased fruit length and diameter with the increase in irrigation and the highest fruit diameter and length were recorded in Kep = 1.0 (full irrigation) in apple trees with drip irrigation. Flesh firmness values decreased as the amount of irrigation water through drip increased for both irrigation intervals of drip irrigation treatments during the study (Fallahi et al., 2010). A four-year study carried out by Fallahi (2018) suggested that trees require full irrigation ETc rates to produce large fruits and any reduction in irrigation (65% ETc and 50% ETc) resulted in smaller fruits.

Many studies in stone fruits (Goldhamer et al., 2002; Girona et al., 2004; Lopez et al., 2006) have also shown that fruit weight at harvest is reduced under water stress conditions. Haulik (1979) irrigated peach plants at 70, 80 and 100 per cent of field capacity and reported that maximum size of fruits was obtained with irrigation at 80 per cent field capacity.

Reduction in irrigation levels during reproductive cell division stage and pit hardening stage apparently reduced weight of fruit in apricot when observed at the end of pit hardening stage; however, a significant increase in fruit size was noticed after the water stress was reassured at the time of final fruit growth stage (Ruiz-Sanchez et al., 2000; Torrecillas et al., 2000). Goldhamer et al. (2002) reported that in peach decreased fruit weight was observed with early water stress during both at the end of the reproductive cell division stage and at harvest.

By application of 100 per cent ETc through drip irrigation method in kiwifruit cv. Allison, Chandel et al. (2004) reported maximum fruit length (69.97 mm), breadth (45.08 mm) and weight (76.22 g). Intrigliolo and Castel (2005) reported that any reduction in water availability during fruit growth reduced average fruit weight of plum. Yldrm and Yldrm (2005) also observed an increase in fruit size of Santa Rosa plum with the application of irrigation @ 20 per cent moisture reduction compared to higher levels. Girona et al. (2006) in their study suggested that fruit growth rate in peach was linearly related to tree water status expressed as midday stem water potential.

Mousavi and Alimohamadi (2006) studied the effects of DI (100% ETc: full irrigation, 80% ETc: DI and 40% ETc: DI) and drought period on 'Mamaei' almond and reported that under DI and drought condition, fruit size, and fresh and dry weight of fruit, was decreased during Stage I, i.e., fruit growth, fresh fruit weight, fresh and dry kernel weight decreased during Stage II, i.e., kernel growth but no significant differences were observed for any parameters when irrigation treatments were applied at Stage III (preharvest stage). Overall DI and drought condition decreased fruit yield, kernel dry weight and kernel percentage significantly.

To determine the effects of DI on 'Marnaei' almond, Ali et al. (2012) carried out an experiment (100% ETc – full irrigation, 80% ETc – DI, 40% ETc – DI and 0% ETc – without irrigation) and reported that during fruit growth stage, dry and fresh weight of fruit, fruit size, was decreased under DI, whereas due to water stress fruit drop percentage was also increased. During kernel development stage under DI fresh fruit weight, fresh and dry kernel weight was decreased. All through three stages (fruit growth, kernel development and preharvest stage) kernel dry weight, kernel percentage and fruit yield were observed to decrease under DI. Mechlia *et al.* (2012) investigated the cumulative effect of different level of irrigation on fruit quality of late peach cultivar 'Carnival' and reported that total dry matter and sugar content showed improvement with increasing irrigation levels.

While studying the effect of irrigation and fruit position on colour, size, sugar content and firmness of fruits on peach, Alcobendas *et al.* (2013) reported that the fruits obtained from plants grown under RDI were more firmer than those fruits which obtained from plants grown under deficit irrigated conditions, whereas weight and diameter of fruit did not differ significantly. Fruit obtained from the plants that are grown under RDI conditions had more TSS, glucose and different acids. Razouk *et al.* (2013) reported that in peach and plum maximum size of fruits, maximum TSS and minimum acidity was obtained with 100 per cent ETc through drip irrigation while conducting an experiment to study the effect of different drip irrigation methods (50% ETc, 75% ETc and 100% ETc) on quality of different stone fruits. Razouk *et al.* (2013) also stated that kernel quality of almond fruits was remained unaffected by restriction of water upto 75 per cent ETc; however, wrinkles on the epidermal layer of almond kernel were more prominent when 50 per cent ETc through drip irrigation was applied which affects the appearance of kernel. Eid *et al.* (2013) investigated the effect of soil moisture regimes on fruit quality of 'Canino' apricot and observed that the increased soil moisture enhanced fruit length which resulted in more fruit size. Hussien *et al.* (2013) studied effect of different irrigation methods on plum and concluded that fruit quality improved significantly with drip irrigation method in comparison to flood irrigation method.

Velardo-Micharet *et al.* (2017) studied the effect of irrigation on postharvest quality of two sweet cherry cultivars 'Lapins' and 'Ambrunés' and reported that under fully irrigated conditions, fruit length and breadth of 'Lapins' cherry was larger and both the two cultivars had higher fruit weight than non-irrigated cherries at the time of harvest.

Durgac *et al.* (2017) irrigate the Ninfa apricot plants at an intervals of 7, 14 and 21 days and obtained maximum weight of fruit (24.4 g), length of fruit (35.5 mm), total soluble solids (12.0°B), fruit pH (3.61) and minimum acidity (0.89%) with the application of irrigation at 21-day interval, whereas maximum fruit index (0.97), flesh/stone ratio (8.27) and flesh firmness (3.88 kg/cm^2) were observed when irrigation was applied at 14-day interval. Increasing irrigation significantly improved fruit weight; however, maximum flesh/seed ratio, TSS and juice pH values were recorded from I_{100}, I_{25} and I_{50}, respectively.

Verma and Chandel (2017) irrigated the peach plants with 100 per cent ETc through drip irrigation and reported highest fruit weight (104.78 g), fruit length (6.12 cm), fruit diameter (5.56 cm), TSS (12.20°B) and minimum acidity (0.61%).

Zhang *et al.* (2017) studied the impact of irrigation scheduling on water status and yield of peach under different irrigation systems for two consecutive years, i.e., 2012–2013 and 2013–2014, and the results showed that mid-day stem water potentials for well-irrigated trees were maintained at a range of −0.5 to −1.2 MPa, while of deficit irrigated trees dropped to lower values. The number of fruits per plant and weight of fruit was observed highest in well-watered trees. However, no statistically significant reduction in fruit size or fruit quality was observed with DI treatments during both years of study.

12.8   FRUIT CHEMICAL PARAMETERS

Bio-chemical parameters of temperate fruits, viz. total soluble solids, acidity, total sugars and ascorbic acid, show a consistent response to irrigation. Numerous authors have reported a significant increase in apple fruit TSS under DI, and in the majority of experiments, the total soluble solids were 0.7–1.5°B higher when there is a reduction in irrigation as compared to full irrigation (Ebel *et al.*, 1993; Leib *et al.*, 2006; Fallahi *et al.*, 2010; Neilsen *et al.*, 2010). Titratable acidity showed highly variable results to DI. In some experiments, acidity showed non-significant results with restrictions in water (Neilsen *et al.*, 2010), whereas in some experiments acidity was significantly lower in DI conditions than full irrigation in one season and higher under DI than full irrigation in the next season (Leib *et al.*, 2006). Irving and Drost (1987) found no influence of any water restriction on fruit acidity, while other authors observed a decrease in acidity under water stress (Drake *et al.*, 1981; Mills *et al.*, 1994). Citing the literature, Behboudian and Mills (1997) concluded that the response of acidity to irrigation was not conclusive. The concentration of four major soluble sugars of apple (sucrose, glucose, fructose and sorbitol) was generally increased with lower levels of irrigation.

Kilili *et al.* (1996) applied four irrigation treatments to 'Braeburn' apple trees and the results indicated that withholding irrigation throughout the growing season and during late season resulted in higher total soluble solids and sugars than those irrigated throughout the season, while no variation was observed for acidity. Mills *et al.* (1996) reported that DI conditions in Braeburn apple plants 55 days after full bloom increased the total sugar concentration of fruits. Miller *et al.* (1998) withheld irrigation water during early and late summer in the growing season of kiwifruit and reported that withholding of irrigation water late in the growing season of kiwifruit was beneficial and the fruit matured six weeks earlier as compared to control, improved TSS (9.3 vs 8.1%) and fruit were more elongated than control. Mpelasoka *et al.* (2001b) reported that fruit sugars increased significantly with decreasing levels of irrigation. Mpelasoka and Behboudian (2002) also reported that titratable acidity was not affected by withholding irrigation during the entire growing period in Braeburn apple. Studies on apples by Remorini and Massai (2003) showed that moderate water stress induced higher fruit sugar contents. DI application in the second stage of fruit growth, in a mid to late-maturing peach cultivar, resulted in lower titratable acidity and higher TSS/acid ratio compared to those in fully irrigated control fruit (Gelly *et al.*, 2004). Sharma and Chandel (2005) recorded significantly lower total sugar content in apricot fruits produced by plants irrigated @ 20 per cent soil moisture depletion of field capacity and highest in fruits produced by unirrigated control.

Leib *et al.* (2006) showed that total soluble solids in fruits from trees receiving restricted water supply were higher than in fruits from trees receiving conventional irrigation. Connell and Goodwin (2007) studied 'Pink Lady' apple for two consecutive years in Victoria, Australia, and reported that TSS tended to be higher in non-irrigated planted (17.6 and 16.9°B) than those with full irrigation (16.2 and 12.2°B), respectively, in both years. In the two-year study carried out by Talluto *et al.* (2008) on apple cultivar Pink Lady, it was shown that titratable acidity in the first year was lower in the fruits from non-irrigated trees than fruits from fully irrigated trees, whereas in the second year there was no difference in titratable acidity between the treatments. Quality characteristics of two strawberry (Machyang and Seolhyang) cultivars were affected by water stress and higher total soluble solids in response to lower water supply was obtained (Kim *et al.*, 2009). Total soluble solids in peach fruit increased under high water restriction as compared to control and light water restriction (Mercier *et al.*, 2009). Fruit TSS significantly increased in Andross cultivar of peach on limitation of water supply during Stage II of fruit growth (Pascual *et al.*, 2010), and fruit total soluble solids also increased in response to withholding irrigation during Stage III of fruit growth observed (Rufat *et al.*, 2010). Kumar *et al.* (2012) reported that total soluble solids, total sugars and ascorbic acid were significantly higher in strawberry fruits, which were obtained from drip irrigation @ 1.0 IW/CPE level than those obtained from other irrigation levels.

Eid *et al.* (2013) investigated the effect of soil moisture regimes on fruit quality of 'Canino' apricot and observed that the increasing soil moisture enhanced fruit size which resulted in reduced TSS/acid ratio and acidity. In a study carried out by Stino *et al.* (2016) on impact of applied irrigation regime during specified phenological stages on pear cv. Le-Conte, it was concluded that fruit TSS increased with decreasing the actual requirement during any studied stages. Fallahi *et al.* (2018) reported that apple trees receiving different levels of irrigation stated that fruits from trees with 50 per cent drip irrigation had significantly higher total soluble solids in three of four years of study.

12.9 FRUIT AND LEAF NUTRIENT AND LEAF CHLOROPHYLL

Romo and Diaz (1985) reported that the nutritional status of peach plants grown under flood and drip irrigation system had similar nutritional status of soil while conducting an experiment on root system and nutritional status. Concentration of manganese differs seasonally which was stable and low under drip irrigation system; however, it was increased under flood irrigation. Dochev (1968) retained soil moisture at field capacity (50, 60 and 70%) in peach plants and observed significantly higher nitrogen and phosphorous content in the tissue of plants kept at 70 per cent of field capacity. Miculka (1983) reported a reduction in leaf nitrogen, phosphorous and potassium content of peach under drip irrigation method @ 20 cm depth or channel irrigation method at a depth of 60 cm. Layne *et al.* (1996) reported higher magnesium content in peach plants maintained under drip irrigation having low fertigation than the other treatments. Rana *et al.* (2005) observed increased leaf nitrogen, phosphorus and potassium content under high irrigation level and more spacing while conducting an experiment on the impact of rootstock and drip irrigation system on leaf nutrient status Flordasun peach under high density plantation. Maximum nitrogen content was observed under plant spacing of 2.0 × 1.25 m receiving 3 l/day water grafted on peach rootstock and minimum nitrogen content was obtained under plant spacing of 2.0×0.5 m receiving 1.0 l/day water grafted plum rootstock. Maximum leaf phosphorus content was recorded in the treatment combination of 2.0×1.25 m plant spacing receiving 3 l/day water in peach-on-peach rootstock. Plants receiving 3.0 l of water per day observed maximum potassium content when planted at spacing of 2.0×1.25 m grafted on peach rootstock. Verma and Chandel (2017) studied nutrient status of peach leaves cv. Redhaven under different irrigation levels reported maximum leaf nitrogen, phosphorous, potassium, calcium, magnesium, iron, manganese, zinc and copper content as 2.63 per cent, 0.12 per cent, 2.10 per cent, 2.24 per cent, 0.38 per cent, 204.94 ppm, 52.03 ppm, 19.94 ppm and 9.69 ppm, respectively, in drip irrigation at 100 per cent ETc.

Ali *et al.* (2012) reported that during kernel growth (Stage II) under DI condition in Marnaei cultivar of almond, nitrogen and boron content in leaves was reduced. Khan *et al.* (2012) observed highest concentration of nitrogen (2.12%), phosphorous (0.19%) and potassium (1.59%) in Shalimar cultivar of almond while receiving irrigation at 100 per cent ETc.

Demirtas and Kirnak (2009) reported highest leaf chlorophyll content in plants irrigated with mini sprinkler while conducting an experiment on four-year-old apricot cv. Hacihaliloglu using mini-sprinkler and surface irrigation methods with intervals of 12, 20 and 25 days. They also observed decrease in leaf chlorophyll content with the increase in irrigation intervals. Highest leaf chlorophyll-a content was obtained in surface irrigation (3.20 mg/ml) followed by mini-sprinkler method (3.19 mg/ml) of irrigation applied at 12-day interval. Maximum leaf chlorophyll-b content, total chlorophyll content and leaf carotene were recorded as 1.81, 5.00 and 0.213 mg/ml in plants irrigated at 12 days of interval with mini-sprinkler method.

12.10 WATER USE EFFICIENCY

Increase in water use efficiency was always observed as a result of reduction in water losses through drip irrigation system and which is efficiently use by the plants. Sivanappan (1998) reported 68 per cent of water savings with the adoption of drip irrigation method as compared to basin irrigation. Rana *et al.* (2000) reported that irrigation in kiwifruit vines @ 20 per cent soil moisture depletion of field capacity is beneficial for higher production of quality fruits in mid hills of north-western Himalayas. Romero *et al.* (2004) studied the effects of RDI under sub-surface drip irrigation conditions on vegetative parameters and yield of mature almond plants and reported that water use efficiency was increased significantly in the SDI (sub-surface drip irrigation) treatments at 100 per cent ETc, except in the kernel filling stage (20% ETc) and postharvest (75% ETc and 50% ETc).

Water use efficiency was found maximum (2.91 q/ha/cm) under drip irrigation at 100 per cent ETc followed by drip irrigation at 80 per cent (2.76 q/ha/cm) and 60 per cent (2.59 q/ha/cm) ETc (Chandel *et al.*, 2004). Maximum water use efficiency (1.27 q/ha/cm) was recorded under the treatment drip irrigation at 60 per cent of evapotranspiration + black polythene mulch which was 269.9 per cent higher than the minimum water use efficiency (0.36 q/ha/cm) (Sharma *et al.*, 2005).

Intrigliolo and Castel (2010) determined the effect of RDI and crop load on Black Gold cultivar of Japanese plum grafted on Mariana GF81 rootstock using RDI during second phase of fruit growth. Thirty per cent of water was saved by RDI, which increased water use efficiency and less impact on fruit growth and fruit yield was recorded, thereby indicating that during the period of study water stress was minimum. However, contrary to above finding, Wang (2011) reported consistently maximum values for crop water stress index in peach plants under DI as compared to those receiving full irrigation.

Maximum water use efficiency of 2:02 kg/m^3 was obtained from pan coefficient: Kp – 1:00 while determining the amount of irrigation water in Redhaven peach when irrigated by drip irrigation system in Aegean region of Turkey (Gunduz *et al.*, 2011). Singh (2013) recorded markedly higher water use efficiency (0.41 t/ha) in apricot irrigated with drip method as compared to surface irrigation where water use efficiency was 0.19 t/ha.

Bozkurt *et al.* (2012) reported that the irrigation level plays an important role for young apricot plant and water use efficiency was decreased with the increasing irrigation intervals.

Khan *et al.* (2012) observed the highest water use efficiency of 0.60 kg/m^3 with 0 per cent ETc level of irrigation in almond cultivar Shalimar followed by 0.46 kg/m^3 in 50 per cent ETc level which was statistically at par with 0.40 kg/m^3 under 75 per cent ETc level of irrigation.

Maximum water use efficiency of peach plants irrigated with drip irrigation system was observed which were irrigated at 80 per cent ETc as compared to other irrigation levels (Verma and Chandel, 2017).

12.11 WATER STRESS

Water stress especially after fruit set and during summer season reduces the fruit size, fruit yield and fruit quality. Xiloyannis *et al.* (1990) reported that fully grown kiwifruit vines require 80–100 l of water for total daily transpiration from 16–17 m^2 canopy area; however, in summer month, the water requirement is more and vine requires 145–180 l of water per day.

12.12 MATURITY TIME

Mpelasoka *et al.* (2001b) observed the effect of reduced irrigation on fruit maturity depends on the timing of application of DI and reported that late DI and whole DI advanced maturity while early DI showed no effect. Ebel *et al.* (1993) in Red Delicious apple and Kilili *et al.* (1996) in Braeburn apple also reported that RDI advance the fruit maturity. Proper harvesting time of any fruit crop had multiple influences on consumer acceptance of fruit (Iglesias and Echeverria, 2009). The time needed to reach berry maturity was related to temperature and precipitation, which is shortened with the rise of temperature and decrease of precipitation (Martinez-Lüscher *et al.*, 2016). Delays in fruit maturity have been confirmed in peach cultivars 'O' Henry' and 'Ryan's Sun' grown under severe water stress conditions (Lopez *et al.*, 2010, 2011), which indicates that the degree of water stress may play a significant role in fruit maturity and medium-to-low water stress appears to advance fruit maturity, whereas severe water stress delays the process of fruit maturity. Background colour of most apple cultivars is an indicator of fruit maturity, and it was observed that with the DI conditions, either the fruit maturity time decreases or remains unchanged (Mpelasoka *et al.*, 2002).

12.13 CONCLUSIONS

Irrigation plays an important role in temperate fruit crop production and is considered an important aspect of management of orchards. With an optimum amount of irrigation supplied, fruit of high quality with better yield can be produced. But under the changing climatic scenario leading to decreasing land and water availability for irrigation, the horticulture industry is facing major challenges for future fruit production. Water is becoming a limiting factor not only in Indian continent but its reduction has also been observed globally. Traditionally, temperate fruit trees are irrigated frequently throughout the growing period and mismanagement of water supply to trees at critical stages leads to fruit drop, reduced fruit size and quality. Though the level of drought tolerance varies from species to species, different rootstocks available in temperate fruits impart drought tolerance. But, under the current scenario, irrigation water should be economized using modern irrigation methods like drip irrigation which has again been reported the most efficient method of irrigation. The productivity of the orchard and quality of the produce while minimizing the irrigation water can be done through techniques like RDI methods.

REFERENCES

Akin, S. and Erdem, T. 2018. Water use of walnut trees under different irrigation regimes. *Journal of Applied Horticulture* 20(1): 62–65.

Alaoui, S.M., Abouatallah, A., Salghi, R., Amahmid, Z., Bettouche, J., Zarrouk, A. and Hammouti, B. 2013. Impact assessment of deficit irrigation on yield and fruit quality in peach orchard. *Der Pharma Chemica* 5(3): 236–243.

Alcobendas, R., Miras-Avalos, J.M., Alarcon, J.J. and Nicolas, E. 2013. Effects of irrigation and fruit position on size, colour, firmness and sugar contents of fruits in a mid-late maturing peach cultivar. *Scientia Horticulturae* 164: 340–347.

Ali, M.R., Mousavi, A., Tatari, M. and Fattahi, A. 2012. Effects of deficit irrigation during different phonological stages of fruit growth and development on mineral elements and almond yield. *Iranian Journal of Water Research in Agriculture* 26(2): 143–129.

Assaf, R., Levin, I. and Bravdo, B. 1975. Effect of irrigation regimes on trunk and fruit growth rates, quality and yield of apple trees. *Journal of Horticultural Science* 50: 481–493.

Barlow, M. and Clarke, T. 2017. *Blue gold: the battle against corporate theft of the world's water.* Routledge.

Behboudian, M.H. and Mills, T.M. 1997. Deficit irrigation in deciduous orchards. *Horticulture Reviews* 21: 105–131.

Bergamini, A., Angelini, S. and Bigaran, F. 1990. Influence of four different rootstocks on the stomatal resistance and leaf water potential of Golden Delicious Clone B (Virus T Ree) subjected to different irrigation regimes. *Societa Orticola Italiana* 533–544

Bignami, C., Natali, S., Cammilli, C. and Sansavini, S. 1995. Post harvest irrigation of early peach cultivars. 21st Convegno. Peschicolo, Lugo, Ravenna, 27–28, August, 235–245.

Bonany, J. and Camps, F. 1998. Effect of different irrigation levels on apple fruit quality. *Acta Horticulturae* 466: 47–52.

Bozkurt, S., Odemis, B. and Durgac, C. 2012. Effects of deficit irrigation treatments on yield and plant growth of young apricot trees. *New Zealand Journal of Crop and Horticultural Science* 43(2): 73–84.

Bryla, D.R., Trout, T.J. and Ayars, J.E. 2003. Growth and production of young peach trees irrigated by furrow, microjet, surface drip or subsurface drip systems. *HortScience* 38(6): 1112–1116.

Brun, C.A., Raese, J.T. and Stahly, E.A. 1985. Seasonal response of 'Anjou' pear trees to different irrigation regimes. II. Mineral composition of fruit and leaves, fruit disorders and fruit set. *American Society for Horticultural Science* 110(6): 835–840.

Cay, S., Tan, A.F., Dine, N., Bitgi, S., Ozbahce, A., Palta, C. and Okur, O. 2009. Effects of different irrigation programmes on yield and quality characteristics of Granny Smith apple variety grafted onto M9 rootstock. *Journal of Agriculture Sciences Researches* 2(2): 73–79.

Chalmers, D.J., Mitchell, P.D. and Jerie, P.H. 1984. The physiology of growth control of peach and pear trees using reduced irrigation. *Acta Horticulturae* 146: 143–149.

Chandel, J.S., Rana, R.K. and Rehalia, A.S. 2004. Comparative performance of drip and surface methods of irrigation in Kiwifruit (*Actinidia deliciosa* Chev.) cv. Allison. *Acta Horticulturae* 662: 205–213.

Chauhan, P.S., Sud, A., Sharma, L.K. and Mankotia, M.S. 2005. Studies on the effect of microirrigation levels on growth, yield, fruit quality and nutrient assimilation of Delicious apple. *Acta Horticulturae* 696: 193–196.

Chenafi, A., Philippe, M., Eva, A., Abderrahmane, B. and Christoph, C. 2016. Influence of irrigation strategies on productivity, fruit quality and soil plant water status of subsurface drip-irrigated apple trees. *Fruits* 71(2): 69–78.

Connell, M.G. and Goodwin, L. 2007. Responses of Pink Lady apple to deficit irrigation and partial rootzone drying: physiology, growth, yield and fruit quality. *Australian Journal of Agricultural and Resource Economics* 58: 1068–1076.

Consoli, S., Stagno, F., Roccuzzo, G., Cirelli, G.L. and Intrigliolo, F. 2014. Sustainable management of limited water resources in a young orange orchard. *Agricultural Water Management* 132: 60–68.

Daniell, J.W. 1984. Effect of glyphosate for weed control in eleven cultivars of peach trees. In *Proceedings of 37th Annual Meeting South Weed Science Society*, Abstr. No. 126.

Demirtas, M.N. and Kirnak, H. 2009. Effects of different irrigation systems and intervals on physiological parameters in apricot. *Journal of Agricultural Sciences and Biotechnology* 19(2): 79–83.

Dharminder, R.K.S., Kumar, V., Devedee, A.K., Mruthyunjaya, M. and Bhardwaj, R. 2019. The clean water: the basic need of human and agriculture. *International Journal of Chemical Studies* 7(2): 1994–1998.

Dochev, D. 1968. A study on the irrigation of young peach trees. *Grdinarska I LozarskaNauka* 5(7): 3–16

Drake, S.R., Proebsting, E.L., Mahan, M.O. and Thompson, J.B. 1981. Influence of trickle and sprinkle irrigation on Golden Delicious apple quality. *Journal of the American Society for Horticultural Science* 106: 255–258.

Du, S., Tong, L., Kang, S., Li, F., Du, S., Li, S. and Ding, R. 2018. Alternate partial root-zone irrigation with high irrigation frequency improves root growth and reduces unproductive water loss by apple trees in arid North-west China. *Frontiers of Agricultural Science and Engineering* 5(2): 188–196.

Durgac, C., Bozkurt, S. and Odemis, B. 2017. Different irrigation intervals and water amount studies in young apricot trees (cv. Ninfa). *Fresenius Environmental Bulletin* 26: 1469–1476.

Ebel, R.C., Procbsting, E. and Patterson, M.E. 1993. Regulated deficit irrigation may alter apple maturity, quality and storage life. *HortScience* 28: 141–143.

Ebel, R.C., Proebsting, E.L. and Evans, R.G. 1995. Deficit irrigation control vegetative growth in apple and monitoring fruit growth to schedule irrigation. *HortScience* 30(6): 1229–1232.

Eid, T.A., Fatama, L., Grah, A. and Hussein, S.M. 2013. Effect of soil moisture regimes and potassium application on growth, yield and fruit quality of Canino apricot (*Prunus armeniaca* L.). *Journal of Plant Production* 4(4): 621–640.

Failla, O., Zocchi, Z., Treccani, C. and Socucci, S. 1992. Growth, development and mineral content of apple fruit different water status conditions. *Journal of Horticultural Science* 67: 265–271.

Fallahi, E. 2018. Long-term influence of irrigation systems on postharvest fruit quality attributes in mature 'Autumn Rose Fuji' apple trees. *International Journal of Fruit Science* 18(2): 177–187.

Fallahi, E., Fallahi, B., Amiri, M. and Shafii, B. 2011. Long-term fruit yield and quality of various Gala apple strain-rootstock combinations under an evapotranspiration-based drip irrigation system. *Fruit, Vegetable and Cereal Science and Biotechnology* 5: 35–39.

Fallahi, E., Fallahi, B. and Kiester, M.J. 2018. Evapotranspiration-based irrigation systems and nitrogen effects on yield and fruit quality at harvest in fully mature 'Fuji' apple trees over four years. *HortScience* 53(1): 38–43.

Fallahi, E., Neilsen, G.H., Peryea, F.J., Neilsen, D. and Fallahi, B. 2010. Effect of mineral nutrition on fruit quality and nutritional disorder in apple. *Acta Horticulturae* 868: 49–59.

Fernandes-Silva, A.A., Ferreira, T.C., Correia, C.M., Malheiro, A.C. and Villalobos, F.J. 2010. Influence of different irrigation regimes on crop yield and water use efficiency of olive. *Plant and Soil* 333: 35–47.

Fernandes-Silva, A.A., López-Bemal, A., Ferreira, T.C. and Villalobos, F.J. 2016. Leaf water relations and gas exchange response to water deficit of olive (cv. Cobrançosa) in field grown conditions in Portugal. *Plant and Soil* 402: 191–209.

Fernandez, C.J., Mcintes, K.J. and Cothren, J.T. 1996. Water status and leaf area production in water and nitrogen stressed cotton. *Crop Science* 36: 1224–1233.

Garjugin, G.A. 1962. The effect of early autumn irrigation to provide reserve moisture on the growth and fruiting of apple trees. *Agrobiologija* 2: 292–293.

Gehrmann, H. and Lenz, F. 1991. Water requirement and effect of water deficiency on strawberry leaf area development and dry matter distribution. *Erwerbsobitbus* 33(1): 14–17.

Gelly, M., Recasens, L., Girona, J., Mata, M., Arbones, A., Rufat, J. and Marsal, I. 2004. Effects of stage 11 and postharvest deficit irrigation on peach quality during maturation and after cold storage. *Journal of the Science of Food and Agriculture* 84: 561–568.

Ghrab, M., Gargouri, K., Bentaher, H., Chartzoulakis, K., Ayadi, M., Ben-Mimoun, M., Masmoudi, M.M., Ben Mechlia, N. and Parras, G. 2013. Water relations and yield of olive tree (cv. Chemlali) in response to partial root-zone drying (PRD) irrigation technique and salinity under arid climate. *Agricultural Water Management* 123: 1–11.

Girona, J., Marsal, J. and Lopez, G. 2006. Establishment of stem water potential thresholds for the response of 'O'Henry peach fruit growth to water stress during stage III of fruit development. *Acta Horticulturae* 713: 197–201.

Girona, J., Marsal, J., Mata, M., Arbones, A. and Delong, T.M. 2004. A comparison of the combined effect of water stress and crop loud on fruit growth during different phenological stages in young peach trees. *Journal of Horticultural Science and Biotechnology* 79: 308–312.

Goldhamer, D.A., Salinas, M., Crisemo, C., Day, K., Soler, M. and Moriana, A. 2002. Effects of regulated deficit irrigation and partial rootzone drying on fate harvest peach tree performance. *Acta Horticulturae* 592: 343–350.

Goldhamer, D.A. and Viveros, M. 2000. Effects of preharvest irrigation cutoff durations and postharvest water deprivation on almond tree performance. *Irrigation Science* 19(3): 125–131.

Goode, E., Higgs, K.H. and Hyrycz, K.J. 1979. Nitrogen and water effects on the nutrition, growth, cmp yield and fruit quality of orchard-grown Cox's Orange Pippin apple trees. *Journal of Horticultural Science* 534: 295–306.

Gunduz, M., Korkmaz, N., Asik, S., Unal, H.B. and Avci, M. 2011. Effects of various irrigation regimes on soil water balance, yield, and fruit quality of drip-irrigated peach trees. *Journal of Irrigation and Drainage Engineering* 137(7): 426–434.

Hamdy, A., Ragab, R. and Scarascia-Mugnozza, E. 2003. Coping with water scarcity: water saving and increasing water productivity. *Irrigation and Drainage* 52(1): 3–20.

Haulik, T.K. 1979. The effect of three irrigation schedules on two peach cultivars. *Crop Production* 8: 207–210.

Henderson, J.C. 2019. Wastewater effluent transport and contamination: a model for groundwater contamination in the Central West Bank. Drexel University.

Hipps, N.A. 1997. Effect of nitrogen, phosphorus, water and pre-planting soil sterilization on growth and yield of Cox Orange on M9 apple trees. *Acta Horticulturae* 448: 125–131.

Hossein, D., Naseri, A., Anyoji, H. and Encji, A.E. 2007. Effects of deficit irrigation and fertilizer use on vegetative growth of drip irrigated cherry trees. *Journal of Plant Nutrition* 30(3): 411–425.

Hsiao, T.C. 1973. Plant responses to water stress. *Annual Review of Plant Physiology* 24: 519–570.

Hussein, A.H.A. 2008. Response of Manzanillo olive (*Olea europaea* L.) cultivar to irrigation regime and potassium fertigation under Tabouk conditions, Saudi Arabia. *Journal of Agronomy* 7(4): 285–296.

Hussien, S.M., Fathi, M.A. and Eid, T.A. 2013. Effect of shifting to drip irrigation on some plum cultivars grown in clay loamy soil. *Egypt Journal of Agricultural Research* 91(1): 217–232.

Hutmacher, R.B., Nightingale, H.I., Rolston, D.E., Biggar, J.W., Dale, F., Vail, S.S. and Peters, D. 1994. Growth and yield responses of almond (*Prunus amygdalus*) to trickle irrigation. *Irrigation Science* 14: 117–126.

Iglesias, I. and Echeverria, G. 2009. Differential effect of cultivar and harvest date on nectarine colour, quality and consumer acceptance. *Scientia Horticulturae* 120: 41–50.

Intrigliolo, D.S. and Castel, J.R. 2005. Effects of regulated deficit irrigation on growth and yield in young Japanese plum. *Journal of Horticultural Science and Biotechnology* 80: 177–182.

Intrigliolo, D.S. and Castel, J.R. 2010. Response of plum trees to deficit irrigation under two crop levels: tree growth, yield and fruit quality. *Irrigation Science* 28: 525–534.

Irving, D.D. and Drost, J.H. 1987. Effects of water deficit on vegetative growth fruit growth and fruit quality in Cox's Orange Pippin apple. *Journal of Horticultural Science* 62: 427–432.

Jain, N., Chauhan, H.S., Singh, P.K. and Shukla, K.N. 2000. Response of tomato under drip irrigation and plastic mulching. In: *Proceeding of 6ᵗʰ International Micro-irrigation Congress, Micro-irrigation Technology for Developing Agriculture*, 22–27 October 2000, South Africa.

Kachwaya, D.S. 2014. Studies on drip irrigation and fertigation in strawberry (*Fragaria* x *ananasa* Duch). Ph.D. Thesis, Dr Yashwant Singh Parmar University of Horticulture and Forestry, Nauni, Solan (H.P).

Kahramanoğlu, I. 2017. Introductory chapter: postharvest physiology and technology of horticultural crops. In *Postharvest handling* (pp. 1–5). InTech Open.

Kaya, S., Evren, S., Dasci, E., Adiguzel, M.C. and Yilmaz, H. 2010. Effects of different irrigation regimes on vegetative growth, fruit yield and quality of drip-irrigated apricot trees. *African Journal of Biotechnology* 9(36): 5902–5907.

Khan, I.A., Wani, M.S., Mir, M.A., Rasool, K. and Simnani, S.A. 2012. Physiological and yield response of almond to different drip irrigation regimes under temperate conditions. *Indian Journal of Horticulture* 72(2): 187–192.

Kilili, A.W., Behboudian, M.H. and Mills, T.M. 1996. Composition and quality of Braeburn apples under reduced irrigation. *Scientia Horticulturae* 67: 1–11.

Kim, S.K., Na, H.Y., Song, J.H. and Kim, M.J. 2009. Influence of water stress on fruit quality and yield of strawberry cvs. 'Machyang' and 'Seolhyang. *Acta Horticulturae* 842: 177–180.

Kireva, R., Petrova-Branicheva, V. and Markov, E. 2017. Drip irrigation of apples at a moderate continental climate. *International Research Journal of Engineering and Technology* 9: 642–645.

Kirnak, H., Kaya, C., Higgs, D. and Gercek, S. 2001. A long-term experiment to study the role of mulches in the physiology and macro-nutrition of strawberry grown under water stress. *Australian Journal of Agricultural Research* 52(9): 937–943.

Klamkowski, K. and Treder, W. 2008. Response to drought stress of three strawberry cultivars grown under greenhouse conditions. *Journal of Fruit and Ornamental Plant Research* 16: 179–188.

Kocaçalışkan, L. 2005. *Plant physiology*. 5th edn. Academic Press. Dumlupinar University.

Kucukyumuk, C. and Kacal, E. 2010. The effects of different irrigation programmes in drip irrigation on leaf properties of Starkrimson delicious apple variety. *Bulletin UASVM Horticulture* 67(2): 297–300.

Kucukyumuk, C., Kacala, E., Ertekb, A., Ozturka, G., Yasemin, S. and Kurttas, K. 2012. Pomological and vegetative changes during transition from flood irrigation to drip irrigation: Starkrimson Delicious apple variety. *Scientia Horticulturae* 136: 17–23.

Kumar, S. 2004. Effect of drip irrigation, fertigation and root stocks on apple under high density planting. Ph.D. Thesis submitted to Dr. Y.S. Parmar University of Horticulture and Forestry, Nauni-Solan (H.P).

Kumar, P.S., Chaudhary, V.K. and Bhagawati, R. 2012. Influence of mulching and irrigation level on water-use efficiency, plant growth and quality of strawberry (*Fragaria* x *ananassa*). *Indian Journal of Agricultural Sciences* 82(2): 127–133.

Layne, D.R., Cox, D.B. and Hitzler, E.J. 2002. Peach systems trail: the influence of training system, tree density, rootstock, irrigation and fertility on growth and yield of young trees in South Carolina. *Acta Horticulturae* 592: 367–375.

Layne, R.E.C., Tan, C.S., Hunter, D.M. and Cline, R.A. 1996. Irrigation and fertilizer application methods affect performance of high density peach orchards. *HortScience* 31(3): 370–375.

Leib, B.G., Caspari, H.W., Redulla, C.A., Andrews, P.K. and Jabro, J.J. 2006. Partial rootzone drying and deficit irrigation of Fuji apples in a semi-arid climate. *Irrigation Science* 24: 85–99.

Li, S.H. and Huguet, J.G. 1989. Production, fruit quality and development of peach trees under different irrigation regimes. *Fruits* 44(4): 225–232.

Li, S.H., Huguet, J.G., Schoch, P.G. and Orlando, P. 1989. Response of peach-tree growth and cropping to soil water deficit at various phenological stages of fruit development. *Journal of Horticultural Science* 64: 541–552.

Lischuk, A.L., Semash, D.P. and Storchous, V.N. 1988. Transpiration intensity of apple and peach leaves with different irrigation methods. *Byulleten Gosudarstvennogo Nikitskogo Botanicheskogo Sada* 65: 89–93.

Litschmann, T. 2004. Význam zavlažování dále poroste. *Zemědělec* 20: 9.

Liu, J., Yang, H., Gosling, S.N., Kummu, M., Florke, M., Pfister, S., Hanasaki, N., Wada, Y., Zhang, X., Zheng, C., Alcamo, J. and Oki, T. 2017. Water scarcity assessments in the past, present, and future. *Earth's Future* 5(6): 545–589.

Lopez, G., Behboudian, M.H., Echeverria, G., Girona, J. and Marsal, J. 2011. Instrumental and sensory evaluation of fruit quality for Ryan's Sun peach grown under deficit irrigation. *HortTechnology* 21: 712–719.

Lopez, G., Behboudian, M.H., Vallverdu, X., Mata, M., Girona, J. and Marsal, J. 2010. Mitigation of severe water stress by fruit thinning in 'O'Henry peach: implications for fruit quality. *Scientia Horticulturae* 125: 294–300.

Lopez, G., Mata, M., Arbones, A., Solans, J.R., Girona, J. and Marsal, J. 2006. Mitigation of effects of extreme drought during stage III of peach fruit development by summer pruning and fruit thinning. *Tree Physiology* 26: 469–477.

Lovelli, S., Perniola, M., Ferrara, A. and Tommaso, T. 2007. Yield response factor to water (ky) and water use efficiency of *Carthamus tinctorius* L. and *Solanum melongena* L. *Agricultural Water Management* 92: 191–201.

Maas, R. and Van, D. 1996. Adjust water application to expected fruit size. *Fruitlet* 86: 14–12.

Malik, R.S., Bhardwaj, S.K., Sharma, I.P. and Bhandari, A.R. 1992. Water application efficiency and nutrient movement in soils of some fruit crops under drip irrigation in mid Himalayas. *National Horticulture Board Technical Bulletin*, pp. 4–5.

Marsal, J., Behboudian, M.H., Mata, M., Basile, B., del-campo, J., Girona, J. and Lopez, G. 2010. Fruit thinning in conference pear grown under deficit irrigation to optimise yield and to improve tree water status. *Journal of Horticultural Science and Biotechnology* 85: 125–130.

Martinez-Lüscher, J., Kizildeniz, T., Vučetić, V., Dai, Z., Luedeling, E. and van Leeuwen, C. 2016. Sensitivity of grapevine phenology to water availability, temperature and CO concentration. *Environmental Science* 4: 1–14.

Martins-de-Oliveira, C.P., Simões, W.I., Bezerra-da-Silva, J.A., Lopes, P.R.C., Jurema-Araújo, E.F. and Cavalcante, B.I. 2017. Flowering, fruiting and physiology of apple tree under different irrigation levels in the Brazilian semiarid region. *Comunicata Scientiae* 8(1): 99–108.

Mechlia, N.B., Ghrab, M., Zitouna, R., Mimoun, M.B. and Masmoudi, M. 2012. Cumulative effect over five years of deficit irrigation on peach yield and quality. *Acta Horticulturae* **592**: 301–307.

Mercier, V., Bussi, C., Lescourret, F. and Genard, M. 2009. Effects of different irrigation regimes applied during the final stage of rapid fruit growth of an early maturing peach cultivar. *Irrigation Science* 27: 297–306.

Miculka, B. 1983. Effect of positioned irrigation on nutrient concentration in peach leaves. *SbornikUvtizZahrgdnictvi* 10(3): 185–194.

Miletic, R., Nikolic, R., Mitic, N., Rakicevic, M. and Blagojevic, M. 2007. Impact of rainfall and irrigation on pomological technological characteristics of the fruit and on the yield of plum cultivars. *Focarstvo* 41: 113–119.

Miller, S.A., Smith, G.S., Boldingh, H.L. and Johansson, A. 1998. Effects of water stress on fruit quality attributes of Kiwifruit. *Annals of Botany* 81: 73–81.

Mills, T.M., Behboudian, M.H. and Clothier, B.E. 1996. Water relations, growth, and the composition of Braeburn apple fruit under deficit irrigation. *Journal of the American Society for Horticultural Science* 121: 286–291.

Mills, T.M., Behboudian, M.H., Tan, P.Y. and Clothier, B.E. 1994. Plant water status and fruit quality in Braeburn apples. *HortScience* **29**: 1274–1278.

Milovankic, M., Vujanic-Varga, D. and Vucic, N. 1981. The fruiting potential of the apple cultivars Jonathan and Golden Delicious under irrigation. *Jugoslovensko Vocarstvo* 12(55–56): 423–427.

Misgar, F.A. and Kumar, A. 2008a. Effect of soil moisture regimes on plant growth, water relations and yield of Red Delicious/M_9 apple trees. *The Horticultural Journal* 21(2): 53–56.

Misgar, F.A. and Kumar, A. 2008b. Effect of different moisture regimes on physico-chemical characters of Red Delicious/M_9 apple. *Himachal Journal of Agricultural Research* 34(2): 58–61.

Misgar, F.A. and Kumar, A. 2010. Stomatal behaviour of Red Delicious apple leaves as influenced by different moisture levels. *Indian Journal of Horticulture* 67(2): 264–266.

Mitobe, M., Asano, S., Sakai, Y., Okuno, T. and Mukai, B. 1991. Studies of increased yield of Japanese pears using the standard model pruning technique. *Bulletin of the Saitama Horticultural Experiment Station* 18: 67–79.

Monastra, F., Avanzato, D., Martelli, S. and Dascanio, R. 1995. Pistachio trial under different volumes of irrigation in Italy. *Acta Horticulturae* 419: 249–252.

Moriana, A., Corell, M., Girón, I.F., Conejero, W., Morales, D., Torrecillas, A. and Moreno, F. 2013. Regulated deficit irrigation based on threshold values of trunk diameter fluctuation indicators in table olive trees. *Scientia Horticulturae* 164: 102–111.

Mousavi, A. and Alimohamadi, R. 2006. Effects of deficit irrigation and drought during different phenological stages of fruit growth and development in almond production. *Acta Horticulturae* **726**: 489–494.

Moussa, A.M.A. 2018. Assessment of sediment deposition in Aswan High Dam Reservoir during 50 years (1964–2014). In *Grand Ethiopian renaissance dam versus Aswan high dam* (pp. 233–253). Cham: Springer.

Mpelasoka, B.S. and Behboudian, M.H. 2002. Production of aroma volatiles in response to deficit irrigation and to crop load in relation to fruit maturity for Braeburn apple. *Postharvest Biology and Technology* 24: 1–11.

Mpelasoka, B.S., Behboudian, M.H. and Green, S.R. 2001b. Water use, yield and fruit quality of lysimeter-grown apple trees: responses to deficit irrigation and to crop load. *Irrigation Science* 20: 107–113.

Mpelasoka, B.S., Behboudian, M.H. and Mills, T.M. 2001a. Water relations, photosynthesis, growth, yield and fruit size of Braeburn apple: Responses to deficit irrigation and to crop load. *Journal of Horticultural Science and Biotechnology* 76: 120–126.

Mpelasoka, B.S., Behboudian, M.H. and Mills, T.M. 2002. Effects of deficit irrigation on fruit maturity and quality of 'Braeburn' apple. *ScientiaHorticulturae* 90: 279–290.

Mpelasoka, B., Behboudian, M.H. and Mills, T. 2012. Water relations photosynthesis, growth, yield and frait size of Braebum apple: responses to deficit irrigation and crop load. *The Journal of Horticultural Science and Biotechnology* 76(2): 120–126.

Naor, A., Klein, L., Doron, L., Gal, Y., Den-David, Z. and Bravdo, B. 1997. Irrigation and crop load interaction in relation to apple yield and fruit size distribution. *Journal of the American Society for Horticultural Science* 122: 411–414.

Naor, A., Naschitz, S., Peres, M. and Gal, Y. 2008. Responses of apple fruit size to tree water status and crop load. *Tree Physiology* 28: 1255–1261.

Naor, A., Peres, M., Greenblat, Y., Gal, Y. and Ben Arie, R. 2004. Effects of pre-harvest irrigation regime and crop level on yield, fruit size distribution and fruit quality of field-grown Black Amber Japanese plum. *Journal of Horticultural Science and Biotechnology* 79: 281–288.

Nasharty, A.H. and Ibrahim, I.M. 1961. Progress report on effect of frequency of irrigation on quality and quantity of plum fruits. *Agricultural Research Review* 39: 100–107.

Neilsen, G.H., Hoyt, P.B. and Neilsen, D. 1995. Soil chemical changes associated with N-P fertigated and drip irrigated high-density apple orchards. *Canadian Journal of Soil Science* 75: 307–310.

Neilsen, D., Neilsen, G.H. and Herbert, L. 2010. Effect of irrigation and crop load management on fruit nutrition and quality for Ambrosia/M.9 apple. *Acta Horticulturae* 868: 63–72.

Packer, W.J., Chalmers, D.J. and Baxter, P. 1963. Supplementary irrigation of Jonathan apple trees. *Agricultural Journal of Victoria* 61: 453–460.

Pascual, M., Villar, J.M., Domingo, X. and Rufat, J. 2010. Water productivity of peach for processing in a soil with low available water holding capacity. *Acta Horticulturae* 889: 189–195.

Passioura, J.B. 2004. Water-use efficiency in farmers' fields. In *Water-use efficiency in plant biology* (pp. 302–321), M. Bacon (Ed.). Blackwell, Oxford.

Perez-Pastor, A., Domingo, R., Torrecillas, A. and Ruiz-Sanchez, M.C. 2009. Response of apricot trees to deficit irrigation strategies. *Irrigation Science* 27: 231–242.

Poddar, R., Qureshi, M.E. and Shi, T. 2014. A comparison of water policies for sustainable irrigation management: the case of India and Australia. *Water Resources Management* 28(4): 1079–1094.

Powell, D.B.B. 1974. Some effects of water stress in late spring on apple trees. *Journal of Horticultural Science* 49: 257–272.

Rana, R.K., Chauhan, J.S. and Chandel, J. 2000. Effect of different soil moisture regimes on growth, cropping, fruit quality and water relations in kiwifruit. *Indian Journal of Agricultural Sciences* 70: 546–549.

Rana, G.S. and Daulta, B.S. 1997. Effect of different rootstocks, spacing and drip irrigation levels on plant height of peach (*Prunus persica* Batsch.) cv. Flordasun. *Crop Research* 14(2): 293–296.

Rana, G.S., Sehrawat, S.K., Daulta, B.S. and Beniwal, B. 2005. Effect of drip irrigation and rootstock on N, P and K leaf content in peach under high density plantation. *Acta Horticulturae* 696: 223–226.

Razouk, R., Lbijbijen, J., Kajji, A. and Mohammed, K. 2013. Response of peach, plum and almond to water restrictions applied during slowdown periods of fruit growth. *American Journal of Plant Sciences* 4(3): 561–570.

Reeder, B.D., Newman, J.S. and Worthington, J.W. 1976. Trickle irrigation on peaches. Publication, Texas Agricultural Experimental Station. PR-3437. p. 2.

Reid, J.B., Brash, D.W., Sorensen, I.B. and Bycroft, B. 1996. Improvement in kiwifruit storage life caused by withholding early-season irrigation. *New Zealand Journal of Crops and Horticultural Science* 24: 21–28.

Remorini, D. and Massai, R. 2003. Comparison of water status indicators for young peach trees. *Irrigation Science* 22: 39–46.

Romero, P., Botia, P. and Garcia, F. 2004. Effects of regulated deficit irrigation under subsurface drip irrigation conditions on vegetative development and yield of mature almond trees. *Plant and Soil* 260(1): 169–181.

Romo, R. and Diaz, D.H. 1985. Root system and nutritional status of peaches under drip or flood irrigation in warm climates. *Acta Horticulturae* 173: 167–175.

Rufat, J., Arbones, A., Villar, P., Domingo, X., Pascual, M. and Villar, J.M. 2010. Effects of irrigation and nitrogen fertilization on growth, yield and fruit quality parameters of peaches for processing. *Acta Horticulturae* 868: 87–93.

Ruiz-Sanchez, M.C., Torrecillas, A., Perez-Pastor, A. and Domingo, R. 2000. Regulated deficit irrigation in apricot trees. *Acta Horticulturae* 537: 759–766.

Rzekanowski, C.Z. and Rolbiecki, S.T. 2000. The influence of drip irrigation on the yield of some cultivars of apple trees in Central Poland under different rainfall conditions during the vegetation season. *Acta Horticulturae* 537: 929–928.

Sabagh, A.S. and Aggag, A.M. 2003. Response of annual apple trees to different water regimes under drip irrigation system. *Alexandria Journal of Agricultural Research* 48(2): 139–147.

Safran, B., Bravdo, B. and Bernstein, Z. 1975. Drip irrigation. *OIV Bulletin* 48: 406–429.

Serrano, L., Carbonell, X., Save, R., Marfa, O. and Penuelas, J. 1992. Effects of irrigation regimes on the yield and water use of strawberry. *Irrigation Science* 13: 45–48.

Shackel, K.A. 2007. Water relations of woody perennial plant species. *Journal International des Sciences de la Vigne et dhe Vin* 41: 121–129.

Sharma, N.C. and Chandel, J.S. 2005. Effect of different levels of irrigation regimes on yield, fruit quality and nutrient status of apricot (*Prunus armeniaca* L.). *Progressive Horticulture* 37(1): 78–81.

Sharma, I.P., Kumar, S. and Kumar, P. 2005. Effect of drip irrigation and mulches on yield, quality and water-use efficiency in strawberry under mid hill conditions. *Acta Horticulturae* 696: 259–264.

Singh, S. 2013. Effect of drip irrigation and mulch on soil hydrothermal regimes, weed incidence, yield and quality of apricot cv. New Castle. M.Sc. Thesis. Department of Soil Sciences, Dr Y S Parmar UHF Nauni Solan, H.P. India.

Singh, R., Bhandari, A.R., Thakur, B.C. and Singh, R. 2002. Effect of drip irrigation regimes and plastic mulch on fruit growth and yield of apricot (*Prunus armeniaca*). *Indian Journal of Agricultural Sciences* 72(6): 355–357.

Sivanappan, R.K. 1994. Prospects of micro irrigation in India. *Irrigation and Drainage Systems* 8(1): 49–58.

Sivanappan, R. K. 1998. Irrigation water management for sugarcane in VSI, pp. II 100–125.

Sotiropoulos, T., Kalfountzo, D., Aleksiou, I., Kotsopoulos, S. and Koutinas, N. 2010. Response of a clingstone peach cultivar to regulated deficit irrigation. *Scientia Agricola* 67(2): 164–169.

Stino, R.G., Abd El-Mohsen, M.A., Shawky, M.E., Yhia, M.M. and Wahab, A.E. 2016. Impact of applied irrigation regime during specified phenological stages on cropping and its attributes on Le-conte pear. *Annals of Agriculture Science* 54(4): 877–890.

Tagi, A. 1984. Effect of different levels of nitrogen, potassium and irrigation on growth cropping and quality of Santa Rosa Plum (*Prunus salicina* Lindl.). Ph.D. Thesis submitted to Himachal Pradesh Krishi Vishva Vidyalaya, Solan (HP),

Talluto, G., Farina, V., Volpe, G. and Lo Bianco, R. 2008. Effects of partial rootzone drying and rootstock vigour on growth and fruit quality of Pink Lady apple trees in mediterranean environments. *Australian Journal of Agriculture Research* 59: 785–794.

Temocico, G., Alecu, I. and Alecu, E. 2011. Drip irrigation affects apple fruits harvest Scientific papers of the R.L.F.G Pitesti, Vol. XXVII.

Torrecillas, A., Domingo, R., Galego, R. and Ruiz-Sanchez, M.C. 2000. Apricot tree response to withholding irrigation at different phenological periods. *Scientia Horticulturare* 85: 201–212.

Treder, W., Grzyb, Z. and Rozpars, E. 1999. The influence of irrigation on growth and yield of plum trees cv. Valor grafted on myrobalan and Wangenheim Prune. *Acta Agrobotanica* 52(1/2): 95–101.

Veihmeyer, F.J. 1972. The availability of soil moisture to plants, results of empherical experiment with fruit trees. *Soil Science* 114(4): 268–294.

Velardo-Micharet, B., Peñas Diaz, L., Tapia Garcia, I.M., Nieto Serrano, E. and Campillo Torres, C. 2017. Effect of irrigation on postharvest quality of two sweet cherry cultivars (*Prunus avium* L.). *Acta Horticulturae* 1161: 667–672.

Vera, J., Abrisqueta, I., Abrisqueta, J.M. and Sanchez, R. 2013. Effect of deficit irrigation on early maturing peach tree performance. *Irrigation Science* 31(4): 747–757.

Verma, P. and Chandel, J.S. 2017. Effect of different levels of drip and basin irrigation on growth, yield, fruit quality and leaf nutrient contents of peach cv. Redhaven. *The Bioscan* 2(2): 1035–1039.

Verma, M.L., Chauhan, P.S., Sharma, L.K. and Bhardwaj, S.P. 2007. Effect of irrigation levels on soil moisture and water-use-efficiency and yield of apple under drip irrigation. *Indian Journal of Soil Conservation* 35(2): 125–128.

Veverka, V. and Pavlačka, R. 2012. The effect of drip irrigation on the yield and quality of apples. *Acta Univ. Agric. et Silvic. Mendel. Brun* 60(8): 247–252.

Wang, D. 2011. Deficit irrigation of peach trees to reduce water consumption. *Transactions on Ecology and the Environment* 145: 497–505.

Xiloyannis, C., Angelini, P. and Galliano, A. 1990. Drip irrigation of Kiwifruit trees. *Acta Horticulturae* 282: 217–225.

Yldrm, M. and Yldrm, O. 2005. Effect of different irrigation programs on plum tree growth, yield and fruit quality under drip irrigation. *Ziraat Fakultesi Dergisi Uludag Universitesi* 19(1): 37–49.

Zhang, H., Wang, D. and Gartung, J.L. 2017. Influence of irrigation scheduling using thermometry on peach tree water status and yield under different irrigation systems. *Agronomy* 7: 1–12.

Zhao, Z., Wang, W., Wu, Y. and Huang, X. 2012. Yield and water use efficiency of pear trees under drip irrigation with different surface wetted percentages. *International Journal of Agriculture and Biology* 14: 887–893.

13 Livestock in Sustainable Watershed Management

Vijayakumar, P.

13.1 INTRODUCTION

India being an agrarian country is blessed with abundant flora and fauna resources. It is one of the largest producers of a large number of agricultural and livestock produce fulfilling the nutritional requirements of India and other countries. The livestock sector is one of the rapidly progressing sub-sectors of Indian agriculture. India is bestowed with large species of livestock resources like Cattle, Buffalo, Sheep, Goat, Pig, Poultry, Camel, Yak, and Mithun; livestock sector contributes significantly toward the livelihood of the farming community and Indian economy. They provide commodities like milk, meat, egg, hide, and manure and services like Fuel (Go bar Gas), draught power for agriculture and transportation for the benefit of humankind. Livestock is an integral part of the watershed management. They utilize the agricultural waste/by-products and convert them into useful animal produce. They also significantly contribute in maintaining the ecological balance and environmental sustainability. Go bar gas (Fig. 13.1) is an alternative source of green energy fulfilling the requirements of households in rural areas. Animal waste like dung, urine, and leftover feeds/fodder can be converted into nutrient-rich organic manure through composting/vermicomposting (Fig. 13.2), which helps in restoring, enriching, and sustaining the soil health.

Watershed is a holistic approach that integrates the different avenues of agricultural and allied sectors by utilizing the natural resources like land, and water, thereby minimizing the risk and maximizing the output. Livestock farming is not season-dependent unlike agricultural cultivation and can be undertaken throughout the year. The demand for the livestock products is also not season dependent. When animal husbandry is amalgamated into the watershed activity, it significantly improves the livelihood opportunities to the watershed beneficiaries resulting in uninterrupted revenue generation and employment leading to upliftment of socio-economic status of the marginal and destitute sections of the society.

13.2 LIVESTOCK WEALTH IN INDIA

India is bestowed with abundant flora and fauna. Livestock is one such area India should be proud of since it is the home to many species of livestock and poultry along with a large number of indigenous breeds. It harbors different livestock species like Cattle, Buffalo, Sheep, Goat, Pig, Horse, Donkey, Yak, Mithun, and Camel. India also has different species under the group "Poultry" like the Chicken, Duck, Guinea Fowl, Emu, Ostrich, Japanese Quail, and Turkey. From this, we can easily assess the importance of livestock in the Indian economy, livelihood development of rural

DOI: 10.1201/9781003351672-13

FIGURE 13.1   Gobar gas plant.

FIGURE 13.2   Vermicomposting pit.

masses, and nutritional security. Livestock is also best suited for Mixed or Integrated farming system (Fig. 13.3), which is vital for watershed development. As per the 20th Livestock census (2019), India possesses 30.38 crore bovines that include Cattle, Buffalo, Mithun, and Yak; 14.9 crore goats, 7.4 crore sheep, 0.9 crore pigs, and 85.2 crore Poultry (DAHD, 2022). According to the National Bureau of Animal Genetic

Duck-Fish   **Pig-Fish**

FIGURE 13.3   Mixed/integrated farming system.

Resources (NBAGR), Karnal, India possesses 50 registered cattle breeds, 19 buffalo breeds, 34 goat breeds, 44 sheep breeds, 10 pig breeds, and 19 chicken breeds.

As far as the economic contribution of the livestock sector is concerned, according to the report of National Statistical Office (NSO), MoSPI, 2021, the Gross Value Added (GVA) of the livestock sector is about Rs. 1,114,249 crores at current prices during the financial year 2020–21, which is about 30.87% of Agricultural and Allied Sector GVA and 6.17% of total GVA (DAHD, 2022). India being the largest milk producer in the world contributes to about 23% of the total global milk production with an estimated record production of 221.06 million metric tonnes during 2021–22. Milk production in India is growing at 5–6% per year. About 50% of the country's total milk production is obtained from Cows, 45% from Buffaloes, and 3% from Goats and other species like Yak, Mithun, Camel, and Sheep. Even poultry production has taken a quantum leap in recent times with an annual egg production of about 129.6 billion eggs during 2021–22 (DAHD, 2022).

Livestock's contribution to the farming community includes nutrition in the form of milk (Cattle, Buffalo, and Goat), egg (Chicken, Duck, Quail, and Turkey), meat (Chicken, Buffalo, Sheep, Goat, Pig, and Poultry), and their products; draught power or energy for ploughing operations in agricultural fields as well as transportation of produce/inputs (draught animals like oxen, bullock, male buffaloes, camel, horse, and donkey); manure or fertilizer and Fuel from animal dung/feces (dung cake, go bar gas); clothing (Hide); leather and wool for protection from cold weather conditions; and many such useful commodities and services (Fig. 13.4). The livestock sector contributes significantly to socio-economic and livelihood development through income and employment generation.

13.3   LIVESTOCK AND ECONOMY

Livestock impacts significantly to the economy of both farmers as well as the nation as a whole. It not only provides nutritious food, but it is also one of the income-generating activity of farming community. It is also ready to cash to the farmers during a sudden crisis where the farmer can sell his/her animal for the monetary benefit and that is the reason why the livestock can be considered "Money on Hoof" or "Live Cash". It can be the main source of income to the farmers during natural

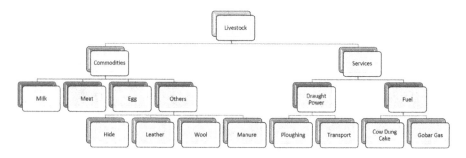

FIGURE 13.4 Commodities and services from livestock.

TABLE 13.1

Reports Reflecting on the Relationship between Livestock and Economy

WASSAN Andhra Pradesh Report (2004)	"Livestock sector is gaining prominence due to higher income elasticity of demand for its products – milk, meat and eggs" and "its contribution of 20% of the state domestic product from agriculture (and allied activities) is increasing".
Maharashtra Report	"Due to the uncertainty of agricultural production, now-a-days livestock is considered an important source of income in most areas of Maharashtra".
Singh *et al.* (2020)	Animal Husbandry sector plays a vital role in the upliftment of socio-economic status of the farming community. Domestic animals and poultry birds are considered a "Mobile Asset" since they can be a ready source of income (sale of animals and birds) to farmers during emergencies and financial crises.
Misra *et al.* (2007)	In areas that are dependent on rainfed farming, livestock contributes immensely to the sustainable livelihood of farming community through income generation and increased economic stability thereby masking the inherent risk of crop failure/loss due to natural calamities. Further, due to hardy nature, small ruminants like sheep and goat can be a savior during drought periods since they can be easily reared in drylands.

calamities like floods, and drought, which had a significant negative impact on the agriculture productivity. Moreover, livestock is not season-dependent and can provide a regular source of income throughout the year. It also utilizes the agriculture waste and by-products and converts them into nutritious animal produce like milk, meat, and egg. The livestock farm waste like dung, urine, and left-over feed/fodder can act as a major source of organic nutrients to the agricultural fields in the form of farmyard manure, vermicompost, compost, etc. The dung and urine can be efficiently converted into green energy sources through generation of biogas/go bar gas for household consumption.

The importance of livestock and the economy has been studied by researchers. The important aspects are presented in Table 13.1.

13.4   LIVESTOCK AND CLIMATE CHANGE

In recent years, global warming is showing an increasing trend that has put ecosystems, habitats, biodiversity, and food security at risk. It is well-established fact that global warming not only affects the animals in tropical countries but also in temperate zones where rise in ambient temperatures is becoming an issue. At the same time, to fulfill the nutritional needs of the rising population, intensive animal agriculture is a must. Current scenario clearly shows that livestock production systems based on grazing, mixed farming, or industrialized systems will be drastically affected (negatively) by the climate change, particularly due to global warming (Bernabucci, 2019).

Livestock is often blamed as one of the major contributors of climate change. It is partially true as livestock are major contributor of some of the greenhouse gases (GHGs) like methane (CH_4), carbon dioxide (CO_2), and nitrous oxide (N_2O). Ruminants like cattle, sheep, and goat are the major contributors of CH_4. At the same time, livestock also has positive impact, which will be discussed later. On one hand, intensive livestock production results in emission of GHGs resulting in climate change, on the other hand, the climate change also impacts the livestock production system through changes in productivity and quality of feed crop and forage, which in turn impacts the milk production and quality; water availability and quality; animal growth, production, and reproduction; biosecurity and animal health.

13.4.1 Impact of Livestock on Climate Change

The impact of livestock on climate change is depicted below (Fig. 13.5).
Livestock impacts the climate change in the following ways:

- Livestock are often linked with the negative impact on the environmental conditions. Livestock is a major contributor of harmful anthropogenic GHGs worldwide like CO_2 (5%), CH_4 (44%), and N_2O (53%).

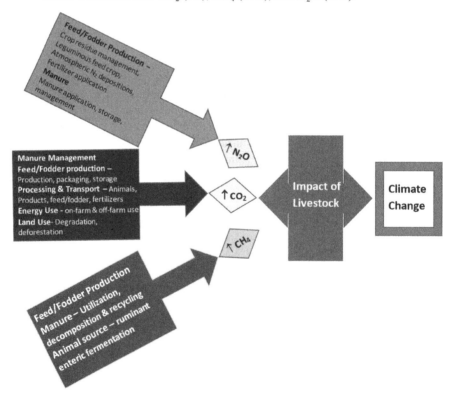

FIGURE 13.5 Impact of livestock on climate change. (Adapted from Rojas-Downing *et al.*, 2017.)

- Direct emission of these harmful gases is through enteric fermentation, respiration, and animal waste excretions (poultry droppings, urine, dung). Whereas, indirect emission is from forage crops, manure, farm operations and machinery, processing, transportation, grazing, deforestation, etc.
- Changes in the land use pattern like conversion of cultivation land to pastures and deforestation due to livestock production are the main sources of CO_2 emissions.
- Feed and fodder production and distribution contribute significantly to GHG emission. Activities resulting in emission of GHGs include production and use of fertilizers/manure for feed/fodder production, processing of feed/fodder, transportation, etc.
- Manure production and utilization results in release of CH_4 and N_2O gases.

13.4.2   IMPACT OF CLIMATE CHANGE ON LIVESTOCK

The impact of climate change on livestock is depicted below (Fig. 13.6).
   Climate change impacts the livestock in the following ways:

- Increase in GHGs and temperature will have negative impact on the quantity and quality of feed/fodder resulting in decreased nutrient availability affecting animal performance.

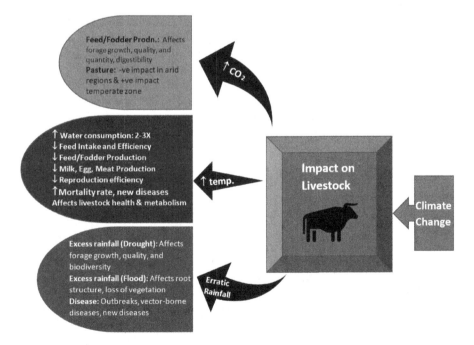

FIGURE 13.6   Impact of climate change on livestock. (Rojas-Downing *et al.*, 2017.)

- Decrease water availability for drinking, feed/fodder production and other activities as well as increase in water consumption by the animals.
- Increase in temperature and GHGs increases the potential for animal diseases and related death. It also results in increase in the growth of microbes and vectors resulting in rapid spreading of diseases, especially vector-borne and food-borne diseases. There are chances of emergence of new diseases and outbreaks.
- Increase in environmental temperature due to climate change may directly lead to heat stress and heat stroke in animals. Indirectly, it will affect the feed/fodder production, increase water intake, decrease feed/fodder intake by animals and altered physiological functions as well as nutrient utilization.

13.5 LIVESTOCK AND ENVIRONMENT

More often than not, livestock is associated with a negative impact on the environment than its positive effect. Livestock is one of the major contributors of GHGs like CH_4, N_2O, and CO_2. CH_4 is a product of enteric fermentation and manure storage, whereas N_2O is generated from manure storage. The release of CO_2 is due to various reasons, including conversion of forest fields and pastures, and degradation due to unsustainable management. N_2O has a higher global warming potential (265 times higher than CO_2) than CH_4 (28 times higher than CO_2) and carbon dioxide. In addition to this, livestock also contributes to global warming through pollution, overgrazing, deforestation, desertification, feed production, and processing.

Amongst the livestock, cattle are the largest contributor of GHGs followed by pigs, buffalo, chicken, and small ruminants. In order to reduce the GHG production from livestock, many manage mental strategies for mitigation have been suggested by the experts which include reducing the number of livestock, particularly the unproductive animals; selection of animals based on their genetic potential to consume less feed/fodder and/or produce less CH_4 per unit of feed/fodder; increasing the efficiency of livestock production through genetic selection of animals and nutritional manipulation; grazing management; and other strategies like better waste management, changes in the consumption pattern toward the products rich in protein, such as meat, milk, egg, and fish (Sejian, 2012).

13.5.1 POSITIVE EFFECTS ON THE ENVIRONMENT

On the other hand, livestock also has positive effects on the environment. Some of the positive impacts of livestock on the environment are presented (Fig. 13.7) and discussed below:

- **Draught Power:** The power generated from livestock, particularly the draught animals like the bullock, oxen, buffalo, horses, donkey, and camel through ploughing, transportation can be considered a source of green energy. In this era of industrialization and mechanization, farmers still are adopting the traditional way of agriculture by using draught animals for different activities *viz.* transportation and agricultural operations like ploughing, tillage, weeding, harvesting, and chaff cutting.

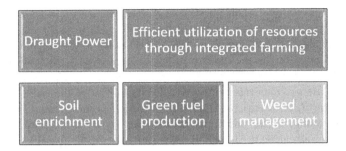

FIGURE 13.7 Positive effects of livestock on the environment.

- **Efficient utilization of resources through Integrated Farming:** Livestock is well suited for mixed farming or integrated farming system. Either livestock can be mixed or integrated with agriculture or other related operations, or different livestock can be reared together (including fish farming). This will not only help in recycling or effective utilization of the waste (as a source of input) but also protect the environment from pollution. For example, agricultural waste like straw or stover/hay can be utilized as a source of dry fodder to the animals thus preventing air pollution due to stubble burning. Some agro-industrial by-products are already widely used as animal/livestock feed like cereal brans/polish, oilcake, meal, molasses, and brewer's yeast. Similarly, the Poultry waste (dropping) can be an input for fish farming. Livestock waste can be used to enrich the soil for agricultural operations resulting in sustainable crop-livestock production systems. Livestock integration with agroforestry contributes significantly to the protection of the environment in many ways such as nitrogen fixation by trees, impeding soil erosion, supplying domestic energy (wood), biogas production, and conserving biological diversity. In addition, integrated farming also helps in the following (Dikshit and Birthal, 2013):

○ Save land through recycling crop residues as animal feed

○ Save land by using dung as a substitute for firewood

○ Save chemical fertilizers by using dung as manure

○ Save fossil fuel by using draught animals as a source of energy in agriculture

- **Soil Nourishment:** Due to recent developments in genetics, new high yielding varieties are available which are used by farmers for getting maximum yield from the agriculture field. This has led to maximum utilization of soil nutrients by the crops resulting in depletion of essential nutrients (NPK) from the soil. Therefore, there is a need to replenish the soil with these essential nutrients from external sources. One of the viable and cost-effective ways of enriching the soil is using farmyard manure. Livestock waste (dung, urine, and farm wastes) can be converted into farmyard manure and can be used to nourish the soil. Other techniques like composting, and vermicomposting can also be used to get nutrient-rich manure.
- **Weed Management:** Weed control can be effectively taken care of livestock and in turn lead to less use of herbicides and prevent water pollution. The livestock species most commonly used in weed control are cattle, sheep, goat, poultry (like duck and geese), and fish.
- **Green Fuel:** In India, cow dung is not only used as fertilizer, it can be used for the production of biogas also called "Go bar Gas". Even today, in villages, cow dung is converted to dung cake and used as fuel for cooking. This will not only be environment-friendly but also help to reduce the dependence on fuelwood for cooking purposes. Biogas can also be used as fuel for generating electricity. Animal waste (dung and urine), wastewater from the farm, and crop residue are some of the organic materials used for biogas production.

13.6   LIVESTOCK IN WATERSHED MANAGEMENT

Watershed is a holistic approach that encompasses livestock as an integral part of it. Livestock has both positive and negative impacts on the watershed. As far as the negative impact is concerned, the major one is soil erosion due to overgrazing. Keeping aside the negative impact, it is pertinent to discuss the positive impacts of livestock in watershed management which are as follows:

- Important contributor to the social, financial, and livelihood improvement of the economically weaker sections of the society.
- Helps in doubling farmer's income particularly through "mixed crop-livestock farming system" which is the predominant system of farming, particularly under rainfed conditions.
- Integrated or mixed farming with livestock components leads to more uniform use of natural resources.
- Livestock rearing facilitates in creating assets and income generation for the poor, marginaland landless farmers.

13.6.1   IMPACT OF LIVESTOCK ON WATERSHED

For the overall development of watershed, livestock is one of the important components. Some of the major areas where livestock has a significant impact on the watershed are:

- **Natural Resource Management:** For the sound health of the ecosystem, one of the critical factors is the ratio of total livestock and human in the ecosystem. Livestock rearing is vital for managing natural resources, particularly the soil. Due to the exploitation of soil to increase food production, the soil is losing its fertility rapidly. Livestock can be a savior to restore the nutrient health of the soil. Manure prepared from livestock waste (FYM or Compost or Vermicompost) itself can supply about two thirds of essential nutrients (nitrogen, potassium, and phosphorus) required by the crops. In the case of concentrated, intensive livestock production system, the environmental problems arising due to increased stocking density can be reduced by proper application of manure produced from the farm. Again, indiscriminate use of manure will also have a drastic negative impact on soil health. Restricted or excessive use of livestock manure can either deprive the soil of nutrients or pollute the environment, respectively. For efficient recycling of nutrients (livestock or agriculture waste), livestock farming should be integrated with crop or agriculture production. Having 8 kg of nitrogen, 4 kg of phosphorus, and 16 kg of potassium per tonne, livestock manure is considered one of the most valuable fertilizers. Manure addition not only provides the essential nutrients (NPK) to the soil, but also improves soil properties like its structure, bulk density, organic matter content, cation exchange capacity, and water retention capacity. This is evident from research studies that concluded that wherever the agricultural production system is disassociated from the animal/poultry production system, the nutrient recycling became difficult.

- **Productivity Enhancement:** In nature, there exists a delicate equilibrium between human-agriculture-livestock. The existence of one is dependent on the other. Any imbalance in the equilibrium will adversely affect the entire ecosystem. For example, livestock rearing is dependent on agriculture and *vice-versa*. Imbalance in the agriculture production system may be detrimental to the stability of the livestock/poultry production system due to its dependent on crop production. Hence, agricultural production in a sound environmental base is a prerequisite for sustained and healthy animal production. Similarly, draught animal power is widely used for ploughing, tillage, transport, water lifting, and powering farm equipment. Animal traction increases the area under cultivation, bringing heavy but potentially very productive soils into production. Animal waste is a cost-effective and healthy organic alternative to costly chemical fertilizers.

- **Socio-Economic and Livelihood Development:** Livestock contributes significantly to the nutritional security of the farming community by providing nutritious food in the form of milk, meat, and egg. It can act as a liquid asset, regular source of income, year-round income source (dairy), instant source of cash (sale of animals) in case of emergency, and generation of employment. Livestock is an important source of supplementary income for rural households, particularly those in the rainfed areas. In watershed Development Projects, effective integration of a comprehensive animal husbandry component would contribute significantly to the sustainable livelihood of the farming community, particularly in the rainfed areas. It is a major instrument for improving rural employment, particularly rural self-employment. For the farmers residing in the drought-prone, hilly, tribal, and other remote areas where crop production on its own may not be capable of engaging them fully, they can take up livestock farming as a subsidiary occupation for earning additional income and also reduce risk of crop failure or loss due to natural disasters.

13.7 ENHANCING LIVELIHOOD SECURITY THROUGH LIVESTOCK

Livestock rearing is an integral part of a farmer's life. Agriculture along with livestock is vital for the success of watershed development. Adopting livestock farming is critical for the socio-economic development of the farming community. Livestock sector not only contributes significantly to the food security but also to the environmental security through draught power and organic manure to the crop sector. It is also a major supplier of many raw materials like hides, skin, bones, blood, and fibers to the industrial sector. Generally, livestock rearing is considered a supplementary source of income. However, in recent times, many small, medium, and large-scale dairy and poultry farms have been established in different parts of the country. This was augmented by the efforts of the central and state governments by promoting livestock rearing through various schemes which provide technical and financial support. Still, this sector supports the farming community in many ways *viz.* supplements income, absorbs income shocks due to crop failure, generates a continuous flow of income and employment, and reduces seasonality in the livelihood patterns of the rural masses.

In India, we generally associate poverty with the small, marginal, and landless farmers who form two-thirds of the rural population. Research studies have revealed that livestock rearing has a significant positive impact on income, employment generation, and poverty alleviation in rural areas. Even the statistics have shown that nearly two-thirds of the rural households in India own livestock predominantly the small, marginal, and landless households. Due to low initial investment and operational cost, the landless farmers rear small animals like sheep, goats, pigs, and poultry for commercial purposes. Moreover, livestock resource distribution is more equitable with 48% marginal farmers having more than half of the cattle population and two-thirds of small ruminants against their share of 24% in land (NABARD, 2018).

13.7.1 CONTRIBUTION OF LIVESTOCK ON LIVELIHOOD SECURITY

The livestock sector contributes significantly to the livelihood security of the farming community through (Fig. 13.8):

- **Financial Security:** Livestock rearing is not season dependent and hence, it can provide the farmers with a continuous source of income. The return from the livestock enterprise depends on the type of enterprise. While dairy farming and layer farming can be a source of income on daily basis

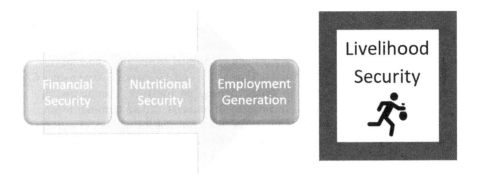

FIGURE 13.8   Livestock and livelihood security.

throughout the year if planned well, broiler farming and other livestock species reared for meat like the sheep, goat, and pig can yield return only after a certain period of time depending on the species reared. However, the success and profit of these farms depend on the availability of feed/fodder which accounts for about 65–75% of the input cost. The livestock/Poultry are an integral part of poverty alleviation programs due to their prevalence in different agro-climatic zones. Livestock farming is slowly transforming itself from subsidiary to the main occupation. Looking at the positive annual growth (5–6%) shown over the past two decades, it is expected that the livestock sector may likely emerge as a powerhouse of agricultural growth in the coming decades. The livestock provides increased economic stability to the farm or household, acting as a cash buffer (small livestock), as a capital reserve (large animals), as well as a deterrent against inflation.

- **Nutritional Security:** In line with financial security, livestock can be a source of daily nutrition (milk, meat, and egg) to the farming community. While farmers get milk from cow and sheep/goat for a period of 305 days and 120 days, respectively; broiler meat can be harvested after rearing birds for a period of 6–8 weeks. Eggs from layer hens can be harvested from 21 weeks of age up to 72 weeks of age. So, unlike the agricultural crops, where the farmer has to wait for a certain period of time to get the nutritious produce, livestock can be a source of nutritious food on a day-to-day basis if the enterprise is well planned and managed.

- **Employment Generation:** One of the important contributions of the livestock sector is employment generation either directly or indirectly. Farmers can take up livestock enterprise as a full-time or part-time occupation. Family members can also be involved in farming activities. Medium and large-scale dairy and poultry farms can also provide employment opportunities to unemployed youth. Indirectly, the livestock sector generates employment opportunities through other allied or related sectors like the veterinary pharmaceutical sector (for production of medicines and vaccines), animal feed manufacturing sector, equipment/utensil manufacturing sector, Transportation and Marketing sector, food processing sector for the production of value-added products from milk, meat and eggs and so on. According to the Department of Animal Husbandry and Dairying, Ministry of Fisheries, Animal Husbandry and Dairying, Government of India. About 20.5 million people depend on livestock farming and are engaged in the activities of farming of animals, mixed farming, fishing, and aquaculture. Recent trends show that an increasing number of workers are looking at various Animal Husbandry activities as a livelihood option. Therefore, Animal Husbandry based livelihoods offer vast scope for rural employment generation, if right kind of ecosystem is built with creation of adequate infrastructure support (NABARD, 2018).

13.8   LIVESTOCK REARING AND WATERSHED HEALTH

Livestock can be a boon or disaster to watershed health depending upon the way we handle them. The impact of livestock on watershed health can be attributed mainly to the grazing management and mixed-farming system. Poor grazing management may be detrimental to the watershed health and water quality, but a mixed-farming system along with agroforestry and silvipasture can have a positive impact on watershed health.

Livestock, especially small animals, can be an integral part of agroforestry or silvipasture systems. Depending on the requirement, they can be managed easily by permanently allowing them to remain on the field or temporarily let in for grazing. Since land is not a constraint in the silvipasture system, the animals can be permitted to graze and remain in the field itself. Overgrazing of the field can be prevented by adjusting the stocking rate. In agroforestry system with annual crops, allowing the animals soon after the harvest of the intercrop will benefit the animals as they get valuable fodder as the interspaces will be littered with residue. At the same time, it will result in the enhancement of the land's fertility through the dung and urine of the grazing animals. Leaves collected from the pruning of the trees can also be used for feeding the animals.

Here are some of the ways to keep watershed healthy:

- Practising seasonal or rotational grazing systems to improve soil health, grass quality, nutrient availability, and water-holding capacity and also to reduce livestock impacts on small or intermittent wetlands.
- Migrating from conventional grazing to "multi-paddock" grazing. It is a process that mimics the natural movement of grazing animals, wherein land was grazed quickly and then rested for long periods. Research studies have shown almost 40% reduction in quick surface runoff, 5% increase in infiltration into the soil, and roughly 30% reduction in immediate stream-flow (Park et al., 2017).
- Preventing the unwanted access by livestock by fencing water bodies like wetlands and streams.

- Strengthening the amalgamation of crop and livestock at farm level to enhance the sustainability of farming systems, improve the efficiency of agricultural systems and enhance ecosystem services. It also allows diversifying income sources and creating activities locally (Moraine *et al.*, 2014).
- Improving water recharge by improved livestock distribution and increased tree cover on pastures.
- Enhancing the water quality through measures such as feed management, proper collection, storage, processing, and utilization of manure.

13.9 CONCLUSIONS

Livestock is one of the largest producers of a large number of agricultural and livestock produce fulfilling the nutritional requirements of India and other countries. Livestock is an integral part of the watershed management. When animal husbandry is amalgamated into the watershed activity, it significantly improves the livelihood opportunities to the watershed beneficiaries resulting in uninterrupted revenue generation and employment. Livestock is best suited to mixed or Integrated farming which is vital for watershed development. Livestock's contribution to the farming community includes nutrition in the form of milk, egg, meat and their products; draught power or energy; Manure or fertilizer and Fuel; clothing (Hide) and leather and wool; and many such useful commodities and services. Livestock is one of the major contributors of GHGs like CH_4, N_2O, and CO_2. The positive effects of livestock on the environment are generation of power, efficient utilization of resources through integrated farming system, soil enrichment, weed control and production of green fuel. Livestock has a significant impact on the watershed through natural resource management, productivity enhancement, and socio-economic development of rural community. Livestock can be a boon or disaster to watershed health depending upon the way we handle them. The impact of livestock on watershed health can be attributed mainly to the grazing management and mixed-farming system. Poor grazing management may be detrimental to the watershed health and water quality, mixed-farming system along with agroforestry and silvipasture can have a positive impact on watershed health.

REFERENCES

Bernabucci, U. 2019. Climate change: Impact on livestock and how can we adapt, *Animal Frontiers* 9(1): 3–5, https://doi.org/10.1093/af/vfy039

DAHD. 2022. Annual Report 2022–23. Department of Animal Husbandry and Dairying, Ministry of Fisheries, Animal Husbandry and Dairying, Government of India.

Dikshit, A. K. and Birthal, P. S. 2013. Positive environmental externalities of livestock in mixed farming systems of India, *Agricultural Economics Research Review* 26(1): 21–30.

http://www.fao.org/3/v8180t/v8180T14.htm, https://texaslivingwaters.org/livestock-management-watershed-health/, https://twitter.com/dept_of_ahd/status/1273920769127399425?lang=en, https://www.ealt.ca/blog/healthy-watershed-livestock

Maharashtra: State Level Report on Watersheds and Livestock. Prepared for the LEAD Study by WOTR, Ahmednagar.

Misra, A. K., Rama Rao, C. A., Subrahmanyam, K. V., Sankar, V., Babu, M., Shivarudrappa, B. and Ramakrishna, Y. S. 2007. Strategies for livestock development in rainfed agroecosystem of India, *Livestock Research for Rural Development* 19(6): Article #83. http://www.lrrd.org/lrrd19/6/misr19083.htm

Moraine, M., Grimaldi, J., Murgue, C., Duru, M. M. and Therond, O. 2014. Integrating crop and livestock activities at territorial level in the watershed of Aveyron river: From current issues to collective innovative solutions. *European IFSA Symposium: Farming Systems Facing Global Challenges: Capacities and Strategies, International Farming Systems Association (IFSA).* AUT., Apr 2014, Berlin, Germany.

NABARD. 2018. Sectoral Paper on Animal Husbandry. Farm Sector Policy Department, NABARD Head Office, Mumbai.

Park, J. Y., Ale, S., Teague, W. R. and Dowhower, S. L. 2017. Simulating hydrologic responses to alternate grazing management practices at the ranch and watershed scales, *Journal of Soil and Water Conservation* 72(2): 102–121.

Rojas-Downing, M. M., Nejadhashemi, A. P., Harrigan, T. and Woznicki, S. A. 2017. Climate change and livestock: Impacts, adaptation, and mitigation, *Climate Risk Management* 16: 145–163, https://doi.org/10.1016/j.crm.2017.02.001

Sejian, V. 2012. CSWRI – Fifty Years of Research Contributions (1962 to 2012). Central Sheepand Wool Research Institute (CSWRI), Avikanagar, Rajasthan, India, pp. 155–178.

Singh, K., Singh, R., Jadoun, Y. S., Deshmukh, B. and Kansal, S. K. 2020. Role of livestock in Indian economy – A review, *International Journal of Current Microbiology and Applied Sciences* 9(8): 432–436. https://doi.org/10.20546/ijcmas.2020.908.050

WASSAN Andhra Pradesh Report. 2004. Andhra Pradesh Policy Report on Watersheds and Livestock, Prepared by WASSAN for the LEAD Study February 2004.Watershed Support Services and Activities Network 12-13-452.

14 Artificial Intelligence Application in Water Resource Management

Khilat Shabir, Rohitashw Kumar, and Munjid Maryam

14.1 INTRODUCTION

Water is essential to life. Water resources must be managed scientifically in order to meet the demands of the expanding population and ensure their availability for future generations. Hydrological activities have a tremendous impact on the earth's ecology. Growth of water resources, especially in coastal plains and river basins, can significantly increase a region's economy, benefit local residents, and assure sustainable development. Managing water resources is a challenging task because of the enormous variations in rainfall patterns, river flow, weather, and climate, as well as the overall hydrological process. It is also influenced by socioeconomic factors like shopping habits, the location of cities and businesses, and festivals. Since many areas of India experience severe water scarcity, the necessity for building an efficient, scientific, and real-time system for managing water resources has become critical. It is necessary to consider all aspects of water resource management in order to meet the increased demand brought on by India's expanding economy and population, while also making sure that the biological cycle is not harmed.

Water resource management requires the creation of new methods that combine, on the one hand, the protection of water resources through new systems and intelligent technology, capable of increasing the efficiency and performance of networks and treatment plants on the territory, and on the other hand, developing new monitoring systems that can be readily accessed and distributed to ensure widespread quality control. In both situations, artificial intelligence (AI) is crucial, especially when there is an abundance of data, which is becoming more common as water and environmental monitoring systems get stronger. The creation of interoperable technologies that can encourage the dissemination and exchange of massive amounts of information between managers, decision-makers, and citizens can result in the creation of a body of knowledge that can be used to power AI systems and support better environmental protection, ultimately having a direct impact on the educational and behavioral aspects. As a result of water's ubiquitous presence in all stages of social and productive declines, it constitutes a natural element for channeling information and consolidating a new culture that combines AI that promotes growth, the sharing of structured expertise, and the sense of belonging to one's territory and its natural

DOI: 10.1201/9781003351672-14

resources. The adoption of a sense of universal (public) ownership and responsibility that must influence both the little everyday decisions and the big planning, managerial, political, and administrative decisions is a prerequisite for the acknowledgment of water as a human right.

14.2 WHAT IS ARTIFICIAL INTELLIGENCE (AI)?

In computer science, AI refers to creating machines and systems that can learn and make decisions similarly to humans. The Association for the Advancement of Artificial Intelligence defines AI as "the study of mechanisms that underpin mind and intelligent behavior, and how these can be implemented in machines." AI involves a wide range of functionalities without any limitation to as (a) learning, which includes a variety of approaches like deep learning (for perceptual tasks), transfer learning, reinforcement learning, and combinations thereof; (b) understanding or deep knowledge representation required for domain-specific tasks like cardiology, accounting, and law; (c) reasoning, which comes in a variety of forms, such as deductive, inductive, temporal, probabilistic, and quantitative; and (d) understanding, or deep

14.3 TYPOLOGY OF AI APPLICATIONS

Most AI applications perform at least one of the following seven tasks: monitoring, discovering, predicting, interpreting, dealing with the physical world, communicating with people, and interacting with other computers.

- Monitoring
  Large data sets may be quickly analyzed by AI to find anomalies and trends. AI is especially well adapted for monitoring applications, such as critical changes in the environment because it can do this much more rapidly and accurately than humans—often in real time.
- Discovering
  Data mining, a term used to describe the process of extracting useful insights from massive databases, is one way through which AI might find new solutions. In particular, AI is highly successful at spotting abstract patterns and exposing unique insights that traditional computer programs cannot since it uses dynamic models that learn and adapt from data.
- Predicting
  AI may predict or model how trends are expected to evolve in the future, allowing systems to anticipate, suggest, and customize solutions. These kinds of applications are probably well known to a large number of users. This application of AI will be advantageous for data-intensive applications like weather forecasting.
- Interpreting
  Up until recently, the majority of data analytics efforts were concentrated on structured data or data that has been carefully organized according to a

predetermined framework, like a spreadsheet of survey results. AI can analyze unstructured data—information that is difficult to categorize, such as photos, video, audio, and text—because it can learn and recognize patterns. Computer systems can now analyze a vastly greater variety of different types of global data as a result.

• Interacting with the physical environment
  AI can enable a wide variety of machine-to-environment connections, enabling autonomous systems to interact with the real world. AI, in particular, makes it possible for robotic systems to navigate and control their environment.

• Interacting with people
  AI can make it easier for people to communicate with computers. When interacting with machines, people generally modify their actions to suit the requirements of the machine, such as when they type on a keyboard, press a button, or turn a dial. With the help of AI, people can communicate with machines, in the same manner, they do with other people since computers can understand voice, gestures, and even facial expressions. For instance, users can converse with AI-powered chat bots to ask questions or signal a robot to approach by nodding or waving.

• Interacting with machines

AI has the ability to autonomously manage challenging machine-to-machine interactions. Additionally, this capability makes it possible for various AI systems to communicate with one another.

14.4 MEASUREMENT WORKS OF THE WATER VARIABLES

To evaluate the quality of the water habitats, monitoring, continuity of measurements, and laboratory work are both challenging and expensive methods. In addition, many of these systems are susceptible to natural disasters. In some circumstances, procedures like calibration, control, and maintenance become challenging. For instance, by measuring flow rates and other variables, streamflow in a river can be determined. After determining the cross-sectional area of the channel, a current meter is used to calculate the flow rate from specific places in sections of a river. A level meter (scales) installed in the stream segment also serves to gauge the river's level. The key curves derived from the regression analysis are afterward commonly applied in practice. This method allows for a reduction in in-situ measurements (Parker, 2003; Murat and Özyildirim, 2008; Humphrey et al., 2008). On the other hand, the measurement intervals gained with respect to time in terms of measuring procedures are one of the most significant challenges in data analysis. It is significantly more challenging to quantify the sediment in a river section than to measure the flow. After in-situ measurements, laboratory work needs to be completed. The key curves derived from the regression analysis are used in practice just like the flow measurements. To assess the level of a river, water level sensors or water level gauges are frequently employed.

Evapotranspirometers are used to measure evapotranspiration and evaporation (lysimeters). Water quality parameters are another metric. The time series are frequently brief and the water quality variables are frequently recorded at nonsystematic intervals. Large numbers of missing values and lengthy periods without measurements can be found in data sets. Water quality data are therefore the most challenging hydrological data, and using traditional hydrological methods to analyze them is likewise challenging. Depending on how the water is used—for irrigation, recreation, energy, and other purposes—the quality variables can change. For establishing the quality of the water, there are both national (Water Pollution and Control Regulation, or WPCR) and international (Environmental Protection Agency [EPA] and World Health Organization [WHO]) rules. About 50 variables, including dissolved oxygen, chemical oxygen demand, biological oxygen demand, pH, temperature, nitrate, nitrite calcium, potassium, phosphate, and fecal coliform, can be used to express it in these documents. It can be claimed that in this instance, there is no such issue with current measures. For instance, a comparison calls for the definition of water quality to include numerous variables, even in a single site. Therefore, when correlations among water quality factors are sought even in a single location, the study gets difficult when considering multiple variables. The analysis is made simpler if every variable is routinely observed at the same frequency.

However, it is very difficult to use the current techniques when multiple variables are recorded at various frequencies in the same place. On the other hand, AI techniques are used as a middle control element, such as in the design of the valve system of the dam's reservoir, the flow and precipitation forecasts based on flood analysis, and the design of irrigation systems. These examples are examples of the middle components of the system. In order to increase the functionality and longevity of many engineering projects, various optimization and modeling schemes coordinated with other scientific fields might be taken into consideration. Thus, AI methods have been suggested as a substitute for overcoming these issues. This emphasizes the requirement for trustworthy regionalization models to calculate the time series of the water variable in an ungauged watershed. For instance, some researchers have examined methods based on AI forecasting from data of hydrologic variables in stations where measurements cannot be made.

Programming languages like MATLAB and FORTRAN have many of the mathematical functions needed to employ these techniques. For the prediction of water variables, such as rainfall-runoff, evaporation, evapotranspiration, streamflow, sediment, dam or lake water levels, water quality factors, artificial neural networks (ANN), fuzzy-based models, and their hybrids are the most often used AI technologies. For instance, AI techniques have been used successfully to forecast water quality variables, evaporation and evapotranspiration, rainfall-runoff, streamflow, silt in a river's width section, and dam or lake water levels.

The AI itself uses early warning indicators to identify and suggest mitigation measures for localized extreme events caused by anthropogenic or natural factors, such as climate change and its ensuing changes in rainfall patterns, in response to what is highly desired by administrators, managers, and citizens (e.g. illicit disposal or accidental spills).

AI has the potential to be crucial in determining and managing adaptation guidelines with respect to the current climate changes. It can be beneficial in particular for:

- water supply management sector:
  - controlling leaks, determining measurement strategies and priorities, and determining the most effective types of interventions;
  - water network and infrastructure investments that support a comprehensive water policy that considers a multitude of technical, managerial, social, and economic factors;
- water resources management sector:
  - the encouragement of water conservation in its natural state by planning the areas in a way that favors both job possibilities and the reduction of hydraulic risk;
  - the consolidation of disparate surveillance activities among the many management and control organizations, also to enhance the quality and effectiveness of information;
  - support capacities for coping with extreme climate occurrences, particularly in relation to drought and flood control;
- transversal sectors, for example, in relation to climate change:
  - the effectiveness of water usage across all industries, as well as ensuring sustainable freshwater extraction and supply to lessen use conflicts and solve water scarcity in the short, medium, and long terms;
  - in order to increase the resilience of water supply, treatment, storage, and transport systems as well as hygienic systems, it is important to support and promote inter-sectoral, regional, national, and subnational policies on water management and quality. This is done by ensuring that adequate knowledge and hygienic practices are followed;
  - support the adoption and application of a risk-based approach in the water and sanitation sector (i.e. water safety plans, sanitation safety plans), including the management of disease data, the design of early warning systems based on projections of pathogen distributions, emerging chemical contaminants, and/or subject to ordinary control;
  - enable the monitoring and modeling of risks, such as the development of toxins and algal blooms in the aquatic environment;
  - stay away from floods' negative impacts on water quality, etc.

AI can also be used effectively in the purification of urban and semi-urban wastewater in urban and semi-urban areas, improving plant efficiency and versatility as well as favoring technologies with low environmental impact, reducing occupied surfaces, minimizing sludge and odor emissions, while maximizing energy recovery and raw material recovery, particularly nutrients and biofuels.

14.5 ARTIFICIAL INTELLIGENCE TECHNIQUES AND THEIR APPLICATIONS TO HYDROLOGICAL MODELING AND WATER RESOURCES MANAGEMENT—SIMULATION

The majority of previous research on mathematical modeling in hydrology has used conventional methods based on process-based (such as deterministic or conceptual models) and/or statistical models (such as Box-Jenkins time series methods, linear/nonlinear regression models). Recent years have seen a significant increase in research in fields that fall under the umbrella of AI. As a result, the technologies discussed go beyond those that are just appropriate for simulating the hydrological cycle to include techniques that could help with the planning and management of larger systems that are closely related to hydrology (e.g. reservoir operation and management, agriculture and irrigation, water quality and environmental policy). Although it is not recommended that AI methods be used in place of traditional hydrological modeling and decision analysis techniques, their incorporation will help to produce more powerful and dependable tools that are better able to deal with hydrological systems.

Following are some of the key AI techniques used in hydrological modeling, such as ANN, evolutionary programming, Bayesian networks (BNs), fuzzy systems, and agent-based models (ABMs). There is no one AI tool or approach that can be used for all tasks (such as prediction, classification, and decision assistance), and no technique is generally ideal for any given activity (Kingston et al., 2008; Langdon et al., 1998; Negnevitsky, 2005; Gibbs et al., 2011; Bharadwaj, 2020). Additionally, the process of hydrological modeling and decision-making can be greatly impacted by selecting the technique that is best suited for a certain task.

14.5.1 ARTIFICIAL NEURAL NETWORKS

ANNs are well adapted to modeling complicated, nonlinear systems with unknown or difficult-to-describe underlying relationships but lots of observed data. Modellers are unable to examine poorly understood systems, especially those that are extremely complex and nonlinear, like those that occur in nature, using traditional statistical methodologies used for hydrological and water resource modeling. By fitting and examining the relationships in big data sets, ANNs, on the other hand, offer a potent method for modeling and enhancing our understanding of such processes. Over the past 10–15 years, ANNs have gained popularity as a "general-purpose" method for modeling water resource variables due to its ability to accurately forecast, categorize, and cluster data. Instead of directly attempting to depict the physical processes taking place within the system, as knowledge-based modeling techniques do, these data-driven models were developed largely to extract information contained in a set of observed data. To put it another way, ANNs are capable of "learning" associations from observed data. ANNs are incredibly adaptable since they can handle continuous and discrete data, linear and nonlinear relationships, and noisy, incomplete, and discontinuous data sets.

ANNs are mathematical models made up of a number of intricately interconnected processing units called "nodes" or "neurons," which are comparable to the biological neurons in the brain (Landahl *et al.*, 1943). ANNs were initially created to replicate how the brain works. These nodes execute relatively simple and constrained computations on their own, but when working together as a network, complex computations may be carried out because of the connectivity between the nodes and the way that data is transmitted through and processed within the network. There are numerous types of ANNs, each with a different structure and operation (Flood and Kartam, 1994). Typically, these are categorized by the type of data used, computations made by each node, and how the network "learns" its function. The two main learning methods that can be used are supervised and unsupervised. Learning a function under supervision entails being provided instances of its inputs and the target value, or desired outcome. On the other hand, unsupervised learning entails discovering patterns in the input data when no target values are given. While supervised ANNs are used for classification and prediction, unsupervised ANNs are typically used for clustering and dimensionality reduction (or regression). Multi-layer perceptrons (MLPs) and self-organizing maps (SOMs) are the most well known and often applied to supervised and unsupervised ANNs, respectively (Islam and Kothari, 2000).

14.5.2 Multi-Layer Perceptrons (MLPs)

As the name implies, MLPs are composed of many layers of nodes. As seen in Fig. 14.1, the nodes are organized into an input layer, an output layer, and one or more intermediate levels, sometimes known as "hidden" layers. The input layer, which has a node for each input variable or characteristic, receives data in the form of a vector representing observed data values. The intensity of the incoming signal is determined by the weight of the connection, which is then used to relay this information to each node in the subsequent layer. Prior to being sent to the nodes in the following layers, the weighted information is combined with a bias value at the nodes and transferred using specified (often nonlinear) activation functions.

Appropriate values of the connection and bias weights, w, which are the network's free parameters, must be estimated in order for the network to carry out the specified function. During a procedure known as "training" (calibration), the model outputs are compared to the goal values, and the weights are subsequently adjusted to reduce

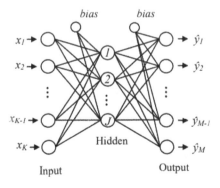

FIGURE 14.1   Layer structure of an MLP.

discrepancies or errors. An independent set of data, not observed during training, must be presented to the network before it can be used to make predictions with any degree of confidence (Maier and Dandy, 2000).

14.5.3 SELF-ORGANIZING MAPS (SOMs)

Visualizing patterns in data sets with more than two dimensions is difficult or impossible. SOMs, sometimes referred to as Kohonen networks, map multidimensional input data to a lower dimensional space to highlight any patterns or clusters that may be present (Kohonen, 1982). There are two levels in these networks: an input layer and an output layer. As depicted in Fig. 14.2, the input layer takes the input data and has a node for each input variable, whereas the output layer is often organized as a two-dimensional grid of rival neurons (processing units). The input variables are linked to the output nodes by weights that can be changed.

The degree to which the weights of each output node match the corresponding values of the input pattern is determined during training by computing the Euclidean distance between the weight and input vectors. The related weight vector is adjusted to further reduce the distance between the associated output unit and the input vector. The output unit with the smallest distance is then deemed to be the "winner." Each output node ultimately serves as a virtual sample, representing typical input data patterns assigned to the node during training or a group of related input patterns (Arab *et al.*, 2004). The SOM still makes an effort to identify clusters in the input data so that any two neighboring clusters in the output grid have comparable weight vectors in the input space. As a result, the neighborhood structure of the underlying distribution is maintained because the weight vectors linked to the winning node's closest neighbors are likewise modified via an SOM learning rule.

14.5.3.1 Applications

Numerous applications, including rainfall-runoff modeling, streamflow prediction, river basin categorization, water quality forecasting, and regionalization of hydrological model parameters, have shown the ability of ANNs, in particular MLPs, to simulate hydrological and water resource variables. SOMs have been used in

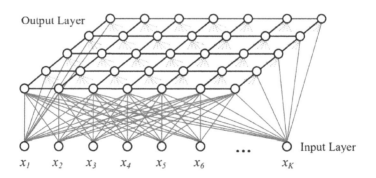

FIGURE 14.2 Grid structure of an SOM.

hydrology-related fields substantially less frequently than MLPs, but these ANNs have also been successfully used in a number of these research. SOMs, for instance, have been used to classify soil moisture profiles, classify river flow events, cluster precipitation fields, forecast runoff volumes and floods (when combined with a linear regression output layer), and cluster ecological assemblage data to evaluate riverine water quality and ecological responses.

In the field of water resource modeling, Kingston *et al.* and Kingston developed a Bayesian framework for ANNs to account for uncertainty in the model and the modeling results. This framework allows for the consideration of uncertainty, the objective selection of model complexity, and the provision of prediction limits. In order to take into account the uncertainty related to ANN estimates of river flows, Khan and Coulibaly also adopted a Bayesian approach. Using weight-based input importance measures and sensitivity analyses, ANNs were evaluated for their explanatory power using input variables, saliency analyses, perturbation analyses, and sensitivity analyses. Hybrid ANN/knowledge-based models, which take advantage of each approach's best qualities, have also shown promise in this area. A popular method for creating a hybrid ANN is to start with a physical model and use the ANN to calculate parameters that are either unknown or impossible to measure. In this method, the known components of the system can be estimated using the ANN, while the unknown components can be characterized by mathematical equations.

14.5.4 Genetic Programming

Genetic programming (GP), like ANNs, can be thought of as a black box, data-driven approach for modeling the relationship between a set of model inputs and outputs. As such, GP is most helpful in hydrological modeling when there isn't a clear mathematical description of the process or when there isn't enough information about specific physical characteristics. However, GP simultaneously optimizes the model structure and parameters, unlike ANNs, which call for the selection of a model structure and the optimization of the related model parameters. This is accomplished by using a search technique based on Darwin's theory of natural selection and evolution. Selection, crossover, and mutation are genetics-inspired operators used here to gradually evolve and enhance a population of solutions. Search techniques like these belong to the branch of AI called evolutionary computation (EC), of which the genetic algorithm (GA) is the most basic.

In GP, each member of the population represents a test computer program that uses a series of functions and parameters to address a specific issue. Tree structures are used to illustrate these many solutions, and children are produced by cutting off "branches" from one parent tree and adding them to another. In order to find better solutions—programs that execute the target function more effectively—genetic operators are used to repeatedly modify the trial programs. In essence, GP imitates how people create computer programs by gradually rewriting and enhancing them (Langdon and Qureshi, 1998). A simulation model of the relationship between a collection of inputs and outcomes can be generated automatically using GP. Since the goal is to identify a function that fits a specific collection of data in symbolic form, this is known as symbolic regression (Liong *et al.*, 2002). When the process being modeled is not well known, GP provides advantages over ANNs because the form of the model does not have to be determined a priori. A model derived using GP is composed of symbolic expressions (a single equation or set of equations), which, depending on the restrictions placed on the representation of individual solutions, may be possible to interpret. As a result, the model may be viewed as a hypothesis about the problem domain rather than simply a predictive model based on data, unlike ANNs, which are frequently criticized for being difficult to interpret (Babovic, 2005).

The terminal set, which includes independent variables, zero-argument functions, and random constants, is part of the tree structures that represent solutions in GP. The functions can be any arithmetic operators (such as +), mathematical functions (such as sin, cos, exp, log), Boolean operators (such as AND, OR, NOT), logical expressions (such as IF-THEN-ELSE), or iterative functions (such as DO-UNT). The normal probability density function is illustrated by an example tree structure in Fig. 14.3.

The tree, as can be seen, is made up of a number of nodes and connections, where the nodes denote which instruction should be executed and the links denote its arguments. The nodes are organized into branches that each explains a sub-component of the model, and the tree's structure groups all of the branches under a root node. The leaf nodes serve as the terminals of the tree, while the root and inner nodes serve as their functions.

14.5.4.1 Applications

GP hasn't seen as many uses in the management and modeling of water resources as other EC techniques, including GAs. It is a far more recent technique, though, and is just now beginning to be considered a competitive alternative to other modeling strategies, such as process-based, conceptual, and empirical black box models (e.g. ANNs). Modeling rainfall-runoff, predicting algal blooms, simulating the spatial distribution of species populations, predicting flow through urban basins, simulating

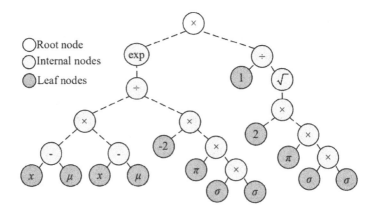

FIGURE 14.3 Example of tree-structured solutions used in GP.

the impact of vegetation on flow conditions, downscaling daily extreme temperatures, and simulating sediment transport processes are just a few of the many applications that GP has already been successfully used for.

14.5.5 BAYESIAN NETWORKS

Water resource management choices must be made despite the uncertainty surrounding them since the modeling and management of these systems frequently rely on noisy, incomplete, and unquantifiable information. Initially, BNs were created to mimic how human experts assess numerous kinds of imperfect information to assign probabilities (Pearl, 1986). They incorporate Bayesian probability theory and offer a valuable method for reasoning under uncertainty. Probability distributions are employed in the Bayesian paradigm to convey the level of belief or knowledge about uncertain variables. The Bayes theorem, which states that the resultant knowledge (posterior probability) is proportional to the product of any a priori knowledge (the prior probability) and the new knowledge obtained from a set of observed data, characterizes the advancement of knowledge gained from new evidence about the uncertain variables (the likelihood). Using all potential values of the uncertain variables and weighing them according to their probabilities makes it possible to make predictions or judgments on a probabilistic basis.

BNs are especially helpful for analyzing the behavior of water resource systems and/or supporting integrated decision-making concerning such systems. Understanding the potential environmental, economic, and social effects of a particular alteration within a watershed is essential for the sustainable management of water resource systems. Modeling such outcomes under numerous alternative management or environmental change scenarios is possible since BNs have the capacity to mix a variety of sources of both factual and subjective information. Stakeholders can see how a specific modification affects all of the components associated with it thanks to the graphical framework, which makes all causal links in the system explicit. Additionally, working with probabilistic descriptions of such complex systems is made simpler by the conditional independencies made apparent in the causal network representation since it enables the use of distinct sub-models to characterize only the direct relationships in the system (Borsuk *et al.*, 2004; Niu and Feng, 2021). When solely employed for system behavior modeling, BNs are frequently referred to as belief networks; however, when paired with a value or utility function, they are typically referred to as decision networks or impact diagrams.

14.5.6 BELIEF NETWORKS

There are two key steps in creating a belief network, and they are as follows:

Development of the causal structure: In order to describe the dependencies inside the system, it is necessary to decide which variables are most pertinent to the issue at hand and how they should be linked. The conceptualization of the issue, the specific model objectives, the spatial and temporal scale, and the outcomes of consultations with various stakeholders should serve as the foundation for the causal structure (Bromley *et al.*, 2005). Any physical, social, economic, or institutional component may be represented by one or more of the variables in a belief network. They may symbolize diverse acts or beliefs as well as the sizes, characteristics, or motion of objects.

The model's goal is to study the most effective management of an irrigation area in terms of environmental, social, and economic considerations, including profit, river water quality and flows, and the dependability of meeting downstream requirements. An example belief network is shown in Fig. 14.4. In this example, the size of the irrigation area, changes in irrigation practices (irrigation efficiency), a change in the price of water, or the availability of river water may have an impact on river extraction and, ultimately, river flow; similarly, the size of the irrigation area, changes in irrigation practices (irrigation efficiency), a change in the price of water, as well as changes in crop type, may have an impact on net profit; and the size of the irrigation area may have an impact on the number of nutrients discharged to each of these variables, shown in Fig. 14.4 as shaded nodes, which can be used to depict a variety of potential scenarios. The availability of water can be used to reflect anticipated future trends, while the size of the area, cost of the water, the effectiveness of irrigation, and crop variety are all controllable factors that represent potential future management options. The belief network can be used to assess changes in any of these factors and explore potential outcomes.

Quantification of conditional relationships: It is necessary to quantify the principal causal links after they have been recognized and represented by the network topology. This is done by specifying the Current Procedural Terminology (CPTs) connected to each node. The states for each of the variables, which are the potential values, or ranges of values, that the

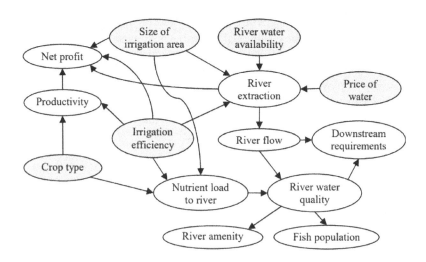

FIGURE 14.4   Example belief network.

variables are likely to acquire, must first be defined before the CPT values can be filled in (de-Santa Olalla *et al.*, 2005). BNs can contain discrete or continuous variables; however, the range of values that continuous variables can take typically consists of sub-ranges that correspond to the variable's many states (Cain, 2001). The user decides how many and what kind of states a variable can have. Consideration should be given to the variable's present state, its expected future state, and any intermediate states when choosing the right states for it. States must effectively capture the effects of each parent node in order to be logical. The CPTs can be filled out once the states have been chosen for each variable. This information may be expert opinion, actual data, simulated output from models, stakeholder perspectives, or observed data. It is possible to obtain probabilistic forecasts of system variables after quantifying each CPT in the network.

14.5.7 INFLUENCE DIAGRAMS

While belief networks are helpful for examining the effects of different management options, the decision of which option to choose is left up to the decision-maker. On the other hand, influence diagrams are a special form of belief networks that have been enhanced to incorporate utility variables and decision variables, which are variables that may be optimized and controlled, in order to identify optimal decisions quantitatively. The sample belief network from Fig. 14.4 is represented as an influence diagram in Fig. 14.5.

Given a variety of management options for the irrigation area, this network can be used to identify the best one. A rectangle is used to represent the decision node

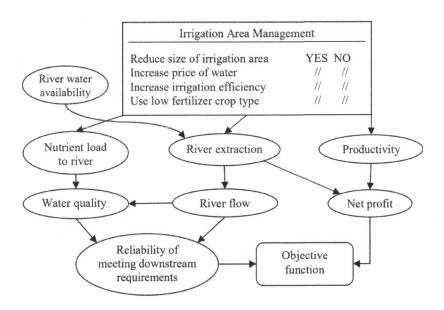

FIGURE 14.5   Example influence diagram.

in the network, a rounded rectangle is used to represent the utility node, and ovals are used to represent the variables that characterize the state of the system. All of the oval nodes receive CPTs, similar to belief networks; however, the decision and utility nodes do not. Utility nodes have associated tables of values that show the utility, or desirability, of each possible combination of their parents' states. Utility nodes are linked to variables defining outcomes that affect utility. The utility values may include data on social, environmental, and economic costs and benefits and may be expressed in any units of measurement. Alternative ways of action are described using decision nodes. To determine the objective function value for each alternative, the state of these nodes (i.e. the action that is executed) can either be manually defined or automatically decided by maximizing the objective function. The result probabilities of the action are coupled with the utilities associated with each outcome to determine the objective function value (i.e. expected utility) associated with the activity.

14.5.7.1 Applications

The application of this technique to various studies dating back to the early 1990s and through subsequent publications has solidly proved the potential of BNs as modeling and decision support tools in integrated water resources management. Many studies have used belief networks to foretell the effects of various land, river, and aquifer planning and management scenarios in recognition that stakeholder participation and uncertainty play essential roles in water resource modeling and management. Applications of this kind include comparing various nitrogen reduction scenarios to lessen eutrophication of an estuary, domestic demand reduction strategies to lessen reliance on water resources in a river catchment, irrigation techniques to sustainably manage an aquifer, watershed development options and their effects on resource availability, river quality management scenarios and their effects on macroinvertebrates, and the assessment of the sustainability of coasts. Influence diagrams have been used in several studies to aid in the determination of the best management strategies, including prioritizing investments in watercourse protection, designing a water quality monitoring system, choosing an irrigation method to improve water management, choosing a vegetation management strategy to manage and prevent dryland salinity in a river catchment, and choosing an irrigation method to improve water management. The majority of these studies present encouraging findings, particularly in terms of improving our comprehension of cause-and-effect linkages and including social factors in modeling and decision-making.

14.5.8 FUZZY SYSTEMS

Similar to BNs, fuzzy systems make an effort to capture subjectivity, ambiguity, and imprecision; however, they do so within a non-probabilistic framework by employing hazy or fuzzy descriptions. It is frequently impractical to precisely define the characteristics or state of an object or system when individuals are speaking to one another. More often than not, ambiguous words like "high," "middle," "low," "wet," "dry," "hot," "warm," and "cold" are employed. These phrases don't have any clear-cut definitions; rather, their meanings depend on how well each individual understands or has personally experienced each condition they refer to. Similar to how people handle concepts, fuzzy systems do so by relying on vague descriptions as opposed to precise ones.

Fuzzy set theory (Zadeh, 1996), which enables the specification of how well an object satisfies a vague description using a mathematical framework, is used in order to do this. In this system, the degree of membership of an object x to a description, or set, I is expressed using membership functions, $m_i(x)$. If the boundaries of the sets were sharp, as they are in classical set theory, then x would either belong to set I (i.e. $m_i(x) = 1$), or it would not (i.e. $m_i(x) = 0$). The amount of "truth" in the assertion that x belongs to I, however, can be expressed by $m_i(x)$, which can take on a continuum of values between 0 and 1. This is made possible by fuzzy set theory. As an illustration, it could be preferable to account for the uncertainty in a set of observed streamflow data by using phrases like "high," "mid," and "low" to characterize the recorded flows rather than exact numerical values. Furthermore, as indicated by the membership functions m_{high}, m_{medium}, and m_{low}, the measurements may be partially represented by each of these terms rather than needing to set precise bounds for what defines a high, medium, or low flow event.

To address subjectivity and uncertainty, fuzzy set theory offers an alternative to probability theory. This is especially helpful when there are unclear problems or little information available (Simonovic, 2005). Due to the data requirements for precisely estimating these distributions, the dependence of probabilistic techniques on distribution functions might be deceptive in such situations. Additionally, not all uncertainty falls under the probabilistic umbrella. For instance, it is only possible to assess variables related to water resources that are not quantifiable (such as the amenity value of a river) using professional experience. Water resource managers and hydrologists frequently have to deal with a mix of quantitative and qualitative data that is difficult to define using probability distributions.

14.5.9 FUZZY RULE-BASED INFERENCE

When something is only partially true, modeling and reasoning are made possible by fuzzy rule-based (FRB) inference. It makes use of fuzzy rules, which translate inputs into outputs by using conditional statements to represent human understanding. Fuzzification, rule evaluation, aggregation of rule outputs, and defuzzification are the four main processes of FRB inference.

14.5.10 FUZZY CLUSTERING

Data is grouped naturally into clusters by clustering algorithms, where data points within a cluster are similar to one another and share specific characteristics. However, it frequently happens that data points cannot be precisely grouped into one cluster because of the ambiguity in the data. In these situations, fuzzy set theory offers a method for fuzzy clustering the data, allowing each data point to partially belong to many clusters as defined by membership functions. The most well-known and often used fuzzy clustering method is the fuzzy c-means (FCM) algorithm, which was first developed by Dunn (1973) and later generalized by Bezdek (2013).

14.5.10.1 Applications

Fuzzy systems have been employed in a variety of hydrological modeling and water resource management applications due to the vast range of applications of the fuzzy methodology. The principal uses include:

FRB models for prediction and decision support: for instance, rainfall-runoff modeling and seasonal runoff forecasting, the prediction of regional droughts, the reconstruction of missing precipitation data, impersonating an expert in the simulation of reservoir operation, modeling nitrogen leaching from agricultural land, and the prediction of algal biomass in eutrophic lakes.

Fuzzy criteria and aggregation methods for multi-criteria decision-making: the greatest water supply system, the most dependable flood protection and management strategies, the most suitable water resource system to accomplish long-term goals, and the finest irrigation management and practices, to name a few.

Fuzzy cluster analysis and classification: for example, classifying river habitat types, clustering hydrological input data to improve their use in predicting future trends, classifying river water quality conditions according to usage mode, and regionalizing watersheds according to flood response.

FRB models can be used to link a set of model outputs to a set of model inputs "model-free" or to (1) model the behavior of a human expert. When there are no universally accepted laws accessible, experts' direct experience is used to generate fuzzy membership functions and rules. When such models are available, existing mathematical models are also used to develop fuzzy membership functions and rules.

14.5.11 Agent-Based Models

Models for the analysis of water resources policy are required to ensure the management of water resources is sustainable. ABMs are helpful for evaluating the so-called soft policies that depend on human interaction to be effective, whereas BNs offer an integrated approach for analyzing the effects of hard management policies. A "bottom-up" approach to complex systems modeling, ABMs emphasize human actions. ABMs make it possible to track the emergence of global, or system-level, features over time and conduct in-depth analysis by mimicking the local behaviors and interactions of several individual agents (such as human actors) and their surroundings.

BMs are made up of an assembly of objects linked together by a common environment (Bousquet and Le Page, 2004). Active objects are termed "agents" and may be used to represent any active entity, from atoms and biological cells to individual people and organizations (Parker *et al.*, 2003). These agents are autonomous and proactive, where autonomy relates to an agent's ability to control its own actions, while proactiveness refers to the goal-directed behavior of agents. Agents also exhibit responsiveness and social ability, which allow them to respond to changes in the environment and to interact and communicate with other agents.

When modeling systems where the incentives of human decision-makers play a significant role (such as challenges with water resource management), the ability of ABMs to directly depict human decision-making is a huge advantage (Parker *et al.*, 2002; Castro and New, 2016). In order to adequately comprehend and simulate these systems, human activity must be taken into account because numerous watersheds and river basins have already undergone and may continue to undergo considerable anthropogenic alterations (Peterson, 2000). ABMs incorporate agent cognition and low-level social-environmental interactions to realistically account for the human component.

14.5.11.1 Applications

ABMs are primarily used as a decision support tool in the management of water resources, especially when soft policies (such as encouragement, education, trade, taxes, and rebates) are used. ABMs can be used to provide answers to issues like what policies are workable and what are their likely consequences on resource development and sustainability. As a result, they are helpful for directing the creation of policies that are viable and sustainable. There has been a move away from management focused on command and control toward policies that emphasize diversity and seek resilience, as resilience is essential to boosting adaptive ability, in many social-environmental systems that are exposed to constant change. The uncertainty surrounding estimating these management policies' real impact, which is mostly dependent on individual actors inside the system, presents one of the challenges in their definition. ABMs are excellent at assessing and analyzing various scenarios and policies, and as a result, they have been used for this purpose in several studies on environmental resource management. Examples include managing groundwater resources, altering the use of land for agriculture, managing irrigation systems, and managing urban water supplies. A number of review papers, specialized journal issues, and books have been published as a result of ABMs' potential for use in these areas.

14.6 ARTIFICIAL INTELLIGENCE TECHNIQUES AND THEIR APPLICATIONS TO HYDROLOGICAL MODELING AND WATER RESOURCES MANAGEMENT—OPTIMIZATION

For a variety of reasons, hydrological modeling and management problems are frequently challenging to resolve, and AI-based optimization methods are typically more suited to handle such issues than conventional optimization or problem-solving approaches. The main reasons for this are that they can handle any type of objective function and constraints and do not require that these functions be continuous or differentiable; they are flexible in their application, and they can be implemented on parallel hardware. Population-based algorithms search from a population of potential solutions rather than a single point. There are several of these optimization techniques, though, and we will go over their benefits, drawbacks, and prior uses in the management of water resources and hydrology.

Numerous shared characteristics of AI-based optimization strategies make them more effective for challenging issues. AI optimization methods are population-based and evolutionary in nature, which means they search for optimal solutions from a number of different locations in many different directions while making use of the information contained in the population to find better solutions. In contrast to traditional methods, which base further exploration of the search space on a single "current best" solution. This makes AI-based strategies significantly less likely to limit the search to local optima than conventional optimization techniques (Simpson *et al.*, 1994). Furthermore, AI approaches can be used to solve any problem as long as it can be simulated because they base their search on evaluations of the objective function rather than information about this function itself (such as gradient information or derivatives). The following are some examples of the many optimization techniques.

14.6.1 EVOLUTIONARY COMPUTATION

Darwin's theory of natural selection and the survival of the fittest, according to which a population evolves over time by selecting and exchanging information among the "fittest" members, served as the basis for the field of AI known as EC. These methods are used to simultaneously explore a solution space from a "population" of solutions, or many solutions. The population's solutions also vie with one another for survival and the benefit of coming generations. Similar to how the genetic makeup of parents is passed on to their offspring, the better, or "fitter," solutions are more likely to do this in nature. As a result, information about better solutions is passed down from one generation to the next. GAs and the University of Arizona-developed shuffling complex evolution algorithm are two notable EC techniques frequently used for optimization.

14.6.2 GENETIC ALGORITHMS

Using GAs, which are a general-purpose stochastic search technique, it is possible to resolve challenging optimization issues. In order to achieve this, they use genetics-inspired operators like selection, crossover, and mutation to shift from one population of synthetic "chromosomes" to a different one. The population that evolves in one generation becomes the population from which another population evolves in the next, and so on. The evolutionary process begins with a population that is entirely random and takes place across a number of subsequent iterations, or "generations."

14.6.3 SHUFFLED COMPLEX EVOLUTION

In recent years, the SCE-UA algorithm (Duan *et al.*, 1992) has gained popularity as an optimization method that is largely employed for calibrating or improving the parameters of conceptual watershed models. While not strictly based on Darwin's theory of evolution and the principle of the fittest, this method combines the advantages of a number of current optimization techniques, such as simplex search, controlled random search, and complex shuffling with competitive evolution, in which communities evolve through a "reproduction" process. The SCE-UA approach, like the GA, starts with a random beginning population of points that represent potential solutions to the problem. Contrary to GAs, decision variables must be continuous, and by setting upper and lower bounds on these variables, the feasible search space must be defined. The population of prospective solutions is separated into a number of communities, or complexes, which are then allowed to reproduce freely. Each member of a complex is a potential "parent" who may take part in the process of reproduction.

The domain independence of EC techniques and the fact that they typically do not necessitate a thorough understanding of the underlying mathematics of the problems they are addressed are two of their key advantages. This means that EC methods can evolve nearly anything if an adequate representation of evolving structures is used, and they are also very quick and inexpensive to implement. GAs are simple to combine with other techniques and can be used in combination with any simulation model to perform optimization (as can the SCE-UA algorithm, given that the decision variables are continuous). Because of their adaptability and broad applicability, EC techniques have been implemented in a variety of disciplines (Gibbs *et al.*, 2011).

14.6.3.1 Applications

In hydrology and water resource management, evolutionary optimization approaches, particularly GAs, have proven to be very successful due to their many benefits over conventional optimization and problem-solving strategies. GAs have been used, for instance, to improve water monitoring networks, irrigation planning and management techniques, reservoir operations, water distribution system design and operation, and management of watersheds, rivers, and aquifers. Additionally, they have been employed in a large number of studies to develop, improve, and calibrate hydrology-related models.

Recently, the SCE-UA algorithm has gained popularity for calibrating theoretical watershed models. It has been demonstrated repeatedly to perform better than alternative calibrating methods for this purpose in different applications. This technique has also been used to solve more broad optimization issues, such as optimizing groundwater management, optimizing urban water supply head works, and programming infrastructure work.

14.6.4 SWARM INTELLIGENCE

A subfield of (AI) known as swarm intelligence (SI) is based on the social behavior of "swarms," which are groups of basic, locally interacting creatures with global adaptive behavior (Millonas, 1994). Ant nests, bird flocks, and schools of fish are a few examples of swarms that can be seen in nature. Such social organisms are intriguing because, despite the fact that each member has their own goals, the group as a whole is extremely well organized, and group collaboration and coordination are mainly self-organized (Bonabeau and Théraulaz, 2000). The group's emergent behavior, which results from the social sharing of information, can also solve challenging issues like determining the shortest path to a food source or spotting promising areas on the landscape when searching for food, whereas simple local interactions between individuals only occur. Consequently, SI techniques offer a practical tool for optimizing intricate, nonlinear functions. Ant colony optimization (ACO), which is motivated by ant foraging behavior, and particle swarm optimization (PSO), which is motivated by flocks of birds' social behavior, are two well-known SI approaches.

14.6.4.1 Applications

The operation of multipurpose reservoirs, the design of water distribution systems, and the scheduling of hydropower plant maintenance have all been successfully optimized using the ACO algorithm due to recognition of the strengths of the algorithm when applied to discrete combinatorial optimization problems. It has also been used to solve a continuous optimization problem, the calibration of a straightforward rainfall-runoff model. When applied to this issue, it was discovered that a GA was more successful and efficient than an ACO. The PSO algorithm has been used in the modeling and management of water resources to optimize selective withdrawal from thermally stratified reservoirs, to train ANNs used in rainfall-runoff modeling and river stage forecasting, and to select size and position hydraulic devices for transient protection in a pipe network.

ACO was created primarily for discrete combinatorial optimization, whereas PSO was created for continuous function optimization. Both methods contain modifications that enable their application to both continuous and discrete optimization problems.

14.6.5 EVOLUTIONARY MULTI-OBJECTIVE OPTIMIZATION

The problems for which there is a single objective, such as cost, demand, and risk, have been explored in relation to the optimization methods that have so far been provided. However, several goals must be achieved at once for the majority of management issues in real-world hydrology. For instance, managing a reservoir may be necessary to achieve goals relating to flood control, hydroelectric power generation, environmental conditions in the reservoir's upstream catchment, and reliability of the water supply. One strategy for dealing with such issues is to reduce them to single-criteria issues by using penalty functions or weighted aggregate objective functions together with conventional optimization techniques. Although this is by far the most typical method for handling multiple objectives, such simplification may produce an optimization problem that is unrelated to the one that needs to be solved (Corne *et al.*, 2003; Massarelli et al., 2021). An alternative is to use multi-objective optimization techniques, which define and optimize a problem in terms of many, sometimes competing, criteria (Michalewicz and Fogel, 2004). These approaches are also known as multi-criteria optimization or vector optimization. Without the need for potentially misleading simplifications, these strategies can handle hydrological modeling and management issues in their full multi-objective form.

14.6.5.1 Applications

Since the mid- to late-1990s, EMOO has been used to solve modeling and management issues related to water resources, but it has only recently started to gain more traction in this area. This is owing to the fact that the discipline of EMOO is still in its infancy and that more scholars and practitioners are just now becoming interested in the field as a result of the effectiveness of new MOEAs. EMOO offers a viable strategy to help with design and decision-making issues pertaining to hydrological systems. The design of groundwater monitoring systems, groundwater remediation systems, calibration of hydrological models, optimization of agricultural land use and management practices within a watershed, optimization of single and multi-reservoir system operation, and determining the best waste load allocation in rivers are a few examples of successful applications of EMOO techniques.

14.7 CONCLUSIONS

The use of AI technology for hydrological modeling and water resource management has increased during the past 15 years. At first, quantitative data-based prediction and forecasting were the main uses of AI approaches. This group includes GP and ANN, the most used AI technique in hydrology and water resource management. In recent years, the inclusion of qualitative and subjective information has become more and more supported by AI technologies like fuzzy systems and BNs. Due to their capacity to comprehensively model complicated hydrological systems, BNs and ABMs have also drawn more interest. The AI-based optimization techniques have a number of characteristics in common that make them suitable for handling modeling and management issues involving water resources, which are frequently characterized by multimodality, constraints, large dimensionality, nonlinearity, noise, and time-varying objective functions. The fact that AI optimization techniques are population-based and that data inside the population is shared in order for one population to "evolve" into another is the most crucial of these traits. With no need for information about the target function that can be impossible or challenging to obtain, they are able to traverse the search space without being stuck in local optima (e.g. gradient information). The output of the models built into the solutions of the nonlinear systems, the stock market efficiency close to the actual system outputs, and the simplicity of model calibration all boost the adoption of these methods, despite the fact that AI techniques are thought of as black box systems. Studies like the behavior of the models, the types of functions, and the number of iterations have an impact on one of the shortcomings of the established mathematical models as it becomes a model for the modeling of a physical occurrence. In this instance, using alternative algorithms and changing the functions are proportionate to a researcher's experience to boost accuracy and reduce the mistake rate. Therefore, it can be argued that compared to traditional methods, AI techniques often produce better modeling outcomes for water resources. Additionally, these methods are combined with fresh ideas and other scientific disciplines, which could result in more accurate predictions.

REFERENCES

Arab, A., Lek, S., Lounaci, A. and Park, Y.S., 2004, December. Spatial and temporal patterns of benthic invertebrate communities in an intermittent river (North Africa). *Annales de Limnologie-International Journal of Limnology* 40(4): 317–327. EDP Sciences.

Babovic, V., 2005. Data mining in hydrology. *Hydrological Processes: An International Journal 19*(7), pp. 1511–1515.

Bezdek, J.C., 2013. *Pattern recognition with fuzzy objective function algorithms*. Springer Science & Business Media.

Bharadwaj, D., 2020. *Integrated water resources management using artificial intelligence.* Department of Research & Publications, Bharat Innovates Universals.

Bonabeau, E. and Théraulaz, G., 2000. Swarm smarts. *Scientific American* 282(3): 72–79.

Borsuk, M.E., Stow, C.A. and Reckhow, K.H., 2004. A Bayesian network of eutrophication models for synthesis, prediction, and uncertainty analysis. *Ecological Modelling* 173(2–3), pp. 219–239.

Bousquet, F. and Le Page, C., 2004. Multi-agent simulations and ecosystem management: a review. *Ecological Modelling* 176(3–4): 313–332.

Bromley, J., Jackson, N.A., Clymer, O.J., Giacomello, A.M. and Jensen, F.V., 2005. The use of Hugin® to develop Bayesian networks as an aid to integrated water resource planning. *Environmental Modelling & Software* 20(2): 231–242.

Cain, J., 2001. Planning improvements in natural resource management. guidelines for using Bayesian networks to support the planning and management of development programmes in the water sector and beyond.

Castro, D. and New, J., 2016. The promise of artificial intelligence. *Center for Data Innovation* 115(10): 32–35.

Corne, D.W., Deb, K., Fleming, P.J. and Knowles, J.D., 2003. The good of the many outweighs the good of the one: Evolutionary multi-objective optimization. *IEEE Connections Newsletter* 1(1): 9–13.

de Santa Olalla, F.M., Domínguez, A., Artigao, A., Fabeiro, C. and Ortega, J.F., 2005. Integrated water resources management of the hydrogeological unit "Eastern Mancha" using Bayesian belief networks. *Agricultural Water Management* 77(1–3): 21–36.

Duan, Q., Sorooshian, S. and Gupta, V., 1992. Effective and efficient global optimization for conceptual rainfall-runoff models. *Water Resources Research* 28(4): 1015–1031.

Dunn, J.C., 1973. A fuzzy relative of the ISODATA process and its use in detecting compact well-separated clusters.

Flood, I. and Kartam, N., 1994. Neural networks in civil engineering. I: Principles and understanding. *Journal of Computing in Civil Engineering* 8(2): 131–148.

Gibbs, M.S., Maier, H.R. and Dandy, G.C., 2011. Relationship between problem characteristics and the optimal number of genetic algorithm generations. *Engineering Optimization*, 43(4): 349–376.

Humphrey, G.B., Maier, H.R., Dandy, G.C., Kingston, G.B., Maier, H.R. and Dandy, G.C., 2008. *Review of artificial intelligence techniques and their applications to hydrological modeling and water resources management. Part 1 – Simulation* (pp. 15–65). Water Resources Research Progress.

Islam, S. and Kothari, R., 2000. Artificial neural networks in remote sensing of hydrologic processes. *Journal of Hydrologic Engineering* 5(2): 138–144.

Kingston, G.B., Dandy, G.C. and Maier, H.R., 2008. *Review of artificial intelligence techniques and their applications to hydrological modeling and water resources management. Part 2 – Optimization* (pp. 67–99). Water Resources Research Progress.

Kohonen, T., 1982. Self-organized formation of topologically correct feature maps. *Biological Cybernetics* 43(1): 59–69.

Landahl, H.D., McCulloch, W.S. and Pitts, W., 1943. A statistical consequence of the logical calculus of nervous nets. *The Bulletin of Mathematical Biophysics* 5(4): 135–137.

Langdon, W.B. and Qureshi, A., 1998. Genetic Programming-Computers Using. In *Natural selection "to Generate Programs," genetic programming and data structures, The Springer International Series in engineering and computer science.* Boston: Springer.

Liong, S.Y., Gautam, T.R., Khu, S.T., Babovic, V., Keijzer, M. and Muttil, N., 2002. Genetic programming: a new paradigm in rainfall runoff modeling 1. *JAWRA Journal of the American Water Resources Association*, 38(3), pp. 705–718.

Maier, H.R. and Dandy, G.C., 2000. Neural networks for the prediction and forecasting of water resources variables: A review of modelling issues and applications. *Environmental Modelling & Software* 15(1): 101–124.

Massarelli, C., Campanale, C. and Uricchio, V.F., 2021. Artificial intelligence and water cycle management. In *IoT Applications Computing*. IntechOpen. DOI: 10.5772/intechopen.97385

Michalewicz, Z. and Fogel, D.B., 2004. Time-varying environments and noise. In *How to solve it: Modern heuristics* (pp. 307–334). Berlin, Heidelberg: Springer.

Millonas, M., 1994. Swarms, phase transitions, and collective intelligence. *Proceedings of the Artificial Life*.

Murat, A.Y. and Özyildirim, S., 2018. Artificial intelligence (AI) studies in water resources. *Natural and Engineering Sciences* 3(2): 187–195.

Negnevitsky, M., 2005. *Artificial intelligence: A guide to intelligent systems*. Harlow: Pearson education.

Niu, W.J. and Feng, Z.K., 2021. Evaluating the performances of several artificial intelligence methods in forecasting daily streamflow time series for sustainable water resources management. *Sustainable Cities and Society*, 64: 102562.

Parker, D.C., Berger, J., Manson, S.M. and McConnell, W.J., 2002. Agent-based models of land-use and land-cover change. In *Proc. of an International Workshop* (pp. 4–7).

Parker, D.C., Manson, S.M., Janssen, M.A., Hoffmann, M.J. and Deadman, P., 2003. Multi-agent systems for the simulation of land-use and land-cover change: a review. *Annals of the Association of American Geographers* 93(2): 314–337.

Pearl, J., 1986. Fusion, propagation, and structuring in belief networks. *Artificial Intelligence* 29(3): 241–288.

Peterson, G., 2000. Political ecology and ecological resilience: an integration of human and ecological dynamics. *Ecological Economics* 35(3): 323–336.

Simonovic, S.P., 2005. A spatial multi-objective decision-making under uncertainty for water resources management. *Journal of Hydroinformatics* 7(2): 117–133.

Simpson, A., Dandy, G. and Murphy, L., 1994. Genetic algorithms compared to other techniques for pipe optimization. *Journal of Water Resources Planning and Management* 120(4): 423–443.

Zadeh, L.A., 1996. Fuzzy sets. In Fuzzy *sets, fuzzy logic, and fuzzy systems*: Selected *papers by Lotfi A Zadeh* (pp. 394–432).

15 Evapotranspiration in Context of Climate Change

Khilat Shabir, Rohitashw Kumar, and Munjid Maryam

15.1 INTRODUCTION

Crop water demand is influenced by evapotranspiration (ET), which is a key component of the hydrologic cycle. As a result, in order to plan irrigation efficiently, it must be quantified. Evaporation and transpiration are combined in ET. During evaporation, water vapor is lost from natural surfaces, including free water, bare land, and living or dead vegetation. Transpiration is the mechanism through which plants lose water as vapor via their leaves. ET, then, is the vaporization of water from the soil (evaporation) or plant surfaces (transpiration). It is measured in depth per unit of time (mm/day, for example).

ET can be directly measured for a certain crop, soil, and climatic circumstances, or it can be calculated using the reference crop ET, which is normally approximated using multiple approaches depending on the availability of data for a given situation. Evaporation, transpiration, and ET are all affected by weather conditions, crop traits, management, and environmental factors. Fig. 15.1 describes the complete ET process.

15.1.1 CLIMATIC FACTORS AFFECTING EVAPOTRANSPIRATION

- **Radiation:** For the ET process, the primary source of energy is radiation. It is determined by the vegetation's albedo and the total solar radiation flux density. Dark-colored plants absorb more solar energy and generate more ET. In order to create accurate estimations or predictions of ET and irrigation water needs, it is necessary to understand the processes that influence the energy balance of cultivated soils. It also makes it possible to more effectively manage irrigation water.
- **Temperature:** As temperature increases over a day, the saturation deficit rises, resulting in greater evaporative demands and higher evaporation rates.
- **Relative humidity:** The temperature and relative humidity of the air interact. The lower the ET, the lower the evaporative requirement, and thus the higher the relative humidity.
- **Wind:** The ET process involves the use of extra energy generated by the horizontal flow of energy from a dry area to a more humid environment. In addition, wind transports water vapor from nearby plants to distant areas.

DOI: 10.1201/9781003351672-15

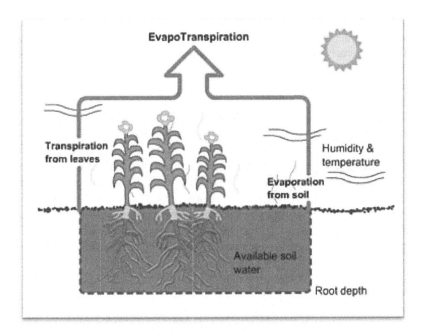

FIGURE 15.1 Evapotranspiration process.

ET plays a significant role in maintaining the balance of surface water in terrestrial ecosystems by accounting for the majority part of the water that leaves the watershed. Surface ET is responsible for two-thirds of worldwide water loss on land. ET is influenced by both external (physical and climatic) and internal (biological) factors, making it one of the most complicated elements in the water cycle. Actual ET and ET ratio (annual actual ET divided by precipitation AET/P) show a significant correlation with temperature, precipitation, and potential ET (which measures energy availability). At continental scales, substantially warmer air temperatures have been reported in recent decades. Consequently, as a result of climate change, particularly global warming, heatwaves, temperature increases, and water vapor deficits are projected to have additional effects on ET and ET ratio. Changes in the ET/ET ratio, as well as their possible causes, have been widely recorded in various places as a result of global warming.

ET is a key factor in balancing the energy and water on the planet's surface. Temperature, rainfall, sunlight, moisture in the air, and wind velocity are all climatic variables that can have a big impact on hydrologic processes. Changes in temperature and precipitation have an impact on evaporation, as well as temporal and regional variability in the runoff. Hulme *et al.* (1994) proposed that climate change would increase evaporation potency, whereas Roderick and Farquhar (2002) revealed that pan evaporation, experimental and physical measurement of potential evaporation

(PET), has dropped in throughout these last several decades in most parts of the world. The supplemental hypothesis predicts that a reduction in actual ET would result in an increase in potential ET. PET was studied in the face of climate change in a Mediterranean location by Chaouche et al. (2010), who discovered that PET is reliant not only on temperature but also on relative humidity, solar radiation, and wind velocity.

The climate is changing. A shift in the average temperature of the planet by one or two degrees can result in potentially hazardous climate and weather changes. Because these observable changes are evident ways that climate change affects the Earth, we call them climate change impacts. Rainfall patterns have shifted in many regions, resulting in greater flooding, droughts, or heavy rain, along with more frequent extreme weather events. Changes in the earth's water sources and glaciers have occurred in addition to climate change, leading water bodies to warm by becoming highly acidic in nature, melting glaciers, and increasing water levels These and other changes are projected to grow increasingly prevalent in the next decades, posing problems to people and the environment.

As a result of rising temperatures, the balance between water and energy in terrestrial systems is systematically changing. Since the 1980s, the climatologically defined 100th meridian in the United States has changed eastward, essentially enlarging the area in Western states where ET is limited by water availability. Water scarcity has hampered forest growth in the twentieth century, woods have experienced increased vapor deficit stress, and the link between temperature and vegetative production has weakened. These are tendencies that are in line with increased water scarcity. The effects of changing land cover on ET have also gotten a lot of attention. Because of differences in leaf area index (LAI), surface albedo, and root depth, different land cover types have varied biophysical impacts. Furthermore, different vegetation species have variable stomatal reactions to rising atmospheric CO_2 concentrations, resulting in different evaporation rates. Despite these findings, it's uncertain how much future warmth and water availability will affect ecosystem function.

Groundwater droughts have increased throughout hot seasons in the twenty-first century (rather than exclusively during dry seasons), and these changes are likely attributable to evaporative alterations with warming. A warmer climate will increase evaporative demand, but the amount of moisture shortfalls and evaporative distress, as opposed to increased ET, is partly determined by the climate. Regional precipitation and moisture levels in soil determine crop water availability, although shallow groundwater may also influence soil moisture and streamflow. Each and every year, 30% of the entire renewable freshwater is returned to the aquifer. Groundwater depths and lateral distribution of water in the subsurface can impact the partition of ET runoff and replenishment, as well as the susceptibility of the land surface to fluctuations in rainfall and temperature. Groundwater is the slowest moving element of the terrestrial water cycle, serving as water storage for plants during dry seasons and causing decadal fluctuations in total availability of terrestrial water. In a warmer environment, increased ET may induce a shift in the proportion of precipitation that flows off as surface water or penetrates to the subsurface as recharge. Water table levels, surface and ground-water interactions, and soil moisture levels can all be impacted by variations in recharge trends over time.

15.2 EVAPOTRANSPIRATION IN CONTEXT OF CLIMATE CHANGE

Increased demand for natural resources is having negative repercussions for sensitive ecosystems and biodiversity, posing a serious global danger to long-term development (Biswas and Tortajada 2009; Oki and Kanae 2006). Changing climate, which now has emerged among the most serious global environmental issues, will become the final threat to mankind in the twenty-first century. Climate change is expected to have a significant negative effect on the world's water resources and sensitive ecosystems. There will be water stress in some areas, which will be exacerbated by increased water demand, putting a growing number of people at risk of water short-ages. Simultaneously, some places would be at risk of flooding, which could lead to the loss of life. Flooding and dry spells will become more common in most areas. Water shortage and floods will become more common in most parts of the world (Bhat and Karim, 2009). The immediate impact of the change in climate on water resources is expected to be mainly a result of ET caused by rising temperatures. Climate change will lead to increased temperatures, shifting rainfall patterns, and changing patterns of land use and cover. Rising temperatures will cause an increase in ET, which will have an impact on the water resources and hydrological cycle (Shahid 2011). For protracted water resource management, quantifying variations in ET caused by climate change is crucial.

As part of both the water flow and the surface energy exchange processes, ET plays a critical role in the management of water resources. Dynamic changes in ET have been studied extensively over the last three decades, finding that global annual ET is on the rise, but the degree of the rise differs by region. Despite the fact that local climate has proven to be a significant driver of ET, vegetation cover can alter ET by affecting canopy interception, canopy transpiration, and ground evaporation, so land cover changes are significant contributors to ET alterations (GE). Quantitatively defining the reasons for ET change is critical for better understanding hydrological process change as a result of climatic and anthropogenic influences, as well as to assist decision-makers in water resource planning and management.

For hydrological and climatological investigations, ET is a crucial quantity. The estimation of ET is a key component in estimating the ER (depth of evaporation from the reservoir), as well as a regulating factor for streamflow and discharge volume. As a result, having an accurate and consistent estimate of ET is critical. Empirically and/or physically based formulae have been established during the last half-century to estimate reference ET (ET_0) for various areas. However, due to large discrepancies in the values produced by different ET models, selecting the right strategy for a broad array of applications can be perplexing and vital for the educated user.

ET_0 is frequently determined using climate variables such as solar radiation, wind velocity, air temperature, and relative humidity. Various approaches for estimating ET_0 are known that use equations varying from the most sophisticated energy balance method which requires extensive climatological data to a simpler way requiring fewer data (Allen *et al.* 1989). Radiation model FAO-24 is a physical model that combines energy budget and mass transfer (Doorenbos and Pruitt 1977; Jensen *et al.* 1990). The Hargreaves approach is straightforward and just requires two meteorological variables: temperature and sun energy. Using high-quality lysimeter data and a broad range of climatic conditions, this method was put to the test (Hargreaves and Samani 1994), and the findings showed that it was virtually as accurate as Penman-Monteith (PM) in estimating ET_o. Irmak *et al.* (2003) reduced the number of input parameters and computations required by the FAO56-PM technique by describing it as a multi-linear regression function. To calculate the ET_o, only radiation from the sun and daily mean temperature are used as inputs.

Studies have looked at how ET has changed in relation to climate change in many instances and the responses of ET to various climatic factors, such as the temperature of air, rainfall, relative humidity, solar radiation, and speed of wind, and these factors are extensively investigated through observations and modeling; however, inconsistent conclusions have been drawn due to different climate zones, vegetation cover characteristics, and land management practices.

15.3 CLIMATE CHANGE AND EVAPOTRANSPIRATION

Climate change is happening all across the world (IPCC 2007), and it could affect factors such as precipitation, air temperature, humidity levels, and radiation from the sun (Haskett *et al.* 2000). It is likely to alter precipitation and evaporation, causing alterations in the hydrological cycle (Bates *et al.* 2008; Huntington 2006). These modifications will have a significant impact on crop ET and irrigation management (Zhang *et al.* 2011). It has been proven that climate variable trends differ from location to place. The average global temperature has risen by around 0.6°C, with two-thirds of this increase occurring since 1975 (WHO 2003). As the temperature rises, food, feed, and fiber production are expected to become more unstable. Climate change projections for crop output are fundamentally uncertain (Asseng et al. 2013). Climate change's negative effects on crop output include increasing temperature and erratic rainfall. Water system stress produced by these two environmental factors reduces agricultural output, and a water shortage might cause plants to wilt.

15.3.1 CLIMATE VARIABILITY

Climate change is the most important determinant of year-on-year variability in social and environmental systems, involving water supply. It has an impact on water resource planning and decision-making, such as flood prevention and management, drought control, and food and fiber production. Furthermore, any changing climate will exacerbate the risk associated with water resource planning. Aside from that, climatic changes will have significant impacts and implications on agricultural and natural ecosystems, as well as society as a whole. Such changes may even affect the position of the world's key crop-growing areas. Food and fiber supply instability caused by climate and weather will affect social and economic sustainability as well as regional competitiveness. As a result, climate change study of hydro-climatic elements such as precipitation, evaporation and transpiration, temperature, air humidity, sunlight hours, pan evaporation, and wind velocity becomes critical.

Short- and long-term variations in meteorological variables are referred to as climatic variability. Changing climate has a long-term and considerable impact on the distribution curve of weather events, spanning decades to thousands of years. It could potentially modify the typical weather patterns or the weather distribution around those conditions (more or fewer extreme weather events). Oceanic events, volcanic activity, variations in solar energy received by the globe, seismic activity, and man-made natural world modification all influence climate change, with the latter elements currently adding to global warming. Climate change is a terminology that is widely used to describe the impact on humans (Tabari *et al.* 2012).

Science has made significant progress in studying climate variability and change. Its origins and consequences aid in the development of a thorough grasp of current and potential impacts on people today and in the coming decades. This knowledge is critical since it allows decision-makers to put climate change in the perspective of the nation's and world's other major concerns. Because climatic fluctuation and variation affect weather patterns in the short and long terms, the water demand, which is a function of weather parameters, varies over time. Changing climate, which is caused by the greenhouse effect, has a significant impact on the hydrological cycle, affecting precipitation, ET, and moisture in the soil distribution. ET_0, being a key element of the hydrologic cycle, will have an impact on crop water requirements and the management of water resources.

15.3.2 Reference Evapotranspiration

The standard crop ET is the rate of ET from a vast area covered by green grass that is 8–15 cm tall grows rapidly, shades the ground entirely, and does not lack moisture. The reference crop, in this case, is grasses. Agricultural management of water resources, irrigation scheduling, and water resource planning all rely heavily on reference ET (ET_0). In research linked to hydrology, climate, and agricultural watershed management, the reference ET (ET_0) flow occurring from cultivated land surfaces is critical. Accurate ET assessment is critical for successful irrigation. Meteorological factors such as temperature, humidity, sunlight hour, and wind speed are used to predict ET. Climate variables greatly influence ET_0 to differ, according to studies. Climatic variables such as temperature, humidity, sunlight hour, and wind velocity are used to forecast ET. According to investigations, climate elements that greatly influence ET_0 differ from one place to the next (Wang *et al.* 2014). ET_0 is important for the worldwide atmosphere system's mass and heat fluxes.

15.4 REFERENCE EVAPOTRANSPIRATION (ET_O) MODELS

ET is a complicated phenomenon that is influenced by a variety of climatic conditions. The climate-based ET_0 estimation methods described in Table 15.1 can be used to compute ET_0. Any ET model's application is restricted by the accessibility of input information. Minimum and maximum daily temperatures (°C), precipitation (mm), pan evaporation (mm), relative humidity (percent), sunshine hours (hr), wind speed (km/hr), potential ET (mm/day), soil heat flux density (MJ/m²/day[1]), and net radiation at the crop surface (MJ/m²/day) are typical observations needed for ET_0 evaluation (Kumar *et al.* 2012).

TABLE 15.1
Reference Evapotranspiration Estimation Methods

S. No.	Method of ET$_o$ estimation	Equations used	Basic reference	Required meteorological data
1	FAO-24 corrected Penman (c=1), (Fc P-Mon)	$ET_o = c\left[\dfrac{\Delta}{\Delta+\gamma}(R_n - G) + \dfrac{\gamma}{\Delta+y}\,2.7W_f(e_a - e_d)\right]$	Doorenbos and Pruitt (1997)	Net radiation, vapor pressure deficit, and wind velocity
2	Priestley-Taylor (P-T)	$ET_o = \alpha\dfrac{\Delta}{\Delta+\gamma}(R^n - G)$	Shuttleworth (1993)	Net radiation, soil heat flux, and vapor pressure deficit
3	FAO-24 Blaney-Criddle (F B-C)	$ET_o = a + b\left[p(0.46\overline{T}+8.13)\right]$	Doorenbos and Pruitt (1977)	Annual daytime hours, temperature, and wind velocity
4	Hargreaves-Samani (H-S)	$ET_o = 0.0135(KT)(R_a)\left(TD^{\frac{1}{2}}\right)(TC+17.8)$ $KT = 0.00185(TD^2) - 0.0433TD + 0.4023$	Hargreaves and Samani (1982, 1985)	Net radiation, min/max temperature
5	FAO pan evaporation (F E-Pan)	$ET_o = k_P E_{Pan}$	Allen et al. (1998)	Pan evaporation
6	Penman Monteith (P-Mon)	$ET_o = \dfrac{0.408\,\Delta(R_n - G) + \gamma\,\left(900/(T+273)\right)\,u_2\left(e_s - e_a\right)}{\Delta+\gamma\left(1+0.34u_2\right)}$	Allen et al. (1998)	Vapor pressure deficit, radiation flux, wind velocity, temperature, and soil heat flux
7	Kimberly-Penman Model (K-M)	$\lambda ET_r = \dfrac{\Delta}{\Delta+\gamma}(R_n - G) + \dfrac{\gamma}{\Delta+\gamma}\,6.43W_f\left(e_z^0 - e_z\right)$	Wright (1982)	Vapor pressure deficit, radiation flux, wind velocity, temperature, and soil heat flux
8	Jensen-Haise Alfalfa Reference Model	$\lambda ET_r = C_T\left(T - T_x\right)R_s$	Jensen and Haise (1963), Jensen et al. (1970)	Solar radiation and mean air temperature
9	SCS Blaney-Criddle Model	$U = KF = \sum kf$	Blaney and Morin (1942), Blaney and Criddle (1962)	Mean air temperature, average relative humidity, and mean percentage of daytime hours
10	Generalized form of ASCE standardized equation	$ET_{SZ} = \dfrac{0.408\,\Delta(R_n - G) + \gamma\,\left(C_n/(T+273)\right)\,u_2\left(e_s - e_a\right)}{\Delta+y\left(1+Cdu_2\right)}$	Penman Monteith by ASCE (Jensen et al. 1990)	Vapor pressure deficit, radiation flux, wind velocity, temperature and soil heat flux

where

n = actual duration of bright sunshine hour

P = atmospheric pressure, kPa

TD = difference between mean monthly maximum and mean monthly minimum temperature, °C;

G = heat flux density to the ground, MJ / m² / d

U_2 = horizontal wind speed at height 2.0 m, m / s

Ta = mean air temperature, °K = $(273 + °C)$

T = mean air temperature, °C

R_n = net radiation, MJ / m² / d

γ = psychometric constant, kPa / °C

ET_0= reference evapotranspiration from well watered grass, MJ / m² / d

α = short wave reflectance (albedo) = 0.23 for green crop

Δ = slope of saturation vapor pressure temperature curve, kPa / °C

Numerous research work has been done in recent times to investigate the consequences of climate change on reference ET (ET_0). These researches revealed that ET_0 trends differ according to climate and area (Rim 2009). In the Tons River Basin in Central India, Darshana and Pandey (2013) found substantial yearly, seasonally, and monthly (for practically all months) declining trends in ET_0, with annual ET_0, decreases ranging from −1.75 to −8.98 mm/year. Similarly, ET_0 has dropped dramatically in the Platte River Basin in central Nebraska, USA (−0.36 mm/year) (Irmak et al. 2012). Considerable (P 0.05) rises in precipitation (0.87 mm/year) led to a considerable reduction in incoming radiation (0.003 2 MJ/(mm/year)), according to these researchers (Irmak et al. 2012). In Iran, considerable decreases in ET_0 have indeed been observed in most locations, although growing tendencies have been observed in others, especially in recent years (Kousari et al. 2013; Tabari et al. 2011; Talaee et al. 2014). Wind velocity was the primary element impacting ET_0 in the southwest and west of Iran, according to Tabari et al. (2012). Espadafor et al. (2011) found statistically significant elevations in ET_0 (up to 3.5 mm/year) in southern Spain, which they attributed to increases in solar radiation and air temperature, as well as declines in ET_0.

Water use efficiency in irrigated lands has been greatly impacted by increased competition for water supplies (Hatfield et al. 1996), particularly in dry and semiarid regions. For assessing the consequences of historical climatic changes on soil moisture distribution and ecosystems in these places with limited water supplies, it is critical to estimate crop ET with the greatest accuracy feasible.

15.5 STUDIES RELATED TO CLIMATE CHANGE AND EVAPORATION

The 2007 survey on changing climate by the United Nations Intergovernmental Panel on Climate Change clearly demonstrates warming tendencies. In many regions, different types of change alleviation and adaptation have begun to emerge. Change in climate is being taken into account in infrastructure designs.

The global water and energy cycle is dominated by ET. A large shift in ET will come from a change in climatic and meteorological factors. Increases in radiation, temperature, and water vapor deficit can all have a direct impact on ET as a result of global warming. Increases of 3–4°C temperature in summer were shown to enhance potential ET by 20% in three Alpine river basins, according to the results of a simulation using a global climate model (GCM) (Calanca et al. 2006). The estimated elevation in solar radiation was five, with a precipitation reduction of 10–20%. Evaporation has grown since 1980 as a result of global warming, according to data from 317 meteorological stations in China (Cong et al. 2008). According to predictions from the Union of Concerned Scientists and the Ecological Society of America, the Gulf Coast of the United States will warm by 3–7°F in the summer and 5–10°F in the winter, based on climate projections (Twilley et al. 2001). ET will increase as the temperature rises. Azam and Farooq (2005) note the importance of developing water-efficient crops and the influence of water vapor production from ET on global warming. Several climate models were used to investigate the influence of global warming on alfalfa ET in California and their relationship with global warming. The data showed a daily mean maximum temperature rise of 4.3°C and a 0.59-mm increase in statewide mean daily ET (Zhang et al. 1996).

A1B, A2, and B1 were simulated using three global circulation models: Canadian Centre for Climate Modeling and Analysis (CCCMA), Geophysical Fluid Dynamics Laboratory (GFDL) in the USA, and Max-Planck-Institute for Meteorology (MPI-M) in Germany (IPCC 2007). The Great River region of Jamaica was downscaled to investigate the effects of climate change on watershed size hydrology. After reviewing global climatological observation network data and remote sensing and adding it into the model, they discovered that ET increased by 7 mm in year 1 between 1982 and 1997. They also came to the conclusion that a shortage of moisture was to blame for the decline in evaporation from 1997 to 2008. According to the watershed scale, the average increase in air temperature from 1980 to 2000 is shown in Fig. 15.2. It is evident from the figure that the air temperature increased by 2.36°C on average. In Fig. 15.3, the monthly increases in ET are plotted based on the outputs of the three models.

Several studies have shown that plant transpiration increases with temperature, but an increase in CO_2 lowers the increase in transpiration related to temperature (California Department of Water Resources 2006; Hatfield et al. 2011). Therefore, as the CO_2 level increases, the evaporation of open water and soil does not decrease. Researchers from the USDA studied the impact of climate change on agriculture by examining CO_2's effect on crop ET without considering the temperature. It was concluded that when there is plenty of nitrogen but not enough water, both C3 and C4 plants' ET will be the same. However, due to a scarcity of nitrogen and water at 550 ppm CO_2, ET is projected to decline (Hatfield et al. 2008). In contrast to these forecasts, ET increases of 6, 13, and 23% are predicted for 2020, 2050, and 2010, respectively, based on modeling implications of changing climate on Finland's water resources (Vehviläinenand Huttunen 1997). A study of the sensitivity of ET to global warming in the dry region of Rajasthan, India, concluded that even a small increase

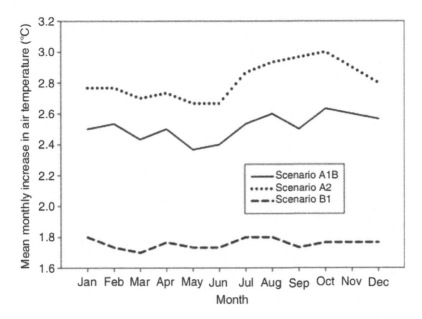

FIGURE 15.2   Expected air temperature increase in the Great River region of Jamaica (2080–2100).

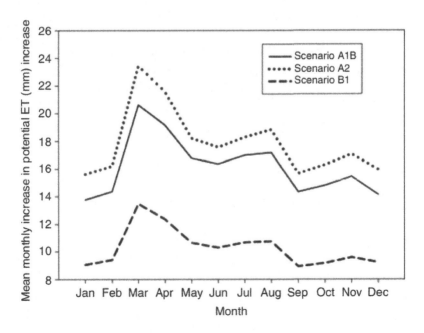

FIGURE 15.3   Expected potential evapotranspiration increase in the Great River region of Jamaica (2080–2100).

in ET in such a climatic area will have a substantial effect (Goyal 2004). In a study of climate change, evaporation, and ET, a study of six global climate models predicted increased ET across India (Chattopadhyay and Hulme 1997). According to a global assessment of the Palmer drought index—which is connected to soil moisture and surface heat effects—the risk of drought will increase as anthropogenic global warming continues, leading to rising temperatures and more drying (Dai *et al.* 2004). Approximately half of the solar energy collected by land surfaces is lost through evaporation, according to Jung *et al.* (2010). They believe that a due to climate change, changes in the hydrologic cycle will cause ET to alter. After reviewing global climatological observation network data and remote sensing and adding it into the model, they discovered that ET increased by 7 mm in year 1 between 1982 and 1997. They also came to the conclusion that a shortage of moisture was to blame for the decline in evaporation from 1997 to 2008.

Despite the overall rise in temperature in recent decades, Ep (pan evaporation) has been reducing in practically all places of India, with the pre-monsoon and monsoon seasons becoming especially remarkable. Stepwise regression was done by Chattopadhyay *et al.* (1997) between Ep and the different meteorological parameters that regulate evaporation: relative humidity (r.h.), wind speed (w.s.), net global radiation (r.n.), maximum (t.x.) and minimum (t.n.) for eight stations (Bikramganj, Varanasi, Raipur, Jabalpur, Hisar, Canning, Akola, Annamalinagar) with 15 years of data (1976–1990) (Table 15.2).

TABLE 15.2
The Number of Times Five Different Meteorological Variables, in Order of Dominance (i–iv), Were Significantly (at 95% Level) Related to Ep and PE in Stepwise Regression Models Established for Stations Over India (1976–1990)

Meteorological Variable	Winter i	Winter ii	Winter iii	Winter iv	Pre-Monsoon i	Pre-Monsoon ii	Pre-Monsoon iii	Pre-Monsoon iv	Monsoon i	Monsoon ii	Monsoon iii	Monsoon iv	Post-Monsoon i	Post-Monsoon ii	Post-Monsoon iii	Post-Monsoon iv
For Ep (eight stations)																
r.h.	3	_	_	_	5	_	_	_	5	_	_	_	4	_	1	_
r.n.	4	1	_	_	1	_	_	_	_	1	1	_	1	1	_	_
t.x.	1	_	1	_	2	1	2	_	2	2	1	_	3	_	_	_
t.n.	_	1	1	_	_	1	_	_	_	_	_	1	_	_	_	_
w.s.	_	1	_	_	_	1	1	_	1	1	_	_	1	2	_	_
For PE (ten stations)																
r.h.	6	2	_	_	7	1	2	_	_	4	1	1	3	2	1	1
r.n.	1	1	1	_	1	6	1	_	9	_	_	_	1	1	_	_
t.x.	_	3	_	2	_	1	2	1	_	_	4	1	_	1	2	_
t.n.	1	_	1	1	2	_	3	_	_	2	_	_	3	1	1	_
w.s.	1	_	_	_	1	_	_	_	_	1	2	_	1	_	_	_

i, ii, iii, and iv denote the stage in the stepwise regression model where the meteorological variable was chosen (i first, i.e. most dominant variable; iv fourth). Only variables that were 95% substantially associated with Ep (pan evaporation) or PE (potential ET h) were included; hence, the number of variables counted in each matrix varies.

The variable most significantly connected with changes in Ep appears to be relative humidity, especially during the pre-monsoon and monsoon seasons. In the winter, radiation is the best predictor of Ep. Increasing relative humidity appears to have been the most crucial variable in more than only counteracting the effect of rising temperatures on Ep and delaying the evaporation process.

15.6 INFLUENCE OF GLOBAL WARMING ON THE WATER CYCLE REVEALED THROUGH SATELLITE DATA

Researchers have traditionally employed approaches such as modeling, remote sensing, and in-situ investigations to quantify variations in ET. While establishing estimations for broad regions or the entire globe, every method has its drawbacks, according to the study. Instead of measuring ET directly, Pascolini-Campbell *et al.* (2021) used satellites to monitor the relevant components of the water cycle in one of their experiments. The authors also employ information from the Gravity Recovery and Climate Experiment (GRACE) and GRACE-Follow On (GRACE-FO) satellite missions to assess change in the status of vast amounts of water, rather than standard satellite imagery.

The authors assume that almost all precipitation is partitioned into ET, river discharge, and groundwater storage to calculate ET. Scientists can determine ET by monitoring the three remaining components of the hydrological cycle separately.

The method is used to create an "ensemble," or collection, of 20 estimates of worldwide land ET for the years 2003–2019.

ET (top), precipitation (second from top), discharge (second from bottom), and change in groundwater levels storage (bottom) have all changed from 2003 to 2019. The average trend is represented in Fig. 15.4 by the black line, while the shade represents the confidence range, with red sections indicating high confidence.

The scientists discover a "significant statistical" rise in ET of 2.3 mm annually from 2003 to 2019, which is around 10% higher than the long-term average. These results "are consistent with the notion that global ET should rise in a warming environment," as per the study.

According to the investigation, precipitation elevated by 3%, and discharge reduced by 6% during the same time period when compared to long-term averages. The scientists also highlight the driving forces behind the observed changes. They discover that global surface temperature is responsible for over half of the rise, with higher temperature leading to increased ET. However, the El Nino Southern Oscillation, a natural climate event, is responsible for 17% of the variability.

As the temperature rises, the amount of rain re-entering the environment through ET increases, whereas "surface runoff" decreases, according to experts. Rainfall that rushes over land and into streams and rivers instead of sinking into the soil is referred to as runoff.

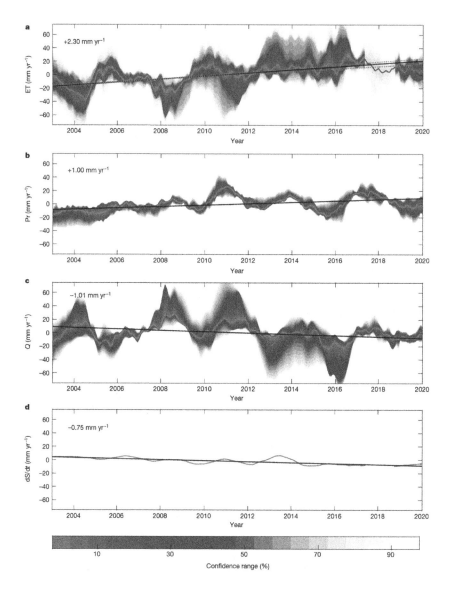

FIGURE 15.4 Time series for evapotranspiration (top), precipitation (second from top), discharge (second from bottom), and change in groundwater storage (bottom) over 2003–2019. The black line shows the average trend and the shading shows the confidence range, where red regions indicate a high confidence (*Pascolini-Campbell et al. 2021*).

15.7 CONCLUSIONS

Temperature-related increases in ET are related to a minor boost in precipitation and a bigger reduction in water storage. This shows that the land surface is drying out, which is likely part of the rationale driving the decrease in humidity levels over land. The quantity of surface water available for irrigation and agricultural supply is affected by increased ET. Increased ET may result in higher water loss from land in certain locations, resulting in water shortages, whereas other areas may see more intense rainfall if the airflow is disrupted. If the current ET current trend continues, the resulting reductions in river flow and increased wildfire danger would pose problems to water management and disaster planners around the world.

The effects of climate change are both seen in evaporation from open water sources and ET from plants. Evaporation and ET have been shown to rise when solar energy and temperature increase. During droughts, a lack of rainfall leads to clearer skies and lower humidity. This situation encourages increased evaporation and ET, leading to faster water loss from lakes, ponds, reservoirs, and land. In research on the impact of climate change on hydrology in all regions, changes in evaporation and ET must be taken into account. If there is enough moisture available, an increase in CO_2 and available energy will boost plant output. The rate and amount of water used by plants in response to climate change are being studied.

Because ET is impacted by climate changes, knowing the trends in variation in crop water demand is essential for planning and managing finite water resources efficiently. The greenhouse effect contributes to climate change, which has a significant impact on the hydrological cycle via rainfall, ET, and soil moisture content. ET_0, being a key component of the hydrologic cycle, will have an impact on crop water requirements and water resource management. ET_0 is the combined effect of four major climatic variables: temperature, relative humidity, sunlight hour, and wind speed, which vary in a predictable way year-round. In terms of variance in climate variables with possible ranges, sensitivity analysis might be used. ET_0's sensitivity can be determined.

Determining the contribution of changing temperature and vegetation covering to ET variation can aid in understanding the causes of hydrological changes and explaining the intricate linkages between vegetation and climate. A number of attribution analyses have been done globally for this aim, and a variety of techniques have been used to study the relationship between ET and various driving variables. Field measurements such as eddy covariance techniques, porometry, lysimeters, and scintillometers are routinely employed to assess the impacts of climate and vegetation on ET. However, these observation methods only cover small areas at the site level with short time periods, leaving the response mechanisms based on various vegetation cover and long-term climatic evolution at the regional scale unaccounted for. Remote sensing has recently been employed to investigate ET changes on a broad spatial scale. Using land surface models and data assimilation based on remote sensing, an increasing variety of ET products have been generated. These products have mostly been used to discover ET trends as well as to statistically regress and correlate ET with environmental parameters such as land cover types, surface temperature, precipitation, and wind speed. Because the link between ET and climatic factors is nonlinear, the correlation values cannot quantitatively describe the contribution of particular elements to ET. Furthermore, assuming linear relationships between ET and climatic factors is inconsistent with reality.

REFERENCES

Allen, R.G., Jensen, M.E., Wright, J.L. and Burman, R.D., 1989. Operational estimates of reference evapotranspiration. *Agronomy Journal* 81(4): 650–662.

Allen, R.G., Pereira, L.S., Raes, D. and Smith, M., 1998. Crop evapotranspiration-guidelines for computing crop water requirements-FAO irrigation and drainage paper 56. *FAO, Rome* 300(9).

Asseng, S., Ewert, F., Rosenzweig, C., Jones, J.W., Hatfield, J.L., Ruane, A.C., Boote, K.J., Thorburn, P.J., Rötter, R.P., Cammarano, D., Brisson, N., Basso, B., Martre, P., Aggarwal, P.K., Angulo, C., Bertuzzi, P., Biernath, C., Challinor, A.J., Doltra, J., Gayler, S., Goldberg, R., Grant, R., Heng, L., Hooker, J., Hunt, L.A., Ingwersen, J., Izaurralde, R.C., Kersebaum, K.C., Müller, C., Naresh Kumar, S., Nendel, C., O'Leary, G., Olesen, J.E., Osborne, T.M., Palosuo, T., Priesack, E., Ripoche, D., Semenov, M.A., Shcherbak, I., Steduto, P., Stöckle, C., Stratonovitch, P., Streck, T., Supit, I., Tao, F., Travasso, M., Waha, K., Wallach, D., White, J.W., Williams, J.R., and Wolf, J., 2013. Uncertainty in simulating wheat yields under climate change. *Nature Climate Change* 3:827–832. DOI:10.1038/nclimate1916.

Azam, F. and Farooq, S., 2005. Agriculture and global warming: Evapotranspiration as an important factor compared to CO_2. *Pakistan Journal of Biological Sciences* 8(11): 1630–1638.

Bates, B., Kundzewicz, Z. and Wu, S., 2008. *Climate change and water. Intergovernmental Panel on Climate Change Secretariat.* Geneva: IPCC Secretariat.

Bhat, R. and Karim, R.R. 2009. Exploring the nutritional potential of wild and underutilized legumes. *Comprehensive Reviews in Food Science and Food Safety* 8:305–333. http://dx.doi.org/10.1111/j.1541-4337.2009.00084.x

Biswas, A.K. and Tortajada, C., 2009. Changing global water management landscape. In *Water Management in 2020 and Beyond* (pp. 1–340). Berlin, Heidelberg: Springer.

Blaney, H.F. and Criddle, W.D., 1962. *Determining consumptive use and irrigation water requirements* (No. 1275). US Department of Agriculture.

Blaney, H.F. and Morin, K.V., 1942. Evaporation and consumptive use of water empirical formulas. *Eos, Transactions American Geophysical Union* 23(1): 76–83.

Calanca, P., Roesch, A., Jasper, K. and Wild, M., 2006. Global warming and the summertime evapotranspiration regime of the Alpine region. *Climatic Change* 79(1): 65–78.

California Department of Water Resources, 2006. *Progress on Incorporating Climate Change Into Planning and Management of California's Water Resources.* Department of Water Resources.

Chaouche, K., Neppel, L., Dieulin, C., Pujol, N., Ladouche, B., Martin, E., Salas, D. and Caballero, Y., 2010. Analyses of precipitation, temperature and evapotranspiration in a French Mediterranean region in the context of climate change. *Comptes Rendus Geoscience* 342(3): 234–243.

Chattopadhyay, N. and Hulme, M., 1997. Evaporation and potential evapotranspiration in India under conditions of recent and future climate change. *Agricultural and Forest Meteorology*: 87(1): 55–73.

Chattopadhyay, S., Singhal, R.S. and Kulkarni, P.R. 1997. Optimisation of conditions of synthesis of oxidized starch from corn and amaranth for use in film-forming applications. *Carbohydrate Polymers* 34: 203–212.

Cong, Z., Yang, D. and Lei, Z., 2008. Did evaporation paradox disappear after the 1980s? A case study for China. *Geophysical Research Abstracts.* Munich: EGU.

Dai, A., Trenberth, K.E. and Qian, T., 2004. A global dataset of Palmer Drought Severity Index for 1870–2002: Relationship with soil moisture and effects of surface warming. *Journal of Hydrometeorology* 5(6): 1117–1130.

Doorenbos, J. and Pruitt, W.O., 1977. Guidelines for predicting crop water requirements. *FAO Irrigation and Drainage Paper* 24: 1–179.

Espadafor, M., Lorite, I.J., Gavilán, P. and Berengena, J., 2011. An analysis of the tendency of reference evapotranspiration estimates and other climate variables during the last 45 years in Southern Spain. *Agricultural Water Management* 98(6): 1045–1061.

Goyal, R.K., 2004. Sensitivity of evapotranspiration to global warming: A case study of arid zone of Rajasthan (India). *Agricultural Water Management* 69(1): 1–11.

Hargreaves, G.H. and Samani, Z.A., 1982. Estimating potential evapotranspiration. *Journal of the Irrigation and Drainage Division* 108(3): 225–230.

Hargreaves, G.H. and Samani, Z.A. 1985. Reference crop evapotranspiration from temperature. *Applied Engineering in Agriculture* 1(2): 96–99.

Hargreaves, G. H. and Samani, Z.A. 1994. Defining and using reference evapotranspiration. *Journal of Irrigation and Drainage Engineering* 120: 1132–1139. https://doi.org/10.1061/(ASCE)0733-9437(1994)120:6(1132)

Haskett, J.D., Pachepsky, Y.A. and Acock, B., 2000. Effect of climate and atmospheric change on soybean water stress: A study of Iowa. *Ecological Modelling* 135(2–3): 265–277.

Hatfield, J.L., Stanley, C.D. and Carlson, R.E. 1996. Evaluation of an electronic foliometer to measure leaf area in corn and soybean. *Agronomy Journal* 68: 434–436.

Hatfield, J.L., Boote, K.J., Kimball, B.A., Ziska, L.H., Izaurralde, R.C., Ort, D., Thomson, A.M. and Wolfe, D. 2011. Climate impacts on agriculture: Implications for crop production. *Agronomy Journal* https://doi.org/10.2134/agronj2010.0303.

Hulme, M., Zhao, Z.C. and Jiang, T., 1994. Recent and future climate change in East Asia. *International Journal of Climatology* 14(6): 637–658.

Huntington, T.G., 2006. Evidence for intensification of the global water cycle: Review and synthesis. *Journal of Hydrology* 319(1–4): 83–95.

IPCC, 2007. Climate Change 2007: Impacts, Adaptation and Vulnerability. Contribution of Working Group II to the Fourth Assessment Report of the Intergovernmental Panel on Climate Change, M.L.

Irmak, S., Irmak, A., Allen, R.G. and Jones, J.W., 2003. Solar and net radiation-based equations to estimate reference evapotranspiration in humid climates. *Journal of Irrigation and Drainage Engineering* 129(5): 336–347.

Irmak, S., Kabenge, I., Skaggs, K.E. and Mutiibwa, D., 2012. Trend and magnitude of changes in climate variables and reference evapotranspiration over 116-yr period in the Platte River Basin, Central Nebraska—USA. *Journal of Hydrology* 420: 228–244.

Jensen, M.E., Burman, R.D. and Allen, R.G.., 1990. Evapotranspiration and irrigation water requirements. *ASCE Manuals and Reports on Engineering Practices, New York* 70: 332.

Jensen, M.E. and Haise, H.R., 1963. Estimating evapotranspiration from solar radiation. *Journal of the Irrigation and Drainage Division* 89(4): 15–41.

Jensen, M.E., Robb, D.C. and Franzoy, C.E., 1970. Scheduling irrigations using climate-crop-soil data. *Journal of the Irrigation and Drainage Division* 96(1): 25–38.

Jung, M., Reichstein, M., Ciais, P., Seneviratne, S.I., Sheffield, J., Goulden, M.L., Bonan, G., Cescatti, A., Chen, J., De Jeu, R. and Dolman, A.J., 2010. Recent decline in the global land evapotranspiration trend due to limited moisture supply. *Nature* 467(7318): 951–954.

Kousari, M.R., Ahani, H. and Hendi-zadeh, R., 2013. Temporal and spatial trend detection of maximum air temperature in Iran during 1960–2005. *Global and Planetary Change* 111: 97–110.

Kumar, R., Jat, M.K. and Shankar, V., 2012. Methods to estimate irrigated reference crop evapotranspiration–A review. *Water Science and Technology* 66(3): 525–535.

Oki, T. and Kanae, S., 2006. Global hydrological cycles and world water resources. *Science* 313(5790): 1068–1072.

Pascolini-Campbell, M., Reager, J.T., Chandanpurkar, H.A. and Rodell, M., 2021. A 10 per cent increase in global land evapotranspiration from 2003 to 2019. *Nature* 593(7860): 543–547.

Rim, C.S., 2009. The effects of urbanization, geographical and topographical conditions on reference evapotranspiration. *Climatic Change* 97(3): 483–514.

Roderick, M.L., and Farquhar, G.D., 2002. The cause of decreased pan evaporation over the past 50 years. *Science* 298(5597): 1410–1411. DOI: 10.1126/science.1075390.

Shahid, S., 2011. Impact of climate change on irrigation water demand of dry season Boro rice in northwest Bangladesh. *Climatic Change* 105(3): 433–453.

Shuttleworth, W.J., 1993. Evaporation. *Handbook of hydrology*. Maidment, DR (Ed.).

Swelam, A., Jomaa, I., Shapland, T., Snyder, R.L. and Moratiel, R., 2010. Evapotranspiration response to climate change. In *XXVIII International horticultural congress on science and horticulture for people*, pp. 91–98. UK: McGraw- Hill Inc.

Tabari, H., Marofi, S., Aeini, A., Talaee, P.H. and Mohammadi, K., 2011. Trend analysis of reference evapotranspiration in the western half of Iran. *Agricultural and Forest Meteorology* 151(2): 128–136.

Tabari, H., Nikbakht, J. and HosseinzadehTalaee, P., 2012. Identification of trend in reference evapotranspiration series with serial dependence in Iran. *Water Resources Management* 26(8): 2219–2232.

Talaee, P.H., Some'e, B.S. and Ardakani, S.S., 2014. Time trend and change point of reference evapotranspiration over Iran. *Theoretical and Applied Climatology* 116(3–4): 639–647.

Twilley, R., Barron, E.J., Gholtz, H.L., Harwell, M.A., Miller, R.L., Reed, D.J., Roser, J.B., Siemann, E.H., Wetzel, R.G. and Zimmerman, R.J. 2001. *Confronting climate change in the gulf coast region*. Cambridge, MA: Union of Concerned Scientists.

Vehviläinen, B. and Huttunen, M. 1997. Climate change and water resources in Finland. *Boreal Environment Research* 2:3–18.

Wang, X.M., Liu, H.J., Zhang, L.W. and Zhang, R.H., 2014. Climate change trend and its effects on reference evapotranspiration at Linhe station, Hetao irrigation district. *Water Science and Engineering* 7(3): 250–266.

World Health Organization (WHO) 2003. *World Health Report 2003. Shaping the future.* Geneva: World Health Organization.

Wright, J.L., 1982. New evapotranspiration crop coefficients. *Journal of the Irrigation and Drainage Division* 108(1): 57–74.

Zhang, X., Chen, S., Sun, H., Shao, L. and Wang, Y., 2011. Changes in evapotranspiration over irrigated winter wheat and maize in North China plain over three decades. *Agricultural Water Management* 98(6): 1097–1104.

Zhang, M., Geng, S., Ransom, M. and Ustin, S., 1996. The effects of global warming on evapotranspiration and alfalfa production in California. In *Department of Land, Air and Water Resources*. Davis: University of California.

Zhang, S., Liu, S., Mo, X., Shu, C., Sun, Y. and Zhang, C., 2011. Assessing the impact of climate change on potential evapotranspiration in Aksu River Basin. *Journal of Geographical Sciences* 21(4): 609–620.

16 Micro-Irrigation
Potential and Opportunities in Hilly and Sloppy Areas for Doubling Farmer Income

Rohitashw Kumar, Munjid Maryamand, and Rishi Richa

16.1 INTRODUCTION

Water is most precious gift of nature, essential for human and animal life, and plays an important role in plant growth. Water resources play an important role as a catalyst for the economic development of a nation. Increases in temperature and changes in rainfall pattern, distribution and intensity pose a serious threat to food security and livelihoods. Climate change holds profound implications for irrigation in India. On macro level, India is a water-stressed country with annual water availability of 1000–1700 m^3 per capita per year. India has total geographical area of 329 Mha, with net sown area of 140 Mha (42.57%). Gross sown area of the country is 194.4 Mha and irrigated area is 68.10 Mha with cropping intensity is 142% (INCID, 1994; Iyer, 2003; CWC, 2005; Kulkarni, 2005). The annual average rainfall of the country for the year 2017 was 1122.9 mm. The overall irrigation efficiency of the country is 38%. Micro irrigation can play a pivotal role in doubling the farmer's income, as at least a 40–50% increase in income levels due to micro irrigation is already proven. The role of micro irrigation system in India has been substantial with significant advances made by the agriculture industry in the country. In the country, National Mission on Micro Irrigation (NMMI) was started in 2010. The future prospect for micro irrigation system industry in India is very optimistic and key growth drivers

16.2 MICRO IRRIGATION

Micro irrigation is the low application of water as discrete or continuous drips, tiny streams or miniature sprays on, above or below the soil by surface drip, subsurface drip, bubbler and micro-sprinkler systems. It is applied through drippers connected to a water delivery pipe line through low-pressure delivery. Agriculture sector consumes 80% of the freshwater in India. Micro irrigation is often promoted by central and state governments as a way to tackle the growing water crisis. It is method of applying sufficient moisture to the roots of the crops to prevent water stress. This is because drip and sprinkler irrigation delivers water to farms in far lesser quantities than conventional flow irrigation. Due to recurring droughts in different years, micro irrigation has become a policy priority in India. The new catch-phrase in one of the central government's schemes, Pradhan Mantri Krishi Sinchai Yojana (PMKSY or Prime Minister's Agriculture Irrigation Programme), is "Per Drop More Crop". Apparently, the shift toward micro irrigation is thought to "save" water and boost crop yields (Kundu, 1998; Singh and Singh, 1978; Kumar *et al.*, 2017; Kumar and Kumar, 2018). The PMKSY was started in 2015 to attract investments in irrigation systems at field level, develop and expand cultivable land in the country (Per drop more crop). Micro irrigation is a quick and efficient tool to achieve the goal of doubling farmers' income by 2022 which slogan given by the Indian Government. The current adoption is just at 10% of the potential. The ultimate potential for micro irrigation in the country is 70 Mha; till now about 9 Mha has been covered, leaving a scope of 60 Mha where 40–50% increase in income levels can be achieved with this solution (Keller and Bliesner, 1990; Rosegrant, 1997; Postal *et al.*, 2001). It is an urgent need to accelerate the adoption rate, a key priority for doubling farmers income by 2022. The adoption rate of micro irrigation is just 1% every year; it will take 100 years to reach the potential.

16.2.1 MICRO IRRIGATION POTENTIALITY IN INDIA

Before the planners and policy makers decide on the various techno economic aspects of introduction of micro irrigation in a state, they should have basic knowledge of its potentiality in the state. It has been assessed that there is potentiality of bringing around 70 Mha under micro irrigation in the country. Out of this 70 Mha, about 45 Mha are suitable for sprinkler irrigation for crops like cereals, pulses and oilseeds in addition to fodder. Out of this drip with a potentiality of around 20 million hectares and the major crops suitable for this system are cotton, sugarcane, fruits and vegetables, spices and condiments. And some pulses crop like red gram. In addition, there is potentiality for bringing an area of about 2.8 million hectares under mini sprinkler for crops like potato, onion, garlic, groundnut, and short stature vegetable crops like cabbage and cauliflower. Water use efficiency under conventional flood methods of irrigation is very low due to substantial conveyance and distribution losses. The fast decline of irrigation water potential and increasing demand for water from different sectors has led to a number of demand management strategies and programs have been introduced to save water and increase the existing water use efficiency in Indian agriculture. Micro irrigation increases water use efficiency as compared to the conventional surface method of irrigation, where water use efficiency is only about 35–40%.

16.3   NEED FOR DRIP IRRIGATION

- To make agriculture productive.
- Environmentally sensitive and capable of preserving the social fabric of rural communities.
- Help produce more from the available land, water and labor resources without either ecological or social harmony.

- Generate higher farm income, on-farm and off-farm employments.
- To use water efficiently.
- No water to runoff or evaporation and soil erosion.
- It reduces water contact with crop leaves, stems and fruits.
- Agricultural chemicals can be applied more efficiently.

16.4   BENEFITS OF MICRO IRRIGATION

Micro irrigation has the following benefits, which is illustrated in Fig. 16.1. The tomato crop grown inside poly house under drip irrigation is also shown in Fig. 16.2.

16.5   MICRO IRRIGATION APPLICATIONS IN HILLS

The cultivation of horticultural crops is more remunerative on the small terraces of upland areas, due to favorable climatic conditions. However, due to the non-availability of irrigation water, farmers grow rain-fed cereals with very low yields. A check basin irrigation method involving high water losses is commonly used in the valley areas. Plant-to-plant hand watering, as used on a small scale in water-scarce upland areas, is commonly used water application method, but it requires a huge amount of labor. Drip irrigation can replace the hand watering system with minimum water losses and labor. Due to topographical advantages, the gravitational head may be used to operate the system, thus eliminating the initial and operational cost of pumping. Different aspects of the design of drip irrigation systems have been discussed in detail by Keller and Bliesner (1990). They assessed the drip irrigation system for the relative effects of hydraulic design, manufacturer's variation, grouping of

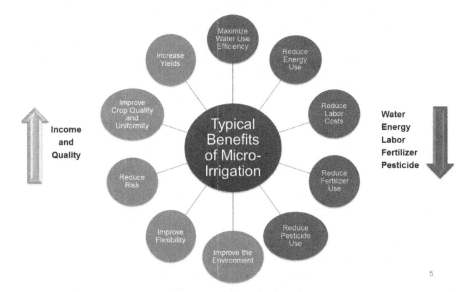

FIGURE 16.1   Benefits of micro irrigation.

FIGURE 16.2   Tomato crop under protected cultivation using drip irrigation.

emitters and plugging. However, these designs are developed for plain areas and the high water pressure is built up by pumping. Some modifications in the design criteria are essential in order to design drip irrigation systems on hilly terraces. Most of the conventionally available long path turbulent flow emitters require an operating water pressure head of 10 m or more for optimum performance. Additional pressure head is required to meet the friction losses in different components of the system, whereas the elevation difference between two adjacent terraces mostly ranges between 0.5 and 5.0 m. This pressure was found to be insufficient to operate the system using turbulent flow emitters. Bhatnagar *et al.* (1998) obtained low emission uniformity (64–72%) for emitter operating at pressure head of 4.0–6.5 m. However, replacing the emitters with micro tubes (1.0 mm diameter) improved the emission uniformity to 94–98% for the same conditions. Another problem encountered is the large variation in emitter discharge, as the system has to be laid on several terraces having varying elevations, sizes and slopes and irregular shapes. Bhatnagar et al. (1998) studied the gravity-fed irrigation system for hilly terraces of the North-West Himalayas. To utilize the scarcely available water, a gravity drip irrigation system was designed for the hilly terraces. The system had the capability to provide uniform emitter discharge throughout the command area spread on several terraces with varying elevations and irregular shapes. The design includes the estimation of the command area of the tank (runoff or low discharge spring-fed) using data on rainfall, evaporation and crop water requirement and the hydraulics of the drip irrigation system. A useful and more pertinent star configuration of micro tube layout with one lateral line feeding four rows was devised, which had some advantages over the traditional layouts in terms of achieving the desired discharge rate, better handling of the system, appropriate water delivery, adjustment in spatial head variation due to friction loss in pipes as well as field slopes and economic factors. Pilot testing of the system showed that the system worked efficiently, with field emission uniformity above 90%. Although the design criteria were developed for the topographical and climatic conditions of the mid-hills of the north-west Himalayas, they can easily be adapted for other locations. Year-wise addition to area under drip irrigation in J&K is shown in Table 16.1.

The horticulture sector is considered the most dynamic and sustainable segment of agriculture all over the world. India is the second largest producer of horticultural crops (240.5 million tones) next to China contributing 10 and 13.8% to world's fruit and vegetable basket respectively. Since horticulture forms an integral part of

TABLE 16.1
Year Wise Addition to Area under Drip Irrigation in J&K

Year of Establishment	Area (ha)
1997–1998	80.00
1998–1999	29.68
1999–2000	18.969
2000–2001	5.00
2000–2002	–
2000–2003	18.33
2000–2004	3.69
2000–2005	–
2000–2006	–
Total area	155.669

nutritional and economic security, it needs serious policy and technological interventions to increase productivity in fruit crops.

Because of wide range of agro-climatic conditions in Jammu and Kashmir state, large number of fruit crops are being grown. Apple is one of the most important fruit crops in Jammu and Kashmir, Fruit production is the back bone of the state economy. In order to optimize the use of resources, hi-tech interventions like precision farming, which comprises temporal and spatial management of resources in apple, is essentially required. Infusion of technology for an efficient utilization of resources is intended for deriving higher crop productivity per unit of inputs. This would be possible only through deployment of modern hi-tech applications and precision farming methods. India is second largest producer of fruit crops in the world, still we are lagging in export potentiality, productivity and quality of fruits, because of low land holdings, rain fed farming, improper utilization of natural resources, over emphasis on chemical fertilizer and pesticides. In order to overcome disadvantages caused by local climatic conditions, one has to go for protected cultivation to produce desired quality apple fruits. It also promises the yield and quality improvement with good benefit cost ratio. The area under different crop in Jammu and Kashmir state is shown in Table 16.2.

TABLE 16.2
Kind-Wise Area and Production of Fruits in J&K State (2015–2016)

Kind	Area (ha)	Production (MT)	Productivity (MT/ha)
Apple	161,773	1,966,417	12.16
Pear	14,475	105,935	7.32
Apricot	6097	14,142	3.32
Peach	2615	5953	2.28
Plum	4279	11,658	2.72
Cherry	2816	10,244	3.64
Grapes	315	1299	4.12
Litchi	872	2264	2.60
Citrus	14,392	33,961	2.36
Olive	707	25	0.04
Mango	12,660	23,856	1.88
Ber	5390	10,752	1.99
Anola	1967	3269	1.66
Guava	2451	8530	3.48
Kiwi	7	–	–
Other	10,376	408	0.04
Total Fresh	241,182	2,217,584	9.19
Walnut	88,900	266,133	2.99
Pecan	415	81	0.20
Almond	7132	7060	0.99
Total dry	96,495	276,415	2.86
Total	337,677	2,493,999	7.39

16.6 GRAVITY-FED DRIP IRRIGATION – A CASE STUDY

The study was conducted to assess the feasibility of gravity-fed drip irrigation system in Kiwi crop under hilly terrace condition during year 2013–2015. The different parameters were measured/estimates during field experiments of the study as (i) discharge rate of different emitters and micro tubes at different pressure heads and different lateral lengths, (ii) estimate the uniformity and distribution coefficients at different water pressures, (iii) soil parameters, such as soil texture, field capacity and hydraulic conductivity and (iv) soil moisture depletion. Gravity-fed drip irrigation was installed in the experimental farm and the system comprises emitters of 4lph, 8lph and micro tubes of different lengths (15, 30, 45 and 60 cm). The gravity system was operated at three different pressure heads (1.6, 1.25 and 1.0 kg/cm^2). The site of water harvesting tank was located at an elevated height 20 m above the field where a drip irrigation system was installed. The water from the tank was supplied through a plastic pipes to the field with a control valve to control the flow of water (Figs. 16.3–16.8).

16.6.1 MOISTURE DEPLETION

Plant roots are the primary pathways for water and nutrient uptake by plants. They connect the soil environment to the atmosphere through water and energy flux exchanges between vegetation canopy and the atmosphere (Feddes *et al.*, 2001).

FIGURE 16.3 Drip irrigation in Kiwi crop through gravity fed drip irrigation.

FIGURE 16.4   Gravity-fed drip irrigation in high density apple.

FIGURE 16.5   Drip irrigation in high density apple.

FIGURE 16.6 Drip irrigation in high density apple.

FIGURE 16.7 Drip irrigation in high density apple.

FIGURE 16.8   Soilless cultivation inside poly house through drip irrigation.

Soil matric potential fluctuations during an irrigation season was measured using watermark sensor at 30, 45, 60 and 90 cm soil depth in gravity-fed drip irrigation system in Kiwi growing field.

16.6.2   SOIL MATRIC POTENTIAL

The soil moisture sensors give the soil water potential at different depths. Soil water potential was recorded between two irrigation or between two rainfalls. The soil water potential reading was zero on the days of applied irrigation and rainfall showing the soil is fully saturated or wet and as the time elasped the water potential reading increased. Irrigation was applied when the water potentialm reading vary in the range 30–40 centibar for study area soil. The highest soil water potential was observed during August 22, 2015 in sensor installed at 30 cm depth, means more moisture was depleted. However, at 45 cm soil depth, highest soil matric potential was observed during September 15, 2015. In 60 cm depth, the soil matric potential was highest during mid-September too as was in 45 cm depth.

The soil water depletion before an irrigation/rainfall at various depths was also studied. Highest soil water depletion was observed on 30th June. The highest soil water depletion was observed at 60 cm soil depth during first week of July. During crop period, the highest water depletion was 10.8% observed at 45 cm soil depth. The soil water depleted during on 30th September was at 45 and 60 cm depth was 11.8%. In the month of August, the water use of Kiwi plant was the largest because by then the canopy is reached the 80–90% of the full shape and the flowering begin to develop very intensively. The variation of moisture depletion at different depths has been determined. It is observed from graph that the maximum depletion of moisture takes place in 45 cm layers of soil and considerably lesser in the lower layers. Similar trend was observed over the entire study period. In the upper part of root zone where root density is high, moisture depletes very fast, whereas in the lower part of the root zone sufficient moisture for the plant is continuously available.

16.7   CONCLUSION

Micro irrigation in India has great risks. Micro irrigation will increase the irrigation potential and water use efficiency and increase productivity of crop. It will change the current land use to more remunerative crop production. Assessment of impact of climate change on water resources also provide information for future strategy for planning of irrigation scheduling and water budgeting of different crops. It is evaluated the effectiveness of better irrigation water management strategies to reduce the impacts of water scarcity for different crops. It was observed in all lengths the discharge rate increased with increasing average distance of emitter from sub-main. The Uniformity Coefficient (UC) of the installed gravity-fed system varied in the range of 58.85–94.5% and Distribution Uniformity (DU) of the system varied in the range of 30.2–95.2%. The discharge rate was found better using micro tubes of 15 cm length. The variation in the uniformity coefficient was found to be inversely proportional to the tube length. The value of discharge rate was decreased with increased micro tube length. It was observed that the maximum depletion of moisture takes place in 45 cm layers of soil and considerably lesser in the lower layers. The matric potential increase than water content in the soil was decreased. The moisture depletion in the root zone showed a non-linear trend during whole crop period.

REFERENCES

Bhatnagar, P.R., Srivastava, R.C., Bhatnagar, V.K. and Prakash, V. 1998. Gravity fed drip irrigation system for cauliflower production in mid-hills of Himalayas. In: Singh, R., Kundu, D.K., James, B.K. and Verma, H.N. (eds), *Progress in micro-irrigation research and development in India* (pp. 115–121). Bhubaneswar: WTCER.

CWC. 2005. Water and related statistics. Ministry of Water Resources, Central Water Commission (CWC), Government of India, New Delhi.

Feddes, R. A., Hoff, H., Bruen, M., Dawson, T., Rosnay, P.D., Dirmeyer, P., Jackson, R.B., Kabat, P., Kleidon, A., Lilly, A. and Pitman, A.J. 2001. Modelling root water uptake in hydrological and climate models. *Bull. Ame. Meteorological Society*, 82(12):2797–2809.

INCID. 1994. Drip irrigation in India. Indian National Committee on Irrigation and Drainage, New Delhi.

Iyer, R.R. 2003. *Water: Perspectives, Issues, Concerns.* New Delhi: Sage Publications.

Keller, J. and Bliesner, R.D.. 1990. *Sprinkler and Trickle Irrigation.* New York, NY: AVI, Van Nostrand Reinhold.

Kulkarni, S.A. 2005. Looking beyond eight sprinklers. In: *Paper Presented at the National Conference on Micro-Irrigation*, G. B. Pant University of Agriculture and Technology, Patnagar, India, June 3–5, 2005.

Kumar, M. and Kumar, R. 2018. Hydraulics of water and nutrient application through drip irrigation – A review. *Journal of Soil and Water Conservation* 17(1): 65–74. DOI: 10.5958/2455-7145.2018.00010.3

Kumar, M., Kumar, R., Rajput, T.B.S. and Patel, N. 2017. Efficient design of drip irrigation system using water and fertilizer application uniformity in different operating pressures at semi-arid region of India. *Irrigation and Drainage (ICID Bulletin)*. DOI: 10.1002/ird.2108

Kundu, D.K., Neue, H.U. and Singh, R. 1998. Comparative effects of flooding and sprinkler irrigation on growth and mineral composition of rice in an Alfisol. In: *Proceedings of the National Seminar on Micro-Irrigation Research in India: Status and Perspective for the 21st Century*, Bhubaneswar, July 27–28, 1998.

Postal, S., Polak, P., Gonzales, F. and Keller, J.. 2001. Drip irrigation for small farmers: A new initiative to alleviate hunger and poverty. *Water International* 26(1): 3–13.

Rosegrant, W.M. 1997. Water resources in the twenty-first century: Challenges and implications for action. In Food and agriculture, and the environment discussion paper 20. International Food Policy Research Institute, Washington, DC.

Singh, S.D. and Singh, P. 1978. Value of drip irrigation compared with conventional irrigation for vegetable production in a hot arid climate. *Agronomy Journal* 70: 945–947.

17 Increase Water Use Efficiency through Micro-Irrigation

Muneeza Farooq, Rohitashw Kumar, and Munjid Maryam

17.1 INTRODUCTION

Micro-irrigation can be defined as the frequent application of slight quantities of water on or below the surface of soil as drop, tiny streams or miniature sprays through emitters or applicators placed along a water delivery lateral line. In India, more than 80% of water that is available is used for irrigation. Micro-irrigation is effective for increasing water use efficiency. Micro-irrigation can be implemented in all types of land, especially where it is impossible to irrigate using flooding method. A land is flooded in flooding technique with water, which results in substantial run-off, anaerobic condition in soil and around the root zone due to which there is not sufficient supply of water to the plants and is thus one of the most inefficient method of irrigation. In undulating topography, hilly areas and areas having shallow depth (less than 22.5cm) micro-irrigation can be used.

Micro-irrigation has become popular because of its low cost and water efficiency. Micro-irrigation systems are immensely popular not only in arid regions and urban settings but also in sub-humid and humid areas where supplies of water are inadequate. Micro-irrigation is extensively used for row crops, orchards, mulched crops, gardens, greenhouses and nurseries.

17.2 OBJECTIVES OF MICRO-IRRIGATION SYSTEM

- Micro-irrigation increases the water use efficiency.
- Using precise water management, it increases the productivity of crops.
- Stimulate micro-irrigation technologies in water-intensive crops that include wheat, sugarcane, rice, cotton etc. and offer adequate focus to spread coverage of field crops under micro-irrigation technologies.
- Using micro-irrigation system can promote fertigation and chemigation.
- Promote micro-irrigation technologies in areas that are water scarce.
- Micro-irrigation technology for agriculture and horticulture improvement with modern scientific knowledge can be improved.

DOI: 10.1201/9781003351672-17

17.3   PREPARATION OF MICRO-IRRIGATION SYSTEM LAYOUT

After taking into consideration the affordability of a farmer, the micro-irrigation system layout for the field is prepared. Its design starts with the selection of emitters, which is dependent on the type of crop, water requirement, operating time, soil type and water quality. Layout consists of alignment of the network of main, sub-main and lateral pipes, and their connection with a water source. The whole area is then distributed into units, reliant on the number of sub-mains to be installed and taking into consideration the pumping capacity of the pump. The main line is planned then for connecting to the sub-mains by allowing for the shortest conceivable route. The main line length is calculated on the basis of the water flow rate so that the frictional head loss is within the indicated limits and the total pressure head needed for the system is within the pump capacity.

17.3.1   Water Sources

Surface and groundwater are the main sources of water for agricultural purposes. One is always required localizing the water source before the installation of a micro-irrigation system. The water source location has to be marked on a map. The following information must be gathered about the water source:

- Depth from the ground surface or height above the ground surface level.
- Details about the pump to be installed.
- Water quality in relations of impurities present (sand, silt, algae etc.).

Tanks, canals, wells, lakes, rivers, ponds, reservoirs, streams etc. are the various sources of surface and underground water, which can be used for the purpose of irrigation. Surface water must be filtered before use because it contains a large amount of impurities.

17.3.2   Tanks

In many States of India like Tamil Nadu, Karnataka, Andhra Pradesh, Telangana and Maharashtra, tanks are the most common source of irrigation. Tanks are built by farmer or groups of farmers by raising bunds across seasonal streams and most of the tanks are usually small in size. Evaporation of water is relatively fast in tanks as effect of large expanse of shallow water. However, they do not make available perennial water supply. In small areas, plastic overhead tanks can also assist the purpose of distributing water to the main and sub-main lines of the micro-irrigation system.

17.3.3   Canals

In India, canals are the second most significant source of irrigation in India. For areas that are extensive like plains and are drained by perennial rivers, such as the northern plains, coastal plains, deltas and broad valleys of the Indian peninsula, canal irrigation is mostly suitable. Plain areas are mostly canal irrigated in India.

States that follow this irrigation system are Andhra Pradesh, Assam, Haryana, West Bengal, Punjab, Rajasthan, Bihar, Karnataka, Tamil Nadu and Uttar Pradesh.

17.3.4 WELLS

Wells are also one of the significant sources of irrigation. The water from the wells is extracted manually using pump or animal power and is obtained from the subsoil. This type of irrigation is most popular in alluvial plain areas, where the high water table is higher. States in which 50% or even more irrigated area is under wells and tube wells are Punjab, Uttar Pradesh, Rajasthan, Gujarat, Maharashtra, Madhya Pradesh and Tamil Nadu.

17.4 CLASSIFICATION OF MICRO-IRRIGATION SYSTEM

Micro-irrigation system is generally classified into the following methods:

1. Drip irrigation system
2. Sprinkler irrigation system
3. Bubbler irrigation
4. Spray irrigation

17.4.1 DRIP IRRIGATION SYSTEM

The most recent method of irrigation is the drip irrigation system. Drip irrigation is also known as trickle irrigation in which water is trickled at very low rates into the soil (2–10 l/h). It is an effective technique of application of water to the plant root at a rate nearly equal to the water requirement rate of plant, thereby reducing the conventional water losses like run-off, percolation and evaporation from soil. Drip irrigation has been found to be extremely successful in various crops like apple, grape, papaya and orchard. Components of drip irrigation consist of filters, mains, sub-mains, laterals, emitters or drippers, control valve, flush valve, air release valve, non-return valve, pressure gauge, gromate and take off, end caps and fertilizing system. These components are defined as follows:

1. **Filters:** Filters are used to clean the suspended contaminations that are present in irrigation water so as to protect the holes and passage of drip nozzles from getting blocked. Filters are known as the heart of drip irrigation system. The different categories of filters used depend on the water quality and type of emitter used.
   a. **Gravel filter:** For inorganic suspended solids, biological substances and other organic materials gravel filters are mostly used. These filters are necessary for open reservoirs and when there is an algae growth. The dust or dirt is stopped and deposited inside the media of the filter. These types of filters usually comprise small basalt gravel or sand (usually 1–2 mm diameter) that is placed in cylindrical tank, which is

made of metal. The water enters from the top and flows through the gravel while leaving behind the dirt in the filter and the clean water is discharged at the bottom. By reversing the flow direction, filter can be cleaned. When the dirt accumulates, the pressure difference between the inlet and outlet is increased and when the pressure difference is greater than 0.5–1.0 kg/cm^2, then filters must be cleaned by adopting back washing. Automatic self-cleaning filters are also available.

    b. **Screen filters:** Screen filters are installed with or without gravel filter, depending upon the water quality. These screens are mostly cylindrical in shape and are made of non-corrosive metal or plastic material.

    c. **Disc filters:** In disc filters, the filtration elements are grooved plastic disc, which are assembled together around a telescopic core, acceding to the preferred degree of filtration. Both the sides of these discs are grooved and grooves cross each other.

2. **Main line:** The main line conveys the water from filtration system to the sub-main. They are usually made of rigid PVC pipes so that to minimize corrosion and clogging. They are placed below the ground surface (60–90 cm), so that they will not hinder cultivation practices. The diameter of main line depends on the system flow capacity. The velocity of flow in mains should not be more than 1.5 m/s and the frictional head loss must be less than 5 ml/1000 m running length of pipeline.

3. **Sub-mains:** The sub-main transfers the water from main line to the laterals. They are also 2–2.5 ft placed below the ground and made of rigid PVC. The diameter of sub-main is generally smaller than main line. There may be a number of sub-mains from one main line based on the crop type and plot size.

4. **Laterals:** These are small-diameter flexible tubing made of low-density polyethylene (LDP) or linear low-density polyethylene (LLDPE) and are of 12 mm, 16 mm and 20 mm in size. The color of lateral is black to avoid the algae growth and effect of ultra-violet radiation. They can resist the maximum pressure of 2.5–4 kg/cm^2. Mains are connected to sub-main at predetermined distance.

5. **Emitters or drippers:** The main component of drip irrigation system is the dripper and is used to discharge water from lateral to the soil. There are various types of drippers available, depending on different operating principles. Drippers are made of plastic, such as polythene or polypropylene. The discharge range of drippers is between 1 and 15 ph. Each dripper has its own characteristics, benefits and drawbacks, which determine its use.

6. **Control valves:** Control valves control the flow of water through particular pipes. Usually, they are installed on filtration system, main line and sub-mains. They are composed of gunmetal, PVC cast iron and their sizes vary from ½″ to more than 5″.

7. **Flush valves:** At the end of each sub-main to flush out, the water and dirt material flush valves are provided.

8. **Air release valve:** At the highest point in the main line, air release valve is provided to release the trapped air. This valve is also provided on sub-main if the length of the sub-main is more.

9. **Non-return valve:** It is commonly used for preventing the damage of the pump from flow of water hammer in rising main line.
10. **Pressure gauge:** Pressure gauge is used for indicating the operating pressure of the drip system.
11. **Gromate and take off:** It is used to attach the lateral to the sub-main. With hand drill of fixed size, a hole is punched in sub-main. It is fixed into the hole. Take off is constrained into the hole that acts as a seal. These are of different sizes, which are 12 mm, 16 mm and 20 mm lateral.
12. **End caps:** The purpose of end caps is to close the lateral ends, sub-main ends or main line ends. Sub-mains and mains are having the flush valve. They are appropriate for flushing the line.
13. **Fertilizing system:** Fertilizing system is used to add the chemical to irrigation water; however, fertigation is not free of vulnerabilities. Chemicals that are added to irrigation water may be toxic to human beings and animals, so prevention must be taken to inhibit the back flow of irrigation water into the water source that might be used for the drinking purpose. To minimize the clogging hazard, only water-soluble fertilizers must be used.

Drip irrigation is further classified into the following types:
a. Surface drip irrigation
b. Sub surface drip irrigation
c. Family drip
d. Online drip
e. In-line drip

a. **Surface drip irrigation**: This system is used for irrigation of perennial and annual crops (Fig. 17.1).
Surface drip irrigation consists of the following components:
   i. **Pump unit:** It consists of a pump and a power unit through which electricity is supplied to the pump. From the source, water is drawn by the pump and provides the accurate pressure for its conveyance into the pipe system.
   ii. **Head control unit:** It comprises shut-off, air and check valves, which control the water pressure and discharge in the system. It also consists of a pressure valve that is installed after the pump unit returns the excess water when the system does not operate at full capacity. For cleaning the water, it may have filters that remove sediment and debris so that it does not clog to the system.
   iii. **Tubings:** It consists of main line, sub-mains and laterals. The main lines convey the water from the source, distribute it to the sub-mains and then carry this water to the laterals, which in turn supply this water to the emitters.
   iv. **Emitters:** Emitters or drippers are used for controlling the discharge of water from the laterals to the plants. Water enters drippers at a pressure of about 1 kg/cm^2. Emitters are usually of two types, which are online emitters and inline emitters.

FIGURE 17.1   Surface drip irrigation.

   b. **Sub-surface drip irrigation:** In this type of drip irrigation, crops are irri-
      gated through buried plastic tubes, containing embedded emitters located
      at regular spacing. A sub-surface drip irrigation system has a same design
      as surface drip irrigation system. But in sub-surface drip irrigation system,
      the drip tubes are typically located 38–84″ (97–213 cm) apart and 6–10″
      (15–25 cm) below the soil surface. In sub-surface drip irrigation, evapo-
      ration is minimized and water is used more proficiently as compared to
      surface irrigation. In sub-surface irrigation, the sound effects of surface
      infiltration like crusting, water losses through evaporation and surface
      run-off are removed. Water is applied straight to the crop root zone as
      opposed to surface irrigation, in which most weed seeds hibernate. Water
      application is effective and uniform in sub-surface irrigation. Sub-surface
      drip irrigation aids in water management in open field agriculture, often

FIGURE 17.2 Sub-surface drip irrigation.

resulting in saving up to 25–50% water as compared to the flood irrigation system (Fig. 17.2).

c. **Family drip irrigation:** This type of drip is also known as gravity-fed drip irrigation system and is a low-cost system established for small family plots. It is fit for house gardening and it can also be used to demonstrate the working of drip irrigation system. Family drip system is suitable for areas measuring 500–1000 m^2. This system comprises five components that are: elevated tank, shut-off valve, filter, main line and drip line. Generally, a family drip irrigation system consists of following components: a drum, control or shut-off valve, filter, main line and drip laterals. The spacing of drip outlet is taken as 30 cm. This system does not require power source; thus, it is cheap and easy to install and operate (Fig. 17.3).

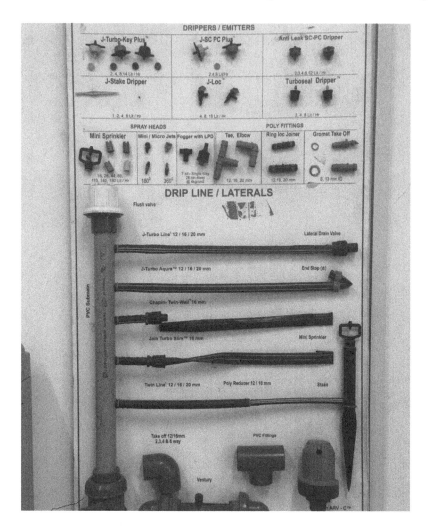

FIGURE 17.3   Detail of drip irrigation.

d. **Online drip irrigation:** In online drip irrigation system, drippers are stationary externally on the laterals at designed spacing (Fig. 17.4). Thus, the drippers can be checked and cleaned effortlessly in case of clogging. To cover the increased root zone of a plant, the dripper spacing can be changed anytime. This system is used in orchards, vineyards, artificial landscapes and nurseries. This system is, generally, used for irrigating horticultural plants like mango, coconut, orange, lemon, banana, grapes, pomegranates and papaya.

e. **In-line drip irrigation:** At the time of manufacturing to meet the prerequisite of various crops, drippers are fixed in the lateral tube at designed spacing in this type of irrigation system. It is operative for row crops like cotton,

FIGURE 17.4   Online drip irrigation.

sugarcane, groundnut, vegetables and flowering crops. The spacing of the drippers depends on water holding capacity of soil and water requirement of crop (Fig. 17.5).

17.4.2   ADVANTAGES OF DRIP IRRIGATION SYSTEM

1. Saves water up to 70% and high water use efficiency of about 95%.
2. Reduced weed growth.
3. Leaching is reduced as water is applied locally.
4. This method can also be applied to the rough areas without leveling.
5. Saves labor and manpower.
6. Yield and plant growth is increased.

FIGURE 17.5   In-line drip irrigation.

17.4.3   Sprinkler Irrigation

In the sprinkler irrigation system, water is sprayed into the air and permitted to fall on the ground surface, which resembles to some extent to rainfall. By the flow of water under pressure through small orifices or nozzles, the spray can be established. With the careful selection of nozzle sizes, operating pressure and sprinkler spacing, the amount of irrigation water necessary to refill the crop root zone can be nearly applied at the uniform rate to resemble the infiltration rate of soil. The application rate should not exceed the maximum allowable infiltration rate for the soil type. Excess application rate will lead to water loss, soil erosion and surface sealing. There may be insufficient moisture in the crop or plant root zone after irrigation and may get damaged. The water flows out of the sprinkle with a certain force and is known as its water pressure and this pressure is measured in pounds per square inch (psi).

FIGURE 17.6   Sprinkler irrigation.

Sprinklers are thus designed to work at certain pressure levels, which are suggested as their operating pressure (Fig. 17.6). The advantages of sprinkler irrigation are as follows:

- Removal of the conveyance channel; therefore, there is no conveyance loss.
- Except heavy clay, it is appropriate for all types of soil.
- Appropriate for irrigating the crops wherever the plant population per unit area is very high. It is mostly applicable for oil seeds and other cereal and vegetable crops.
- Saving of water.
- Closer control of water application is suitable for giving light and frequent irrigation and water application efficiency is higher.
- Yield can be increased.

- Mobility of system.
- It may also be applicable to undulating areas.
- It protects the land as there is no need for bunds etc.
- Effects greater conducive micro-climate.
- Areas that are located at a higher elevation than the source can also be irrigated using sprinkler system.
- Soluble fertilizers and chemicals can also be used.
- Less problem of sprinkler nozzles clogging because of sediment-loaded water.

17.4.4 Components of Sprinkler Irrigation System

This system consists of pump unit and valves, filtration unit, pipelines and sprinklers:

a. **Pump unit and valves:** For developing, the required pressure pump is needed. The pump is required when the land is undulating, porous, erodible and impermeable. Pumps are used where the flow rate is very low.
b. **Filtration unit:** The impurities that are present in irrigation water can be removed by installing the filtration unit. Different types of filters are used, which include hydro-cyclone, media and screen filter. The filter is selected depending on the quality of water and if the quality of irrigation water is low, then a higher mesh-sized filter is used.
c. **Pipeline:** The design of mains, sub-mains and laterals depends on soil characteristics, source of water and topography. The mains should be laid along the gradient and the laterals across the gradient.
d. **Sprinklers:** The sprinklers can be selected based on their nozzle size and the pressure, which is used to discharge water. It should be noted that the water discharged does not cause any damage to the crops. Water must be supplied in a uniform manner and should meet the water requirement of crop.

17.4.4.1 Types of Sprinkler Irrigation

i. Centre pivot
ii. Rain gun
iii. Towable pivot
iv. Impact sprinkler
v. Pop up sprinkler
vi. Linear move sprinkler

17.4.4.1.1 Center Pivot

It is applicable for irrigating mostly field crops. It comprises a single sprinkler lateral which is supported by a series of towers and rotates around a fixed central point called pivot point and is fixed at one end. A drive unit touches the ground surface that contains essential components for moving the machine. It comprises a base beam, drive train, wheels and other structural support equipment. The towers are self-driven so that the lateral rotates around the pivot point, which is installed in the center of the irrigated area. The long pipes between the drive units are known as spans. Spans contain the main water line, sprinklers and a supporting structure that holds the weight between the towers. The drive unit components are controlled by a tower box, with respect to the direction and duration.

17.4.4.1.2 Rain Gun

It is generally applicable as a water spray mist. Water is discharged at less than 175 lph. To irrigate trees and other crops that are widely separated, rain gun is used. Fruits, such as citric fruits, mango, guava and avocado, can be irrigated using this method. The rain gun has a small diameter. Thus, the discharge of filtered water is necessary, amounting up to a necessity of 60–80 mesh. The minimum operating pressure of 1.5–2 kg/cm^2 is suitable for rain gun method. The head is fixed to plastic wedges that are 20–30 cm above the ground level. It is applicable to field crops, such as onion, groundnut and cotton, and plantation crops like coffee and tea.

17.4.4.1.3 Towable Pivot

It resembles to the center pivot. The only difference is that in towable pivot, pivot is towed away by a tractor. In the center of the pivot, there are three-to-four wheels that make it possible to move the pivot from one place to another with the help of the tractor. This helps the farmers to carry out mechanized irrigation in an efficient manner. It can effortlessly irrigate the fields as the machine can be towed away from one field to another in a minimum time.

17.4.4.1.4 Impact Sprinkler

The sprinkler is driven in a circular motion by the force of the outgoing water and one of its arms extends from the head. This sprinkler arm in the system is continuously pushed back into water stream by a spring. When the water stream is stroked by the arm, it disseminates the stream and re-orients the flow, permitting a uniform watering area around the sprinkler. For closely spaced field crops impact sprinkler is usually recommended like leafy vegetables, potato cotton, oilseeds, pulses, cereals and fodder crops.

17.4.4.1.5 Pop Up Sprinkler

This system comprises an inlet, cap, body, wiper seal, riser, radius adjustment screw and nozzle. It is easy to install and is a portable system and thus is suitable for irrigating lawns, planting beds and seasonal flowers.

17.4.4.1.6 Linear Move Sprinkler

This system is similar in construction as that of center pivot system with the only exception that no end of lateral pipeline is fixed. This system consists of series of towers that move in the direction of rows laterally. The whole line moves down the field at right angles to the lateral. Water is delivered to the continuously moving lateral by a flexible hose or open ditch pickup. Both the center pivot and linear systems are proficient of high-efficiency water application. Water efficiency means reducing water wastage by determining the amount of water required for a particular purpose and the amount of water delivered or used. This system is not labor-intensive but requires high capital investment.

17.4.4.1.7 Micro-Sprinklers

Micro-sprinklers are the emitters, which are usually known as spray heads or sprinkler. They operate commonly in scheduled patterns by spreading the water through air. Based on the water flow patterns, micro-sprinklers are categorized as 'mini-sprays'', jets', micro-sprays' or 'spinners'. The external emitters are sprinkler heads that are exclusively connected with the lateral pipes, normally, using 'micro-tubes', which is also known as small-diameter tubing. The sprinkler heads are fixed on a support stake, the height of which is 25–30 cm, and are connected to the supply pipe. Less energy is required by micro-sprinklers, and generally, it works at a pressure range of 1–3 kg/cm² and a discharge of 40–75 lph. Micro-sprinklers are appropriate because only fewer sprinkler heads are required to cover a large area. This system is appropriate for shallow rooting crop pattern, such as garlic and onion.

17.5 BUBBLER IRRIGATION

In this irrigation system, water is applied to the surface of soil as a little stream, from a small diameter tube, which is 1–13 mm or a commercially available emitter. To control the water distribution, small basins or furrows are required as the application rates generally surpass the soil infiltration rates. The use of this system in agriculture is limited but bubbler application is extensively utilized in landscape irrigation systems. The low head and pressurized systems are the two major types of bubbler irrigation system, which are usually available. This irrigation is a localized, low pressure, solid permanent installation system and is suitable for tree groves. The water is infiltrated into the soil surface and wets the root zone. The water is applied through bubblers. These are small emitters that discharge water at flow rates of 100–250 l/h placed in the basins. Each basin can have one or two bubblers as needed. The bubblers are small plastic head emitters with a threaded joint. They were basically designed for use on risers above the ground surface for flood irrigation of small ornamental areas. They have been successfully used in recent decades in several countries for the irrigation of fruit trees. They perform better under a wide range of pressures delivering water in the form of a fountain or small stream.

17.6 SPRAY IRRIGATION

Jets, foggers or spitters are used in a spray irrigation system. Only a certain fraction of ground surface water is applied. Series of nozzles on micro-sprayers are used in this system to eject the fine jet of water. Each nozzle can deliver water to an area of several square meters, which tends to much greater than individual areas wetted by a single drip emitter.

17.7 ADVANTAGES OF MICRO-IRRIGATION SYSTEM

1. **Saving of water:** Micro-irrigation is proven to be an effective technique in saving water and improving water use efficiency in comparison to the conventional surface method of irrigation, where there is only water use efficiency of about 35–40%.

2. **Increase in irrigation efficiency:** In correctly designed and managed drip irrigation systems, the on-farm irrigation efficiency is determined to about 90%. Farmers using a pumping system to irrigate their fields should certify that the pump and pipe size are appropriate for their requirements, thus eluding overuse of water and energy and consequent leakages.

3. **Higher yields:** The yields are higher in micro-irrigation than traditional flood irrigation. The gain in productivity due to use of micro-irrigation is determined to be in the range of 20–90% for different crops. There is an increase in yields of crops 45% in wheat, 20% in gram and 40% in soybean.

4. **Less water loss:** There is also lower loss of water in this system due to reduction in loss of water in conveyance and also decrease in loss of water through evaporation, run-off and by deep percolation.

5. **Energy efficient:** Due to the decrease in water consumption in micro-irrigation system, energy use is also decreased that is needed to lift water from irrigation wells.

6. **Lower consumption of fertilizers:** An efficient drip irrigation system lowers the consumption of fertilizer through fertigation.

7. **Reduction in diseases and weeds:** It helps in constraining growth of weeds as it keeps limited wet areas. Under this condition, the occurrence of disease is also decreased.

8. **Cost savings:** There are substantial reductions in irrigation costs and savings of electricity and fertilizers.

9. **Precision farming:** Developing computerized GPS-based precision irrigation technologies for micro-irrigation systems will permit farmers to apply water and agrochemicals more precisely and site specifically to meet soil and plant status and requirements as provided by wireless sensor networks.

17.8 CONCLUSION

In Indian agriculture, micro-irrigation has achieved greater importance. To the farmers, micro-irrigation has provided alternatives it saves water and increases yield. Micro-irrigation has materialized the concept of 'more crop per drop' by ensuring the availability of adequate quality and quantity of water, especially in the areas where there is water scarcity for crop production. Sub-surface drip irrigation gives results that are valuable under dry climatic conditions and the adverse effect of water shortage is mitigated and results in good yield. When drip irrigation is properly designed, managed and installed, it achieves conservation of water by decreasing evaporation and deep drainage in comparison to other types of irrigation such as flood or other irrigation methods and in drip irrigation system water can be applied more precisely to the roots of a plant. Micro-irrigation system can reduce many kinds of diseases that are spread via contact of water with the foliage. The future revolution in Indian agriculture can be obtained from precision farming and micro-irrigation can be the stepping stone for obtaining the goal of making agriculture profitable and productive.

REFERENCES

AGRICOOP (2005). Policy initiative, Micro-irrigation. Dept. of Agriculture and Cooperation, Ministry of Agriculture, Govt. of India. (http://agricoop.nic.in/PolicyIncentives/microirrigation.htm).

Anonymous (2004). Report of the Task Force on Micro-irrigation, Ministry of Agriculture, Dept. of Agriculture and Cooperation, Govt. of India, New Delhi, Jan, 2004.

Batchelor, C., Lovell, C. and Murrata, M. (1996). Simple microirrigation techniques for improving irrigation efficiency on vegetable gardens. *Agricultural Water Management* 32: 37–48.

Camp, C.R. (1998). Subsurface drip irrigation: A review. *Trans ASAE* 41(5): 1353–1367.

Casey, P., Lake, A., Falvey, C., Ross, J.A. and Frame, K. (1999). *Spray and drip irrigation for wastewater reuse, disposal. National small flows clearing house.* Morgantown. W.Va: West Virginia University,.

Mane, M.S. and Ayare, B.L. (2007). *Principles of sprinkler irrigation.* New Delhi: Jain Brothers.

McCauley, G.N. (1990). Sprinkler vs. flood irrigation in traditional rice production regions of southeast Texas. *Agronomy Journal* 82(4): 677–683.

Michael, A.M. (2010). *Irrigation theory and practice* (p. 578). Delhi: Vikas Publishing House PVT Ltd,.

Rao, G.C. and Shankar, S.M. (2008). Training manual on Sprinkler Irrigation (Design and Maintenance), Precision farming Development centre, Department of Agricultural Engineering. College of Agriculture, Rajendranagar, Acharya NG Ranga Agricultural University, Hyderabad, Andhra Pradesh, India. p. 1

Sivanappan, R.K. (1994). Prospects of micro-irrigation in India. *Irrigation and Drainage Systems* 8(1): 49–58.

Westcott, M.P. and Vines, K.W. (1986). A comparison of sprinkler and flood irrigation for rice. *Agronomy Journal* 78(4): 637–640.

18 Groundwater Conservation and Management by Artificial Recharge of Aquifer

Karan Singh, Rohitashw Kumar, and B. A. Pandit

18.1 INTRODUCTION

Groundwater is one of the valuable resources for daily sustenance as it provides fresh water. From a usage perspective, it is extensively used for agriculture, industries, and the supply of drinking water. Due to its limited existence and ever-growing population, India's groundwater resources are being destroyed at a concerning rate. According to Seckler *et al.* (1998), the International Water Management Institute (IWMI) research indicated that in the first quarter of the next century, nearly 25% of the earth's population (1.4 billion people) would face the issue of water scarcity. If corrective solution is not followed, India will experience water stress before the year 2025 and severe water scarcity by 2050, according to a World Bank assessment (World Bank document, 2006). Due to the whims of the monsoon and a shortage of surface water, the dependence on groundwater has grown significantly in several areas of the country, particularly in arid and semi-arid areas, over the past few decades (Kannan and Mathew, 2008; Singh and Ravichandran, 2011). With industrialization and improvements in living standards, demand for water has increased significantly—this is causing a decline in groundwater availability. The significant land use changes and rise in paved surfaces in urban areas are also having a negative impact on groundwater table recharge. It is urgently necessary to increase groundwater supplies through appropriate management interventions in order to address the risks of de-saturation of aquifer zones. As we know, the groundwater is replenished through both artificial and natural methods. Natural recharge occurs when rainfall water percolates into the aquifer's bed through the process of infiltration. But as already stated, agricultural intensification, recent rapid urbanization, industrialization, and staggering population increase have reduced areas for natural infiltration, which has reduced the possibilities for natural recharge of the groundwater. Unlike natural recharge (resulting from natural causes), artificial recharge techniques have been shown to be beneficial in various studies (Alston and Watkins, 1973).

The term "artificial recharge" describes the human-induced transport of surface water to the aquifer. The percolation of surface water that is either stored or flowing and would not otherwise percolate into the aquifer speeds up the natural process of recharging. The method under which groundwater is increased at a rate greater than that under normal replenishment conditions is known as artificial recharge (commonly referred to as planned recharge). It describes how water moves from the surface of the earth to underground strata where it can be stored for use in the future. Therefore, any man-made facility that recharges an aquifer with water may be termed an artificial facility (Central Ground Water Board [CGWB], 1994).

The goal of "artificial recharge" is to increase the natural refilling of groundwater through construction, water dispersion, or artificial changes to the natural conditions. It helps to decrease overdraft, conserve surface runoff, and expand the amount of groundwater that is accessible. Depending on whether or not recharge results from routine water use, it may be accidental or intentional.

The technology principle of artificial groundwater recharge is illustrated in the schemes in Fig. 18.1 (Heviánková *et al.*, 2016). A natural stream or other above-ground water source is pumped into the pretreatment technology with the use of a pumping station. The water that is pumped from the natural stream is treated using pretreatment if necessary. The water is either pumped or lifted by gravity from the

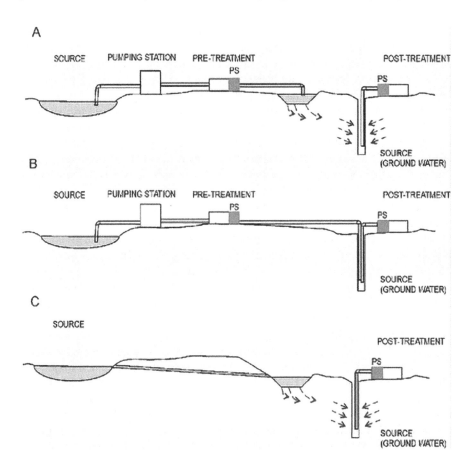

FIGURE 18.1 Scheme of artificial recharge from a surface stream into an underground source with pretreatment and infiltration basin (A), without an infiltration basin with direct charging of the underground source (B), with infiltration by means of gravity recharge with no pumping or pretreatment (C) (Heviánková *et al.*, 2016).

technical pretreatment into the infiltration point like infiltration basin and finally the recharged water combines with the subsurface source's water.

18.2 NEED FOR ARTIFICIAL RECHARGE

The nation's groundwater resources continue to be over-exploited in many areas, and natural replenishment of the reservoirs is inadequate and unable to keep up. As a result, a sizable area of the country today has diminished groundwater resources and falling groundwater levels. Artificial recharge to groundwater has developed into a significant and purposeful management approach in order to supplement the groundwater's natural supply. Through appropriate civil infrastructure, the efforts primarily consist of enhancing the natural flow of water into the groundwater reservoirs. The methods used for artificial aquifer recharge relate to and incorporate the groundwater reservoir and source water, and they depend on the regional hydrogeological conditions.

The country's rainfall is monsoon-dependent, and in a significant portion of the country, it only occurs for 20–30 days to 3 months. This is the only time that the groundwater reservoir will receive natural recharge. The aim of artificial recharge techniques is to extend the post-monsoon recharge period by around three extra months, hence supplying more recharge. Artificial recharge techniques seek to increase the sustainability of groundwater sources throughout the dry season by prolonging the recharging period in the post-monsoon period by three or even more months. In the country's arid regions, annual precipitation ranges from 150 to 600 mm with fewer than ten wet days. A significant portion of the precipitation is brought on by three-to-five intense storms that last a few hours. These regions have very high rates of potential evapotranspiration (PET), which frequently range from 300 to 1300 mm.

The annual water resource management must be done by preserving the rainfall and by conserving the water in surface and subsurface reservoirs when the average yearly PET is significantly higher than the rainfall. Artificial recharge techniques must be used in locations where the climate is unfavorable for producing surface storage in order to quickly shift the majority of the surface storage to subsurface water reservoirs. Even if rainfall in hilly locations is relatively substantial, post-monsoon water shortages are common as the majority of the available water is wasted to surface runoff. During the post-monsoon season, springs, the main supply of water in such terrains, are also reduced. In these regions, rainwater harvesting and surface storages placed strategically in spring's recharge zones can give long-term yields and improve recharging both during and after the rainy season.

18.3 BENEFITS OF ARTIFICIAL RECHARGE

The need for artificial recharge is rising as a result of the need for sustainable groundwater source to meet the demands of a rising population. Artificial recharging might have both intangible and tangible advantages. The key benefits of the artificial recharge are as follows:

- Free subsurface storage space is available, and overflow is prevented.
- It improves the sustainable production in the regions where excessive development has drained the aquifer.

- Minimal evaporation losses occur.
- It can help in storage and preservation of extra surface water in order to meet future needs.
- Quality improvement by permeable media infiltration. Due to the diluting effect, the quality of groundwater has improved (changes in concentrations of dissolved ions, individually or collectively).Furthermore, biological purity is very good.
- It has no serious negative effects, like population relocation or the loss of agricultural land.
- There are relatively few temperature changes.
- It is eco-friendly, prevents flooding and soil erosion, and supplies enough soil moisture even in the summer.
- Underground water is comparatively less prone to both natural and man-made disasters.
- It offers a naturally occurring system of distribution between recharge and discharge points.
- Leads to energy savings because of a decrease in suction and delivery head caused on by an increase in water levels.
- Increased vegetation cover in the surrounding areas.
- Maintaining well yield, water level, and command area during years of low rainfall.

However, there are several problems related with the usage of artificial recharge techniques such as the low recovery efficiency (i.e. not all the injected water may be recovered), pollution hazards from the injection of recharge water of low quality, aquifer's clogging, high cost of operation, etc. Therefore, selecting an acceptable site for artificial recharge in a specific area should be done with caution.

18.4   SOURCES OF ARTIFICIAL RECHARGE

One of the essential conditions required for groundwater recharge is the availability of source water, which is primarily measured in terms of surplus monsoon runoff that has not been committed and is now being wasted. This element can be evaluated by looking at the rainfall pattern, including a number of rainy days, its frequency, the maximum amount of precipitation per day, and fluctuation across time and space. For determining the availability of surplus surface water, it is possible to take into account the differences in rainfall patterns across time and space and their significance to the potential for artificial recharge of subsurface reserves. The following forms of source water may be used for artificial recharge:

- *In situ* rainfall in the watershed or region
- Aquifer system, stream, or spring in vicinity
- Supply of water (canals) from large reservoirs located within the watershed or basin

- Trans-basin water transport for surface water supply
- Treated industrial and municipal wastewaters
- Additional particular source(s)

From all of these sources, water that is available for artificial recharge varies greatly from location to location. The main water sources for the artificial recharge of groundwater are rainfall and runoff. The main source of replenishment for the underground water reserve is rainfall. Seepage from streams, tanks, and canals, as well as the return flow from applied irrigation, are additional important sources of recharge. It is essential to have a firm understanding of rainfall and runoff in order to properly evaluate the availability of source water. Hydro meteorological and hydrological data collection and analysis are crucial in determining the water availability for the planning of artificial recharge schemes. The following sections go into more detail on each of these.

18.4.1 RAINFALL

The monsoon season brings the majority of India's annual precipitation. In June, the southwest monsoon arrives in India and often lasts until September. It rains in the northeast from October through December. The northern and eastern states and Kerala benefit from both of monsoons; therefore, they get rain for seven to eight months. Except for Tamil Nadu, this has significantly more rain during the northeast monsoon; the rest of the nation experiences rain during the southwest monsoon. In these places, the rainfall is confined to three to four months. Also different across the nation is the average annual rainfall, which ranges from 200 to 250 mm in Rajasthan to >3000 mm in north-eastern states and Kerala.

Rainwater is free of organic matter, soft in nature, and bacteriologically safe. It is feasible to collect and use rainwater runoff, which would otherwise go through storm drains and sewers and be wasted. The majority of the rain that falls on the earth's surface runs off into rivers, streams, and eventually the sea. It is estimated that rainfall contributes 8–12% of the total recharge to the aquifers. The process of rainwater harvesting involves gathering rain from tiny catchment areas, such as rooftops and flat or sloping surfaces, depending on local conditions, either for direct consumption or to supplement groundwater resources. In order to manage and store the running water, little streams must have minor obstacles built across them. It contributes to decreasing the frequent drainage backups and flooding that occur during rainfall events in urban locations when there aren't many open places and there's a lot of surface runoff. In the years to come, a catastrophic water problem in the majority of our country can be greatly avoided by reviving the old techniques of rainwater harvesting under scientific lines.

18.4.2 RUNOFF

Runoff is the portion of precipitation that travels as a surface or subsurface flow toward rivers or oceans. The excess rainwater drains out through the smaller natural channels on the earth's surface to the larger drainage channels after infiltration and other losses from the precipitation (rainfall) occur. These flows are referred to as surface flows. A portion of the infiltrated precipitation flows beneath the surface of the ground in a direction parallel to it before resurfacing at specific locations. These are referred to as interflows. Another portion of the infiltrated water flows laterally to emerge in depressions and rivers and join the surface flow after percolating downward to groundwater. Subsurface flow or groundwater flow is the term used to describe this type of movement.

The fundamental and most important input required for the construction of recharge structures with the best capacity is precise runoff estimation. Oversized or undersized constructions, which must be avoided in any case, are frequently the result of unrealistic runoff projections of catchments' output.

18.5   QUALITY OF SOURCE WATER

Planning and method selection for recharging are also influenced by the physical, biological, and chemical quality of the recharge water. Chemical quality of recharge water refers to the type and amount of dissolved particles and gases, whereas physical quality means the type and amount of suspended solids, temperature, and the concentration of trapped air. The type and quantity of living things are referred to as biological quality. Any or all of these traits may reduce recharge rates in particular conditions.

   i. **Physical quality:** Surface application methods are more effective than subsurface methods if suspended particulates are there in the recharge water. Even though suspended particles could lead to clogging, it is possible to remediate the infiltration surfaces. When indirect recharge techniques are employed, suspended particles hardly ever cause issues. Induced recharging would definitely be one of the best options in these situations.
   ii. **Chemical quality:** In order to prevent chemical interactions that would decrease effective porosity and recharge capacity, recharge water should be chemically compatible with the aquifer material through which it flows as well as with the native groundwater. Concerns include chemical precipitation, unfavorable exchange reactions, and the existence of dissolved gases. Clay particles may inflate or disperse as a result of sodium-based cation exchange reactions in recharge water, which will reduce infiltration rates and aquifer permeability. Gases that have been dissolved in an aquifer may change their pH or try to escape from solution, creating gas pockets that fill pore space and reduce aquifer permeability.
   iii. **Biological quality:** Additionally, the recharge water may contain bacteria or algae. Organic wastes may contain hazardous bacteria, or they may be encouraged to seek conditions to grow and decompose, or they may produce an excessive amount of nitrate or other by products when broken down. Growth of algae and bacteria at the time of recharging can clog surfaces for infiltration and produce gases that make the process even

more difficult. Surface blockage can significantly lower the penetration rate, even if surface spreading filters out the majority of algae and bacteria before the recharge water reaches aquifer. It is generally not advised to inject bacterial and algae-containing water through wells since it can clog the screens or materials in the aquifer, which is expensive and difficult to fix.

18.6   FACTORS AFFECTING ARTIFICIAL RECHARGE PHENOMENON

For artificial recharge, major factors to keep in mind are as follows (Hare et al., 1986):

- The aquifer's transmission characteristics
- The volume and quality of water sources that are accessible
- The depth of underground storage space
- The availability of underground storage
- Availability of wastewater
- Potential for clogging
- Cultural/social considerations
- Costs
- Legal/institutional constraints

18.7   DIFFERENT METHODS FOR ARTIFICIAL RECHARGE

Many different methods are popular for recharging groundwater reservoirs. The broad variability in artificial recharge techniques is comparable to those in the hydrogeological framework (Fig. 18.2). Artificial recharge techniques can be divided into three categories—

  i. Direct methods
 ii. Indirect methods
iii. Combination methods

18.7.1   DIRECT METHODS

18.7.1.1   Surface Spreading Techniques

In order to improve infiltration and increase groundwater storage in phreatic aquifers, this form of artificial groundwater recharge uses a variety of approaches to increase the contact area and residence duration of surface water over the soil. The vadose zone should be permeable and devoid of clay lenses and the area should have gradually sloping ground without gullies or ridges.

  i. **Flooding:** Flooding is a very effective method for recharging the unconfined aquifer in specific locations with good hydrogeological conditions by dispersing excess surface water from canals and streams over a wide area

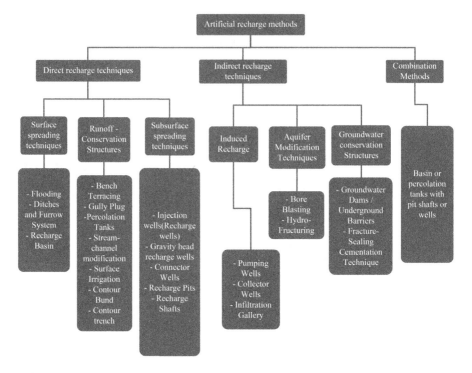

FIGURE 18.2 Hydrogeological framework recharge techniques.

for a sufficient amount of time to replenish the groundwater body. This method can be applied on gently sloping ground without gullies and ridges that has a slope of 1–3% (Fig. 18.3).

ii. **Ditches and Furrow system:** Shallow, flat-bottomed, and closely spaced ditches and furrows can be found in locations with irregular topography. This relates to locating the basin or impermeable aquifer layers that prevent recharge to subsurface aquifers. Concerns about the chemical mixture of surface waters and local groundwater as well as hydrological variability within the aquifers are also significant. For the purpose of replenishing water from the source stream or canal, prepare the necessary feasibility and offer the greatest amount of water contact area. Compared to the recharge basin technique, this method requires less soil preparation and is less susceptible to silting.

iii. **Recharge Basins:** Excavated artificial recharge basins or those surrounded by levees or dykes are both options. They are frequently constructed parallel to transient or sporadic stream courses. This approach has a very high water contact area, which typically accounts for 75–90% of the total recharge area. With this technique, space is used effectively, and basin shapes can be changed to suit the topography and the available area (Fig. 18.4).

FIGURE 18.3 Flooding method (Veeranna and Jeet, 2020).

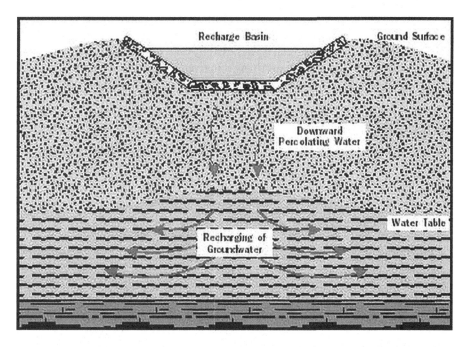

FIGURE 18.4 Recharge basin (http://www.waterencyclopedia.com/A-Bi/Artificial-Recharge. html).

18.7.1.2 Runoff Conservation Structures

They are suited for places with little to no potential for water transfer from adjacent areas and low to moderate rainfall, often during a single monsoon season.

i. **Bench Terracing:** By retaining runoff water for longer periods of time on the terraced area, it helps to conserve soil by increasing infiltration and groundwater recharge.

ii. **Gully plug:** Gully plugs are the smaller runoff conservation structures constructed across streams and small gullies that rush downhill slopes bringing drainage from smaller catchments in the rainy season. The barrier is typically built out of local stones, soil, brushwood, weathered rock, and other similar elements.

iii. **Contour bunds:** Increased soil moisture storages are accomplished by the use of contour bunds as a watershed management technique. This method is typically used in regions with little rainfall.

iv. **Contour trenches:** These are rainwater-gathering structures that can be built on the sides of hills and on ruined and arid wastelands in both high and low rainfall regions.

v. **Percolation pond/tank:** In order to force surface runoff to percolate and refuel the groundwater store, these tanks are artificially formed surface water bodies that immerse a particularly permeable land region. In the Indian setting, a percolation tank must not really hold onto water beyond February month. It must be situated downwind of a runoff area (Fig. 18.5).

vi. **Stream-channel Modification:** These techniques are frequently used in alluvial regions, but they can be successfully used in hard rock places where good phreatic aquifers are overlying thin river alluvium or if the rocks have

FIGURE 18.5 Percolation pond/tank (Mati, 2012).

nel. If there are surface storage dams upstream of the recharge sites, which allow for the regulated release of waters, artificial recharge by stream channel alterations may be more successful.

vii. **Surface Irrigation:** It aims to boost agricultural production by reliably watering crops in monsoon breaks and other times when there isn't a monsoon. Surface irrigation should be prioritized above other uses of source water in areas where adequate drainage is guaranteed since it has the added benefit of boosting groundwater supplies.

18.7.1.3   Subsurface Techniques

These are intended to recharge deep aquifers that are covered by hard layers in order to prevent infiltration of water from surface sources that would otherwise recharge them naturally.

i. **Injection Wells (Recharge Wells):** Injection wells are tube well-like constructions used to "pump in" treated water with pressure to increase the groundwater storage capacity of restricted aquifers. The aquifer that has to be recharged is typically one that has already been excessively pumped by tube wells, and a downward trend in water levels has begun. To stop the infiltration of seawater and address the issues of land subsidence in constrained regions, artificial recharge via injection wells is carried out in coastal zones (Fig. 18.6).

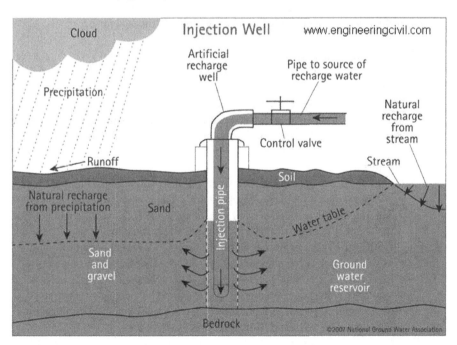

FIGURE 18.6 Injection well (Malu, 2017).

  ii. **Gravity-Head Recharge Wells:** When source water becomes available, regular bore-wells and dug-wells that are used for pumping can also be used as recharge wells in addition to injection wells. When using existing wells for recharge, the source water used should be properly filtered and disinfected. To prevent scouring of the bottom and air bubble entrapment in the aquifer, the recharge water should be directed to the bottom of the well below the water level.

  iii. **Connector Wells:** A special kind of recharge well called a connector well allows water to move from aquifer to aquifer without any the need for pumping. The upper head aquifer horizons begin replenishing the lower head aquifer horizons.

  iv. **Recharge pits:** Structures called recharge pits help with the challenging task of artificially replenishing the phreatic aquifer from surface water sources. The sole distinction between them and recharge basins is that they have a smaller bottom area and are deeper.

  v. **Recharge Shafts:** A shaft is used to generate artificial recharge in situations where impermeable strata cover the aquifer that is deep below the land surface. The cross-section of a recharge shaft is substantially smaller than that of a recharge pit.

18.7.2 INDIRECT METHODS

The indirect approaches aim to recharge aquifers using indirect methods rather than by providing aquifers with direct water supplies. Induced recharge and aquifer modification techniques are the most popular approaches in this category.

  1. **Induced Recharge**

The indirect way of artificially recharging the groundwater reservoir that entails pumping water, from an aquifer that is hydraulically linked to surface water. The biggest benefit of this method is that, in the presence of good hydrogeological conditions, surface water quality usually increases as a result of passing through aquifer materials before being discharged from the pumping well (Fig. 18.7).

    i. **Pumping Wells:** Perennial streams that are hydraulically linked to an aquifer via the permeable rock material of stream channel are installed with an induced recharge system nearby. The outside of a stream bend is a suitable place. Most crucial factors during induced recharge include the chemical composition of the surface water source.

    ii. **Collector Wells:** Collector wells are built to extract enormous amount of water supplies from lake-bed deposits, riverbed deposits, or swampy locations. Horizontal wells may be preferable than vertical wells in locations where the phreatic aquifer next to the river has a thin layer. More induced recharge from the stream is possible for collectors with horizontal laterals and infiltration tunnels.

    iii. **Infiltration Gallery:** Other structures for accessing groundwater reserves beneath the strata of riverbeds include infiltration galleries.

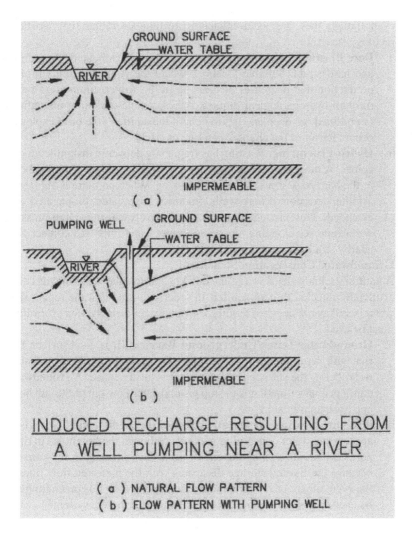

FIGURE 18.7 Induced recharge (PHED, 2019).

The gallery is an open-jointed, horizontally perforated or porous structure (pipe) that is encircled by a gravel filter envelop that is buried in permeable and saturated strata with a shallow water table and a year-round source of recharge. The galleries are typically installed between 3 and 6 m below the surface to collect water by gravity. As a result, the choice should be taken first taking into account the desired yield, then economic factors.

2. **Aquifer Modification Techniques**
These methods alter the properties of aquifer to boost the ability to transmit and hold water. However, because of the resulting increase in groundwater storage in the aquifers, they are also being regarded as recharge structures

even though they are yield-enhancing techniques rather than artificial recharge structures.

    i. **Bore Blasting:** These methods work well on consolidated, crystalline, and hard strata. Suitable locations where the aquifer exhibits limited production that decreases or dries up in the winter/summer are found through hydrogeological inquiry. All blast holes, whether unconfined or confined, go deep enough to benefit the aquifer. One once explosion occurs for all of the charges in a line or circle.

    ii. **Hydro-Fracturing:** Blasting has frequently delivered insignificant outcomes. A more modern method for increasing secondary permeability in the hard rock strata is hydro-fracturing. When pressure is applied to a solitary portion of bore-wells, fractures are created, propagated, and extended. This process is known as hydro-fracturing. High-pressure water removes clogging, opens up fissures, and improves contact with nearby strata that contain water.

3. **Groundwater Conservation Structures**

A natural groundwater flow regime quickly controls the water that has been artificially injected into an aquifer. In order to ensure that the replenished water is still available when required, groundwater conservation procedures must be used.

    i. **Groundwater Dams/Underground Barriers:** It is a subsurface barrier built on a stream to slow down the groundwater flow and store water below the surface for use when it is most needed. A groundwater dam's primary function is to stop groundwater flow out of the sub basin and raise aquifer storage.

    ii. **Fracture-Sealing Cementation Technique:** In dry conditions, it is an appropriate water conserving method. Although the boreholes in these zones are productive, by the end of the winter or summer, they are dry because the limited storage dissipates with the preferred flow lines in the event of a poor topographical scenario. Further, this precaution can be used to stop dirty or saline water from entering the system.

Under ideal hydrogeological circumstances, a variety of combinations of surface and subsurface recharge techniques can be used together for the best possible recharging of groundwater reservoirs.

18.7.3 Combination Methods

Under ideal hydrogeological conditions, a variety of combinations of surface and subsurface recharge techniques may be used in conjunction for the best replenishment of groundwater reservoirs. The combination of approaches in these situations should be site-specific. Recharge basins are one of the most widely used combination techniques. Water can move from an upper to a lower aquifer zone through the annular area between the walls and casing (connector wells) in percolation ponds with recharging pits or shafts, induced recharge with wells tapping various aquifers, etc.

18.8   OPERATION AND MAINTENANCE

Before choosing the site and manner of recharging, a thorough hydrogeological analysis must be carried out to ensure the effective functioning of an artificial recharge system. Artificial recharge structures need to be maintained on a regular basis because silting, chemical precipitation, and organic matter buildup quickly diminish infiltration capability. The period between one cleaning and the next can be increased by making the injection or connector wells dual-purpose wells, but for spreading structures aside from subsurface dykes designed with an overflow or outlet-annual de-silting is required. Due to the fact that the buildings were put in place to alleviate drought, it is unfortunate that routine maintenance is sometimes neglected until a drought comes, at which point the structures must be repaired. Maintenance of buildings is typically done by a number of organizations and people.

18.9   GROUNDWATER MANAGEMENT THROUGH AQUIFER MAPPING

Because the strata that contain groundwater are hidden below the surface, it is challenging to directly assess the amount of water that is presented there and available for use. Therefore, it is essential to use indirect approaches to estimate the same, which frequently serves as an eye-opener and gives decision-makers critical information for sustainable groundwater management. The problem of groundwater assessment is somewhat challenging due to the complexity of the processes governing the occurrence and movement of water, primarily because not only large amounts of data should be collected, but also a multidisciplinary scientific approach must be used to determine where groundwater is located in space and time and its quantity and quality. Thus, estimating groundwater supplies is crucial to the development, planning, and management of water resources. Aquifer mapping is a scientific procedure that assesses the groundwater quantity, quality, and sustainability in aquifers using a combination of hydrologic, geology, geophysical and chemical fields, and laboratory analyses. Systematic mapping is anticipated to advance our knowledge of the hydrologic makeup of aquifers, their geologic conceptual frameworks, their water levels and temporal variations, and the incidence of anthropogenic and natural contaminants that effect groundwater portability.

18.10   NEED FOR AQUIFER MAPPING

The groundwater regime in many regions of the country has suffered as a result of numerous development activities throughout the years. In order to create efficient management methods with community involvement and improve groundwater governance, scientific planning is required for the conservation of groundwater under various hydrogeological circumstances. Although the Central Ground Water Board (CGWB) and other central and state agencies have generated a significant amount of hydrological as well as hydrogeological data through scientific investigations, these mostly relate to administrative units and have only rarely stated the problems of the entire aquifer system. A process of systematic data collection, data generation, compilation, analysis as well as synthesis is required in order to provide accurate and comprehensive information about the various factors of groundwater resources existing in different hydrogeological settings given the country's emerging challenges.

18.11   OBJECTIVES OF AQUIFER MAPPING

The following are objectives of aquifer mapping given as

- The lateral and vertical placements of aquifers, as well as their characterization, are defined on a 1:50,000 scale generally and are further detailed up to a 1:10,000 scale in areas that have been designated as priority areas.
- To develop aquifer management plans and enable the sustainable management of groundwater resources, quantification of availability and evaluation of groundwater quality are necessary.
- Explains the aquifer geometry, aquifer type, behaviors of the groundwater regime, hydraulic features, and geochemistry of multi-layered aquifers on a scale of 1:50,000.
- It makes use of novel geophysical methods and establishes the usefulness, effectiveness, and applicability of these methods in various hydrogeological contexts.
- It completes the methodology and approach that will be used to implement the national aquifer mapping program across the entire nation.
- To create an aquifer information and management system based on the prepared aquifer maps for the sustainable management.
- Applying the learned lessons, the activities to prepare micro-level aquifer mapping can be scaled up.

18.12   GROUNDWATER STUDIES AND AQUIFER MAPPING IN INDIA

Groundwater was regarded as a source for enhancing surface water supplies and expanding India's potential for irrigation when thorough studies of aquifers and groundwater resources started in the 1950s. At that time, the Geological Survey of India's Ground Water Wing conducted groundwater research across the nation, while the Exploratory Tube Well Organization (ETO) was established to find suitable locations and drill tubes or bore-wells for developmental purposes. With the mission to "develop and disseminate technologies and monitor and implement national policies for the scientific and sustainable development and management of India's ground water resources, including their exploration, assessment, conservation, augmentation, protection from pollution, and distribution, based on principles of economic and ecological efficiency", these two organizations were combined into a single organization known as the CGWB in 1972.

The CGWB has been conducting hydrogeological research over time in an effort to understand the scope, traits, yield potentials, and development prospects of significant hydrogeological units throughout the nation. Exploratory drilling operations, surface and subsurface geophysical investigations, and groundwater quality assessments were used to aid and augment these studies. As a result of these actions, numerous hydro-geological units that may be established to complement the surface water sources were identified and demarcated. In conjunction with international organizations, a number of initiatives were started in significant Indian basins to conduct water balance analyses on aquifer groundwater regimes and development prospects.

The focus of the CGWB's hydrological research increasingly switched from development to management beginning in the early 1990s as a result of the increased development of groundwater resources to fulfill increasing demands and the consequent detrimental environmental repercussions. As a result, studies on groundwater augmentation, regulation, and integrated water resources management through the joint use of surface and groundwater resources were added to the CGWB's scope of work. The Board's Information, Education & Communication (IEC) operations increased as a result of the realization of the value of stakeholder involvement in effective groundwater management. The CGWB has focused a number of its operations during the past ten or so years on "sustainable ground water resource management".

In terms of maps that show aquifers, the Geological Survey of India released the first hydrogeology map in 1969 under the title "Geohydrological Map of India" with a scale of 1:2 million. Following that, CGWB released an updated "Hydrogeological Map of India" at a scale of 1:5 million based on its study. The CGWB produced its first edition of the hydrogeological map of India at 1:2 million scales in 1985 and its second edition in 2001 based on surveys, exploration, and special research carried out. Nine significant aquifers (hydrogeological units) were included on the map based on stratigraphy and the data that were available at the time.

The necessity to delineate aquifer units on a broader level and establish approaches for their long-term development was then becoming increasingly obvious as a result of the country's increasing reliance on groundwater as a source of fresh water and the increasing strain on the available resources. This resulted in the CGWB's database being updated with information obtained from its own and numerous other organizations' scientific investigations, allowing the Board to produce the very first aquifer map of the nation at a scale of 1:125,000 with 14 major aquifer systems that were further divided into 42 main aquifers in the first phase. On the other hand, the current program of national aquifer mapping has been begun in order to map these aquifer systems at 1:50,000 scale or bigger, which is seen to be necessary for planning their sustainable development with stakeholder participation. In addition, the Board has begun five pilot aquifer mapping projects in conjunction with the nationwide aquifer mapping in a variety of hydrogeological settings across the country, including hard rock, coastal, basaltic, desert, and alluvial terrains, to assess the efficacy of using various geophysical tools and to complete the protocol for using various geophysical methods to assist hydrogeological research in the nationwide aquifer mapping.

18.13 AQUIFER MANAGEMENT

The CGWB undertakes a follow-up program for aquifer management, concentrating on participatory groundwater management plans, based on the findings of aquifer mapping and modeling. Stakeholders are involved in the definition of management plans, which also take socioeconomic factors and local water challenges into account. Finding a good solution is done at the village or watershed level. The stakeholders are given a thorough image of the aquifers in each location, including aquifer-wise water budgeting, water availability, and water quality in the pre-monsoon and post-monsoon periods. Aquifer management plans, for instance, identify areas for artificial recharge, provide for groundwater abstraction to improve the water balance in the watershed, and identify alternate sources of water supply, particularly in geologically contaminated areas where aquifers that are free of contamination recognized through aquifer mapping could be tapped.

18.14 GROUNDWATER MANAGEMENT USING GIS

Information on the properties and possible occurrences of groundwater resources is available using a variety of methodologies. Groundwater conditions and their occurrences can be learned via remote sensing (RS) data, such as satellite photography that can be processed with sophisticated software (Abdelkareem and El-Baz, 2015). Thus, groundwater exploration and mapping using satellite RS and geographic information system (GIS) approaches are effective, especially in arid areas (Abdalla, 2012; Moubark and Abdelkareem, 2018). The knowledge of the spatial distribution of groundwater can be improved with the use of advances in image processing techniques (Abdelkareem et al., 2012; Lentswe and Molwalefhe, 2020; Saraf and Choudhary, 1998; Yeh et al., 2016). Delineating potential groundwater areas using GIS is advantageous since the technique allows for the analysis and integration of a wide range of spatially distributed data under numerous logical conditions. GIS techniques have been effectively used in several research studies to map groundwater availability zones.

The mainstay of conventional approaches used to identify, define, and map the groundwater potential zones are ground surveys using geophysical, geological, and hydrogeological techniques. These surveys are typically expensive and time-consuming (Israil et al., 2006). On the other hand, geospatial technologies can quickly and affordably produce and model useful data in a variety of geosciences domains (Moghaddam et al., 2015; Russo et al., 2015). In order to determine and map groundwater potential regions, researchers have employed a variety of techniques, including probabilistic models like frequency ratio (Ozdemir, 2011; Razandi et al., 2015), multi criteria decision analysis (Mukherjee et al., 2013; Rahmati et al., 2015), weights of evidence (Lee et al., 2012; Pourghasemi and Beheshtirad, 2015), logistic regression (Pourtaghi and Pourghasemi, 2014), evidential-belief function (Mogaji et al., 2015; Pourtaghi and Pourghasemi, 2014), certainty factor (Razandi et al., 2015), decision tree (Chenini et al., 2010), ANN model (Lee et al., 2012), Shannon's entropy (Naghibi et al., 2015), and machine learning techniques (Rahmati et al., 2016) like random forest. The RS and GIS are powerful tools that can be used to quickly estimate natural resources out of the various techniques. Before using intricate and expensive surveying procedures, the method is affordable and can be used to successfully explore groundwater (Faust et al., 1991; Hinton, 1996). This issue has already been the subject of numerous studies, which confirm the application of RS and GIS methods for mapping groundwater potential zones throughout the world (Dar et al., 2010).

The creation of maps showing groundwater availability depends on a number of variables. Lineaments, topography, lithology, stream networks, and thematic layers related to the climate are some of these (Ganapuram *et al.*, 2009; Srivastava *et al.*, 2012). For their development, GIS tools are used. Such methods can be applied in a variety of settings, including arid regions. The spatial analyst tool of the GIS uses a weighted overlay approach to merge the thematic layers (Sikdar *et al.*, 2004).

18.15 CONCLUSION

Groundwater resources around the world are under extreme stress due to overexploitation and significant changes in climate over time. The need to assess groundwater potential and aquifer productivity rises along with the global need for potable water for agricultural works, human consumption, and industrial applications. Systematic aquifer mapping is planned to advance our knowledge of the hydrologic makeup of aquifers, their geologic underpinnings, their water levels and temporal variations, and the incidence of anthropogenic and natural contaminants that are affecting groundwater potability. It is necessary to determine and map the aquifers at the micro scale, accurately measure the aquifer-wise groundwater resource, evaluate aquifer water quality, characterize the aquifer yield and storage capabilities, identify the aquifer recharge and discharge zones, and make institutional arrangements in order to manage the groundwater resource through aquifer-based management against rising demand.

REFERENCES

Abdalla, F. 2012. Mapping of groundwater prospective zones using remote sensing and GIS techniques: A case study from the Central Eastern desert, Egypt. *J Afr Earth Sci.* 70:8–17.

Abdelkareem, M., El-Baz, F. and Askalany, M. 2012. Groundwater prospect map of Egypt's Qena Valley using data fusion. *Int J Image Data Fus.* 3(2):169–189. DOI:10.1080/19479832.2011.569510

Abdelkareem, M. and El-Baz, F. 2015. Analyses of optical images and radar data reveal structural features and predict groundwater accumulations in the central Eastern Desert of Egypt. *Arab J Geosci.* 8:2653–2666.

Alston, P.H. and Watkins, R. 1973. Apple breeding at East Malling. *Proc Eucarpiafruit Breed Symp.* 14–29.

Central Ground Water Board (CGWB). 1994. Manual on Artificial Recharge of Ground Water. Technical Series M, No. 3, Central Ground Water Board, Faridabad, March 1994, 215 p.

Chenini, I. and Mammou, A.B. 2010. Groundwater recharge study in arid region: An approach using GIS techniques and numerical modeling. *Comput Geosci.* 36:801–817.

Dar, I.A., Sankar, K. and Dar, M.A. 2010. Remote sensing technology and geographic information system modeling: An integrated approach towards the mapping of groundwater potential zones in Hardrock terrain, Mamundiyar basin. *J Hydrol.* 394:285–295.

Faust, N.L., Anderson, W.H. and Star, J.L. 1991. Geographic information systems and remote sensing future computing environment. *Photogramm Eng Remote Sens; (United States).* 57.

Ganapuram, S., Kumar, G., Krishna, V.I.V., Kumar, M., Kahya, E.M. and Demirel, C. 2009. Mapping of groundwater potential zones in the Musi basin using remote sensing data and GIS. *Adv Eng Softw.* 40(7):506–518.

Heviánková, S., Marschalko, M., Chromíková, J., Kyncl, M. and Korabík, M. 2016. Artificial ground water recharge with surface water. *IOP Conf Ser: Earth Environ Sci.* 44:022036. http://iopscience.iop.org/1755-1315/44/2/022036.

Hinton, J.C. 1996. GIS and remote sensing integration for environmental applications. *Int J Geogr Inf Syst.* 10:877–890.

India: India's Water Economy, Bracing for a Turbulent Future – 2006; Report No. 34750-IN, Agriculture and Rural Development Unit South Asia Region, Document of the World Bank.

Israil, M., Al-Hadithi, M. and Singhal, D.C. 2006. Application of a resistivity survey and geographical information system (GIS) analysis for hydrogeological zoning of a piedmont area, Himalayan foothill region, India. *Hydrogeol J.* 14:753–759.

Kannan, S. and Mathew, 2008. GIS and remote sensing for artificial recharge study in a degraded Western Ghat terrain. *J Centre Stud Res Eng.* 1–15.

Lee, S., Kim, Y.S. and Oh, H.J. 2012. Application of a weights-of-evidence method and GIS to regional groundwater productivity potential mapping. *J Environ Manage.* 96:91–105.

Lentswe, G.B. and Molwalefhe, L. 2020. Delineation of potential groundwater recharge zones using analytic hierarchy process-guided GIS in the semi-arid Motloutse watershed, eastern Botswana. *J Hydrol Reg Stud.* 28:100674.

Malu, S. 2017. Artificial Recharge of groundwater. https://www.engineeringcivil.com/artificial-recharge-of-groundwater.html

Mati, B. 2012. Developing Ground Water and Pumped Irrigation Systems: Training Manual 6.

Mogaji, K.A., Lim, H.S. and Abdullah, K. 2015. Regional prediction of groundwater potential mapping in a multifaceted geology terrain using GIS-based Dempster–Shafer model. *Arab J Geosci.* 8:3235–3258.

Moghaddam, D.D., Rezaei, M., Pourghasemi, H.R., Pourtaghie, Z.S. and Pradhan, B. 2015. Groundwater spring potential mapping using bivariate statistical model and GIS in the Taleghan watershed, Iran. *Arab J Geosci.* 8:913–929.

Moubark, K. and Abdelkareem, M. 2018. Characterization and assessment of groundwater resources using hydrogeochemical analysis, GIS and field data. *Arab J Geosci.* 11:598.

Mukherjee, S., Mukherjee, S., Garg, R.D., Bhardwaj, A. and Raju, P.L. N. 2013. Evaluation of topographic index in relation to terrain roughness and DEM grid spacing. *J Earth Syst Sci.* 122:869–886.

Naghibi, S.A., Pourghasemi, H.R., Pourtaghi, Z.S. and Rezaei, A. 2015. Groundwater qanat potential mapping using frequency ratio and Shannon's entropy models in the Moghan watershed, Iran. *Earth Sci Inf.* 8:171–186.

O'Hare, M.P., Fairchild, D.M., Hajali, P.A. and Canter, L.W. 1986. Artificial recharge of groundwater. *Proceedings of the Second International Symposium on Artificial Recharge of Groundwater.*

Ozdemir, A. 2011. GIS-based groundwater spring potential mapping in the Sultan Mountains (Konya, Turkey) using frequency ratio, weights of evidence and logistic regression methods and their comparison. *J Hydrol.* 411:290–308.

PHED. 2019. Retrieved from https://megphed.gov.in.

Pourghasemi, H.R. and Beheshtirad, M. 2015. Assessment of a data-driven evidential belief function model and GIS for groundwater potential mapping in the Koohrang Watershed, Iran. *Geocarto Int.* 30:662–685.

Pourtaghi, Z.S. and Pourghasemi, H.R. 2014. GIS-based groundwater spring potential assessment and mapping in the Birjand Township, southern Khorasan Province, Iran. *Hydrogeol J.* 22:643–662.

Rahmati, O., Pourghasemi, H.R. and Melesse, A.M. 2016. Application of GIS-based data driven random forest and maximum entropy models for groundwater potential mapping: A case study at Mehran Region, Iran. *Catena*. 137:360–372.

Rahmati, O., Samani, A.N., Mahdavi, M., Pourghasemi, H.R. and Zeinivand, H. 2015. Groundwater potential mapping at Kurdistan region of Iran using analytic hierarchy process and GIS. *Arab J Geosci*. 8:7059–7071.

Razandi, Y., Pourghasemi, H.R., Neisani, N.S. and Rahmati, O. 2015. Application of analytical hierarchy process, frequency ratio, and certainty factor models for groundwater potential mapping using GIS. *Earth Sci Inf*. 8:867–883.

Russo, T.A., Fisher, A.T. and Lockwood, B.S. 2015. Assessment of managed aquifer recharge site suitability using a GIS and modeling. *Groundwater*. 53:389–400.

Saraf, A. and Choudhary, P. 1998. Integrated remote sensing and GIS for ground water exploration and identification of artificial recharge site. *Int J Remote Sens*. 19:1825–1841.

Seckler, D., Upali, A., David, M., Radhika de, S. and Randolph, B. 1998. World Water Demand and Supply, 1990 to 2025: Scenarios and Issues, Research Report 19. Colombo, Sri Lanka: International Water Management Institute.

Sikdar, P., Chakraborty, S., Enakshi, A. and Paul, P.K. 2004. Land use/land cover changes and groundwater potential zoning in and around Raniganj coal mining area, Bardhaman District, West Bengal – A GIS and remote sensing approach. *J Spatial Hydrol*. 4(2):1–24.

Singh, L. and Ravichandran, S. 2011. Studies on Estimative Methods and their Role in Artificial Ground Water Recharge. 3.

Srivastava, S., Ghosh, S. and Kumar, A. 2012. Spatial and temporal investment pattern in irrigation development and its impact on Indian agriculture. Research Bulletin No. 55. Bhubaneswar: Directorate of Water Management (ICAR).

Veeranna, J. and Jeet, P. 2020. Groundwater recharges technology for water resource management: A case study, in B. Kalantar (ed.), *Groundwater Management and Resources*, IntechOpen, London. 10.5772/intechopen.93946

Yeh, H., Cheng, Y. and Lin, H. 2016. Mapping groundwater recharge potential zone using a GIS approach in Hualian River, Taiwan. *Sustain Environ Res*. 26(1):33–43.

Index

Printed in the United States
by Baker & Taylor Publisher Services